Time in Eternity

Time in Eternity

Pannenberg, Physics, and Eschatology in
Creative Mutual Interaction

ROBERT JOHN RUSSELL

University of Notre Dame Press

Notre Dame, Indiana

University of Notre Dame Press
Notre Dame, Indiana 46556
www.undpress.nd.edu
All Rights Reserved

Published in the United States of America
Copyright © 2012 by University of Notre Dame

Library of Congress Cataloging-in-Publication Data

Russell, Robert J.
Time in eternity : Pannenberg, physics, and eschatology
in creative mutual interaction / Robert John Russell.
p.   cm.
Includes bibliographical references (p.   ) and index.
ISBN-13: 978-0-268-04059-8 (pbk. : alk. paper)
ISBN-10: 0-268-04059-1 (pbk. : alk. paper)
ISBN-13: 978-0-268-15839-2 (web pdf)
1. Religion and science.   2. Pannenberg, Wolfhart, 1928–
3. Space and time—Religious aspects—Christianity.   4. Eternity.
5. Eschatology.   6. Cosmology.   I. Title.
BL240.3.R877   2012
261.5'5—dc23

2012015584

I dedicate this book to Wolfhart Pannenberg

for his extraordinary contributions to

the constructive dialogue between the natural sciences

and systematic theology. I thank him for the Christian hope

in the Resurrection of Jesus that he has helped me

more deeply grasp professionally as well as personally.

I am honored to call him both theological mentor

and friend.

CONTENTS

PART 1. SRP → TRP

*Revising Pannenberg's Trinitarian Conception of Eternity
and Omnipresence and Its Role in Eschatology in Light of
Mathematics, Physics, and Cosmology*

PART 2. TRP → SRP

The Theological Reconstruction of Pannenberg's Views in Light of Mathematics, Physics, and Cosmology as Offering Suggestions for New Research Programs in the Philosophy of Time and in Physics

ACKNOWLEDGMENTS

This project has been wonderfully and critically influenced by numerous friends and colleagues around the world and at the Graduate Theological Union (GTU) in Berkeley, California. I took advantage of the opportunity afforded by participating in a variety of international conferences over the past decade to develop materials for this book. I also have had several occasions to try out portions of this material in doctoral and seminary courses at the GTU. There are thus many more people than I can possibly thank here. Nevertheless I wish to mention some to whom I am most indebted.

I am particularly grateful for the opportunity to teach a doctoral seminar in the spring of 2010 with Ted Peters based on drafts of this volume as well as on the historical and contemporary material in physics, cosmology, mathematics, philosophy, and theology that serve as background to this volume. Feedback from students in the seminar was very helpful, particularly from Fady El Chidiac, Junghyung Kim, John King, and Oliver Putz. I am also very grateful for extensive written comments from, and many detailed conversations with, Johanne Stubbe Teglbjaerg, a postdoctoral fellow from Copenhagen University in residence at the GTU during the summer and fall of 2011. And I appreciate the written responses to earlier versions of the book from Philip Clayton, LeRon Shults, and Dean Zimmerman.

I want to offer particular thanks to Ted Peters, whose prolific writings on theology and science have taught and inspired me over the past three decades. I am grateful for the many conversations we have had and for the experience of teaching a variety of doctoral seminars with him on theology and science in general and on Pannenberg in particular.

I am especially grateful to Joshua Moritz, for editing the entire manuscript with extraordinary care, for his many valuable suggestions about the text, and for our extensive and fruitful conversations over the past decade. He is also the gifted artist who created most of the volume's figures and diagrams; I am particularly delighted with those in chapters 2 and 4.

Finally I want to thank my daughters Christie Lavigne and Lisa Galicia for their constant encouragement of and interest in my work. Most of all I want to thank my wife Charlotte for her abiding faith, love, and support during the—seemingly endless!—process of writing this volume.

This project was funded in part by a grant from the Metanexus Institute. Their funding in turn came from the John Templeton Foundation. I am grateful for the financial support of both of these organizations.

ABBREVIATIONS

Frequently cited works of Wolfhart Pannenberg are indicated by the following abbreviations.

ETS   "Eternity, Time and Space." In *The Historicity of Nature: Essays on Science and Theology,* ed. Niels Henrik Gregersen. Philadelphia: Templeton Foundation Press, 2007. 163–74.

ETTG   "Eternity, Time and the Trinitarian God." *Dialog: A Journal of Theology* 39, no. 1 (2000): 9–14.

JGM   *Jesus—God and Man.* Trans. Lewis L. Wilkins and Duane A. Priebe. 2nd ed. Philadelphia: Westminster Press, 1977; originally published in German as *Grundzuge der Christologie,* 1964.

M   *Metaphysics and the Idea of God.* Trans. Philip Clayton. Grand Rapids, MI: Eerdmans, 1990.

ST1   *Systematic Theology.* Vol. 1. Trans. G. W. Bromiley. Grand Rapids, MI: Eerdmans, 1991.

ST2   *Systematic Theology.* Vol. 2. Trans. G. W. Bromiley. Grand Rapids, MI: Eerdmans, 1994.

ST3   *Systematic Theology.* Vol. 3. Trans. G. W. Bromiley. Grand Rapids, MI: Eerdmans, 1998.

TKG   *Theology and the Kingdom of God.* Ed. Richard John Neuhaus. Philadelphia: Westminster Press, 1969.

Introduction

Thy reign come
—Matthew 6:10

*Is there conceivable any positive relation between the con-
cept of eternity and the spatio-temporal structure of the
physical universe? . . . This is one of the most arduous, but
also one of the most important questions in the dialogue
between theology and natural science. . . . Without an answer to the question regarding time and eternity, the relation of God to this world remains inconceivable.*

—Wolfhart Pannenberg,
"Theological Questions to Scientists"

A. THE TOPIC AND MOTIVATION FOR "TIME IN ETERNITY" IN THEOLOGY AND SCIENCE

The topic of this volume, generically described as "time and eternity,"
arises in and is shaped specifically by two distinct but interrelated contexts in philosophical and systematic theology. The first context involves
the relation between God and the world, specifically the relation between
the eternity of God and the temporal character of the world. Traditionally,

1

the divine eternity was viewed either as timeless or as unending time. Some twentieth-century Trinitarian theologians such as Wolfhart Pannenberg, however, go beyond both of these views to depict eternity as fully temporal and the source of creaturely time. The second context is the relation between the eternity of the eschatological New Creation and time in the present creation. This second context is made more complex when the New Creation is understood as grounded in God's radical act at Easter. In this case, according to many New Testament scholars and systematic theologians, the New Creation arises out of the transformation of the present creation in a way that is analogous to the bodily resurrection of Jesus.[1] In this volume I will treat the topic of time and eternity in both of these contexts, drawing out the intrinsic relations between them.

My motivation for exploring the topic of time and eternity in both contexts is multilayered. The primary motivation is to respond aggressively to the direct challenge from physics to the topic of time and eternity in its first context just cited. Here Albert Einstein's special theory of relativity seems to most physicists and philosophers to challenge the temporal character of the world, and thus indirectly to challenge the temporal character of eternity when relativity is imported into the theological conversation about God. According to these scholars, relativity's most convincing interpretation is that the world is not dynamic, one in which time is "flowing." Instead, they argue that the world is static, a timeless world, or "block universe," where the flow of time is a subjective illusion. Physics and cosmology even more severely challenge the topic of time and eternity in its second context—the eternity of the eschatological New Creation in relation to time in the present creation—when the scope of eschatology expands beyond individual human life, beyond collective human history, and beyond the evolutionary history of life on earth to embrace the universe as a whole—as God's creation—and therefore as the subject of God's radical transformation into the New Creation. Here, in discussing "time and eternity," we must inevitably face the challenge that scientific cosmology poses to Christian eschatology, one in which the scientific predictions of a cosmic future of "freeze" or "fry" undercut, even render meaningless, an eschatology based by analogy on the bodily resurrection of Jesus.

Over the past decades additional issues have arisen in the dialogue between theology and science and motivate this volume. 1) The beginning of our universe at $t = 0$ (the "big bang") and the anthropic principle as a response to the universe's fine-tuning for life are deeply consonant with the doctrine of creation *ex nihilo* (creation out of nothing). In the background, though, has been a clear "dissonance" that would eventually need to be addressed thoroughly: if we welcome t = 0 and the anthropic principle into the dialogue with theology we cannot ignore the challenge raised by the "freeze or fry" scenarios to Christian eschatology.

2) Granted that we interpret biological evolution in rather general terms as the means by which God creates the diversity of life on earth, following the doctrine of *creatio continua* (continuous creation). But can we do this in a way that goes beyond what I call "statistical deism," in which God sets up the general laws and the initial conditions at the beginning of the universe and then simply allows the natural processes to unfold entirely on their own? Instead, can we deliver on a robust form of theistic evolution—one in which God is understood to act objectively in the temporal development of natural processes but without intervening in, subordinating, and violating these natural processes? And if we are successful in developing such an account of non-interventionist objective divine action (or what I refer to as "NIODA"), will this in turn help us address the challenge raised by science to Christian eschatology?

3) Finally, if we can develop a robust version of theistic evolution by employing NIODA, how do we understand God's action in nature in relation to the problem of "natural evil," such as suffering, disease, death, and extinction, when natural evil is constitutive of evolutionary biology and not a consequence of an historical Fall? The best response is clearly to start with the cross in which Jesus takes on the suffering of humanity, and extend his embrace of suffering to the suffering of all life on earth. But the theology of the cross demands that we move ahead to the resurrection of Christ, and this in turn leads directly to eschatology and thus the inevitable challenge from physics and cosmology (for details, see appendix to the introduction, section A).

Progress in the discussion of these issues has required long periods of intense scholarly research by individuals and international conferences in theology and science. New summits of understanding have been reached, and from them have arisen both a rewarding sense of accomplishment and a growing sense of impending problems that increasingly require attention. Some of these problems have been long known in general terms but are now seen in a clearer and deeper way, while others were unexpected and have yet to be explored in any depth. Both by the inner logic of each of the issues discussed above and by their mutual entailment as a sequence of issues, the overall result has been a conceptual vector leading inexorably to the "eye of the hurricane," the mutual challenge between theology and science. On the one hand, at the centerpiece of Christian faith we encounter the bodily resurrection of Jesus of Nazareth on Easter as the normative instance of God's now-and-coming eschatological transformation of the creation in all its dimensions into the New Creation. Since these dimensions include what science understands as the physical universe, eschatology obviously challenges the scientific predictions for the future of the universe as "freeze or fry." On the other hand, physics and cosmology challenge the truth and meaningfulness of any Christian eschatology that affirms the physical/biological transformation of this world into the New Creation, since all forms of biological life will become impossible long before the universe recollapses and fries, let alone if it expands and freezes forever. In essence, we seem to be at a fundamental impasse between theology and science.

At the same time, progress in biblical studies has led to a deepening reaffirmation of the centerpiece of Christian faith, the bodily resurrection of Jesus. A growing wave of New Testament scholars have made an increasingly compelling case for the bodily resurrection of Jesus in contrast to those who continue to demythologize the empty tomb accounts and reduce the appearance traditions to the psychological projections and existential experiences of the disciples. With the ascendency of the former group of biblical scholars, the connection between the physical dimension of the bodily resurrection of Jesus and its cosmic significance for the eschatological future of the universe becomes vastly more pronounced, requiring that we now take scientific cosmology explicitly into the conversation. In addition, many theologians writing specifically on Christian eschatology acknowledge that any eschatology whose scope includes not only personal salvation and social transformation but also environmental justice must engage the natural sciences. This engagement with the natural sciences should lead once again to the theological encounter with cosmology. Unfortunately, most scholars in both New Testament studies and Christian eschatology ignore the overwhelming challenge of science to eschatology (see appendix to the introduction, sections B and C).

The most fruitful way to explore all of these issues, I believe, is to inquire into the relation between time and eternity in both of the contexts described above: the eternity of God and the temporal character of the world; and the eternity of the eschatological New Creation and the temporal character of the present world, where the term "present" serves to emphasize its difference from the eschatological New Creation that arises from it. First, however, I will clarify some basic terms that will hold for this entire volume. By "time" I mean our ordinary, daily experience of time—with its fleeting present moment seemingly lost forever as it vanishes immediately into the past, only to be replaced by what was before a future of uncertain possibilities. By the "eternity of God," I mean something much richer and more complex than its two traditional alternatives: 1) Eternity as timeless on the one hand, in which all the distinctions between the rushing moments of our life, each with their unique pasts and futures, are lost as they are conflated into the dimensionless eternal "*nunc,*" a single structureless and unchanging "now." 2) Eternity as unending flowing time on the other hand, in which we are again imprisoned in a momentary present that immediately vanishes into a lost past only to be replaced by an event emerging contingently from an ever inaccessible future. Here the only difference between eternity and ordinary time is that the temporal process of this world simply continues forever.

Instead, following most twentieth-century Trinitarian theologians,[2] I understand eternity to be the boundless temporality of the Trinitarian God, a lavishly rich "supra-temporality" that is both the source and fulfillment of the temporality of creation: the temporality we experience in nature, in our lives, and in history. This is an eternity that flows out of the endless perichoretic dance of the divine persons ceaselessly taking place within the unity of Trinitarian community. By the "eternity of the New Creation," I refer to the gift of true temporality of the Trinity to our world, both as it is to be and as it is being transformed into the New Creation by God's radically new act beginning with the bodily resurrection of Jesus at Easter. It is an eternity of renewed and transformed creaturely life in which creatures retain their distinctive personal and social histories along with the specific temporal events of past, present, and future underlying them and contributing to their intrinsic identity, but without the separation of times into a past that is forever gone and a future that is never available in the lived moment such as in the kind of temporality we now experience. The eternity bequeathed to the New Creation by the Trinity is a form of true temporality, a structured duration of diversity in unity. It is an eternity that holds all the events of the creation in an over-arching and differentiated unity, a unity that brings together our lived experience of the flow of fragmentary present moments without subsuming their distinctions or separations into one timeless moment. It is an eternity in which we will experience everlasting life with all of our present life available to enjoy endlessly in an ever-widening and deepening experience.

The temporality of the New Creation is also a time of the redemptive purging and healing of our lives. Each event of sin that you and I commit, personally and through our participation in broken family and social, economic, and political institutions, must finally be confessed and, then mercifully, forgiven and forgotten (Jer. 31:31). Yes, each act of virtue (through the activity in us of God's grace, our brokenness notwithstanding) is forever celebrated as a redemption given us freely and undeservedly by Jesus Christ. This forgiveness takes our own deserved judgment up into the suffering that Jesus bore and replaces our brokenness with Christ's wholeness and healing balm. I believe it extends to all creatures—wherever there is biological death even if there is no moral sin. Thus eternity is not only a time of endless rejoicing in all that is true and good and beautiful, it is also a time of leaving off and destroying of all that is wrong and false and ugly in this creation through the amazing grace of justice and mercy bequeathed to all of life in the Incarnation, life, ministry, death, and Resurrection of Christ. In short, it is a mystery in which all events in our lifetime—including life's intrinsically unnecessary but, in this world inevitable, death—is taken up into the endless eternity of the New Creation through the power of Christ the eschaton. Each individual life is purged, forgiven, restored, and fulfilled just as the whole history of the world is purged, forgiven, restored, and fulfilled in the first Easter—and this even while all life and history await their full consummation in the eschatological future. As Christ's resurrection has both revealed the future of the cosmos and redeemed the entirety of its history—and therefore every moment before and since the Easter event—so with Christ's future return "all will be well and all will be well and every kind of thing will be well."[3] This redemptive aspect of Christian eschatology requires a careful exposition and analysis of a number of deeply connected theological doctrines and themes that lie beyond the scope of this volume. Nevertheless it should be understood that it lies within the wider theological context presupposed for this book.

In sum, then, the theme of "time and eternity" engages both (1) the relation between the ordinary time of our experienced life and the eternity of God, and (2) the transformation of ordinary time into the eschatological eternity of God's New Creation based by analogy with the bodily resurrection of Jesus as transformation. By situating these vital theological questions within the heart of the growing dialogue and interaction between theology and science, I hope to address what is arguably the most difficult and the most promising topic for this interaction as emphasized by Pannenberg in the epigraph to this introduction.

To undertake this interaction, however, requires that we face a profound challenge to the assumption of "flowing time" that is both ubiquitous to theology in general and that clearly pervades the specific theme of this volume, "time and eternity." As indicated above, this challenge is posed by the view of the majority of physicists and Anglo-American philosophers of time who unabashedly defend a static, "block universe" view of nature as timeless. According to the block universe view, the flow of time is strictly *subjective;* in the *objective* world, the past, present, and future are equally real. For anyone working on the interaction between theology and science, such a challenge is unavoidable because the strongest case for the block universe view arises from one of the two fundamental theories in contemporary physics as it is most often interpreted philosophically: Einstein's special relativity (SR).[4] Until the twentieth century, scholars supporting timelessness waged a pitched battle on basically philosophical terms with those defending flowing time. Since the birth of relativity in 1905, however, an immense new arsenal of seemingly overwhelming force has been supplied to the defenders of timelessness, and their position has been reinforced by such outstanding scientists and mathematicians as Bertrand Russell, Hermann Minkowski, and, of course, Einstein himself. The depth of their commitment to timelessness is enshrined in the words of Einstein's remarkable letter to the widow of his recently deceased friend: "Michele has preceded me a little in leaving this strange world. This is not important. For us who are convinced physicists, the distinction between past, present, and future is only an illusion, however persistent."[5]

This book takes up the challenge posed by the block universe understanding of time by articulating and defending a new argument in support of a flowing time interpretation of special relativity, an issue about which I have thought for many years and, more recently, which I have vigorously studied. Such an argument for flowing time in light of SR can be of value not only to the theology and science community but also to philosophers and scientists engaged in the flowing time/block universe debate. More broadly, it is my hope that the overall developments in this volume will both address the specific challenges outlined above and more generally provide a fruitful example of how Christian theology and contemporary philosophy and physics, cosmology, and mathematics can constructively interact and inform one another. To accomplish this task I will rely on a method that I have developed over the past decade for the productive interaction between theology and science, one that I call Creative Mutual Interaction (CMI).

According to CMI, a robust philosophical interpretation of scientific theories can lead to a creative reformulation of theological doctrines. But in what might be considered a startling move, a theology that is so reformulated in light of science can also lead to suggestions for creative new research programs in science and in the philosophy of science. I see this book as an opportunity to set out a fully developed example of CMI, one that will test the viability of this method as it constructively engages a matter at the very heart of the interaction between theology and science: time and eternity and its implications for Christian eschatology in light of physics and scientific cosmology. I will say a bit more about CMI shortly, but first I want to discuss the theological focus of this volume: the writings of Wolfhart Pannenberg.

B. THE FOCUS ON WOLFHART PANNENBERG

Many theologians have been influential in persuading me that the doctrine of the Trinity provides the most appropriate context for exploring the theme of "time and eternity." In particular, I have been moved by Karl Barth's beautiful conceptualization of creaturely time as "embedded" in God's eternal life, by Elizabeth Johnson's compelling imagery of the perichoretic life of the divine Persons, by Catherine Mowry La-

Cugna's celebration of the retrieval of the Trinity in the twentieth century, by Jürgen Moltmann's insistence on the economic Trinity *in* creation and creation *in* the immanent trinity,[6] by Ted Peters' development of the theme of "temporal holism" in his writings on eschatology and his claim that we find the unambiguous goodness of creation in the eschatological future, and by Karl Rahner's pivotal insight that the economic and the immanent Trinity are one and the same (i.e., Rahner's Rule). In light of these visionaries, and intending to reflect the overarching point upon which they all seem to agree—namely that eternity is fully temporal and the source of creaturely time—the title of this volume is "time in eternity." Nevertheless, the theological content is based explicitly on the work of Wolfhart Pannenberg.

Pannenberg's profound insights on "time and eternity" draw from classical as well as contemporary philosophy and from the entire history of Christian theology as it is woven together by him with astonishing clarity, depth, and fresh insight. He uniquely portrays the endless richness of the temporality of the Trinitarian divine life as a "differentiated unity." This eternity is unique in offering the gift of structured duration to human experience and to the natural world, a duration that preserves the distinctions between the past and future of every moment without separating them into irretrievable realms such as we experience in ordinary time. Pannenberg understands God as Trinity to be at work in the world, both continually appearing in history as the "arrival" of the immediate future and as reaching back from the eschatological future to the Easter event in order to transform the world into the New Creation. In a breathtaking move, Pannenberg thematizes the latter as "prolepsis": although the New Creation still lies in our future, or more correctly in the "future of our future," the Easter event is already and normatively a manifestation in our time and history of what is the not-yet still-future eschatological-apocalyptic destiny for all the world. In the Easter event, the New Creation, having been transformed by God out of the original creation, reaches back over and into, and is manifested proleptically in, world history. Then, through this proleptic character of Easter, all of history prior to it and following it is filled with the promise of New Creation.[7]

Throughout his writings, Pannenberg engages both the natural sciences and the philosophy of science. To deepen that engagement on the theme of "time and eternity," much of his work used in this volume

must first be reconstructed within the framework of contemporary physics, cosmology, and mathematics. This is because in Pannenberg's theology his language tends to presuppose the way both ordinary language
and classical physics treat time and space. I hope that what I bring to the
table, as both a theologian and a physicist, will help in this process of
"friendly reconstruction" in light of science. But let me be clear here. The
purpose of this book is not what has become routine in the academy:
first a critical engagement of Pannenberg's work, then an exposition of
his theological "weaknesses," and then an advance of one's own theological agenda. Instead, the goal of this book is to "take him at his word,"
and to explore the "what if" question: What if much of what Pannenberg claims is deeply *right* about "time and eternity," even though some
reconstruction is first needed to take into account contemporary physics,
cosmology, and mathematics? What can we learn from Pannenberg about
time and eternity both as a thematization of God's present relation to the
world and of God's eschatological relation to the New Creation and thus
proleptically to the world now? From here, I pursue a novel way of assessing and demonstrating the fruitfulness of Pannenberg's reconstructed
theology by exploring what it has to offer for a dialogue with scholars
in the philosophy of time and with theoretical physicists and cosmologists. And this, in turn, reflects the method of CMI that structures this
volume.

C. VOLUME OVERVIEW

This section will present a detailed outline of the volume, noting the division of chapters as parts one and two based on CMI and orienting the
reader to the topics of each chapter. First, however, I would like to make
some general comments about CMI, the method which structures the
entire volume. I would also like to summarize the three key concepts that
arise in discussing Pannenberg's work on time and eternity—duration,
co-presence and prolepsis—since they are so pervasive in the volume
and so important to its overall argument.

    As mentioned above, over the past decade I have developed and
tested a new and very general interdisciplinary method for relating theology and science that I call Creative Mutual Interaction (CMI). In CMI,

theology is first reformulated in light of contemporary science. Once reformulated, such a theology is then available as a source of insights for potential research directions in contemporary science and the philosophy of science. Through this double move it is my hope that Christian theology can articulate the New Testament *kerygma* forcefully and evocatively to today's scientifically informed culture as well as be an engaging voice offering suggestions for potential research in the dialogue with scientists and philosophers of science. CMI includes five paths for importing the discoveries of the natural sciences into constructive theology where, as with any theological source, they are incorporated into theology through critical reflection and philosophical analysis. These paths represent the traditional theological method of *fides quaerens intellectum* (faith seeking understanding). But CMI goes on to offer something quite new: it includes three additional paths in which such a reformulated theology can inspire new worldviews for science as well as new research directions in science, including new theories and new criteria of choice between existing theories. (For readers unfamiliar with CMI, see the appendix to this introduction, section D.)[8]

Over the past decade I have used CMI in articles that relate various specific topics in theology and science, such as the doctrine of creation and big bang cosmology or non-interventionist divine action and quantum mechanics. This volume is the first time that CMI plays an explicit and ubiquitous role in structuring a book-length work. Accordingly, it is organized into two parts to reflect the structure CMI: part one utilizes the first five paths to reformulate theology in light of science and part two, the second three paths to suggest research ideas to science and the philosophy of science in light of this reformulation.

During the past decade I have also constructed a very specific set of guidelines to monitor the interaction between theology and science as shaped by CMI when the theological topic is Christian eschatology and the scientific topic is cosmology. These guidelines are meant to keep theological reconstruction from going in directions that are clearly unhelpful and to lead such reconstruction to areas that seem most likely to bear fruit. Used extensively throughout this volume, these guidelines play a crucial role in overcoming the challenge from cosmology to eschatology and in helping pave the road for a continuing constructive conversation between such a revised theology and the natural sciences. I invite the reader to explore the details regarding these guidelines in the appendix to the introduction (section E).[9]

It is appropriate and important to note explicitly here how Pannenberg's writings on the bodily resurrection of Jesus have influenced me in the formation of these guidelines for relating theology and science. Guideline 1, in particular, plays a crucial role in taking us beyond the challenge posed by scientific cosmology to eschatology, with its predictions of "freeze or fry." Guideline 1 entails the rejection of two philosophical assumptions about science: the argument from analogy and its representation as nomological universality. The guideline draws on Pannenberg's insight that we can consider the resurrection of Jesus as an historical event if we set aside the Enlightenment assumption about the uniformity of history. On the one hand, if the historian dogmatically assumes that "the dead stay dead," then the Resurrection is ruled out of bounds a priori. But if, on the other hand, we do not insist on this assumption, then the Resurrection is open to historical research (see appendix to the introduction, section F). I have made a similar argument about cosmological predictions based on science. If we make the philosophical assumption that the predictions of well-winnowed scientific theories, such as big bang cosmology, must come to pass then of course science challenges eschatology. But if, instead, we assume that the regularities of nature studied by science have their ultimate basis in the faithful action of God as continuous creator, then—if God has acted in a radically new way at Easter and will continue to do so—the predictions of science for the cosmic future will not hold. In this way, belief in the eschatological New Creation does not conflict with science (for details, see appendix to the introduction, section E, guideline 1). Finally, if Christian eschatology is based on the view of the resurrection of Jesus as a transformation, a view held by many New Testament scholars and theologians including Pannenberg, then we may be able to find physical features of the universe as it is now that will eternally be a part of it, and science can return as a friend in studying these features (see appendix to the introduction, section B; and section D, guideline 6). All of these ideas will be present in the background of this volume. They provide a wider sense of the way Pannenberg's body of writings has implicitly influenced my research here. With this we are prepared to explore three key concepts in my interpretation of Pannenberg's work underlying the interaction between his theology and mathematics, physics, and cosmology.

1. Three Key Concepts in Pannenberg's Theology

Three concepts arise repeatedly throughout this volume and derive from my interpretation of Pannenberg's theology of eternity and omnipresence: duration, co-presence, and prolepsis. Although I have already touched on them, it will be helpful to describe them again here at the outset of the volume.

Duration

In my reading of Pannenberg, time in nature is not point-like, the terminus or limit of the infinite division of a one-dimensional continuum. Instead, time involves duration, or temporal thickness, not only in our conscious experience of memory and anticipation but also in nature, including its fundamental processes as studied by mathematics, physics, and cosmology. The basis for duration both in consciousness and in the physical world, according to Pannenberg, is the temporal structure of eternity. Here eternity as a divine attribute takes up the times of our lives and unifies them via duration, even if we only experience this unity briefly.

Co-presence

It is important to note that Pannenberg's concept of eternity is more subtle than the two "garden-variety" options so often found in philosophical and theological literature. Thus, in Pannenberg's view, eternity is not timelessness, the conflation of all moments of time into a single timeless "now" in which all temporal thickness is lost. Nor is eternity endless ordinary time, a continuing succession of separate temporal moments each of which exists only for an instant as the "present" and then is gone forever. Instead, the divine eternity is one of duration, but according to my read of Pannenberg it is a duration that includes an internal structure that I will call "co-presence." In this novel concept, duration is a differentiated unity that holds together as co-present all events in the history of the universe both now and in the eschatological New Creation. Within the duration of eternity, each event retains its unique past and future. I call this time's "past-present-future structure," or "ppf structure." All events, in turn, each with their own ppf structure, are held together without conflation and without separation in the duration of eternity: that is, they are held together "simultaneously." And in perhaps his greatest *ansatz* in this topic, Pannenberg makes the bold claim that this understanding of eternity is possible only because God is Trinity. To paraphrase Pannenberg, the divine eternity as a differentiated unity is the eternity of the differentiated unity of the Trinitarian God.

Let me spell this out in more detail. For Pannenberg, the relation of time and eternity involves the following claim: the *distinction* between events in time will be sustained in eternity while the *separation* between events in time will be overcome in eternity. By the "distinction" between events, I mean the unique character of every event as present, a unique character reflected in its unique ppf structure. This ppf structure will be preserved in the unity of the divine eternity. This part of the claim argues against eternity being a mere conflation of all events into a single, timeless, unstructured present. By the "separation" of every event as present from all other events as present, I mean the fact that each event in creaturely time can only be experienced as present once, and, for each event, its future events and its past events are never available as present. But in eternity, this separation is overcome. All events are equally available to be re-experienced, forgiven, and savored endlessly. This part of the claim argues against eternity being merely an endless sequence of time as we now know it. Finally, this relation between time as we know it and the divine eternity is true not only in the "now" of our creaturely lives, although experienced only partially and by way of anticipation, but even more so as a relation between this lived "now" and the endless life that we will have eschatologically in the New Creation, when we shall forever experience God's eternity immediately. As Pannenberg writes, the distinction between God and creation will remain in the New Creation, but the distinction between the holy and the secular will be overcome.

Prolepsis

I also argue that Pannenberg's view of the relation between the eschatological future and the present includes two distinct concepts. I refer to the first as the "immediate causality of the future": in every present moment in life and in nature the factors predisposing, but not predetermining, the character of that present moment include not only the immediate efficient causality of the past, as represented by forces and interactions in physics. These factors also include what I call the immediate causality of the future in which the future is manifested (or, as Pannenberg phrases it, "appears") in the present as an additional contributing factor in making the present concretely what it is. The second concept is prolepsis: a strikingly topological view of the relation between creation and the New Creation in which the eschatological future "reaches back" and is revealed in the event of the resurrection of Jesus. This reaching back is not within the topology, or spatial structure, of the universe as we know it, that is, as the creation with its past and future as described by special relativity, general relativity, big bang cosmology, etc. Instead, it is a more extensive topology, one that connects the universe as creation with the New Creation where the New Creation is thought of as emerging through God's radically new action starting at Easter and continuing until its consummation in the global eschatological future of creation. On the one hand, both creation and New Creation are part of a single divine act of creation *ex nihilo,* including this topological connection. Yet, on the other hand, this connection is not so much a part of the present creation as it is a proleptic act originating in the New Creation and reaching back into our world and its history.

2. Summary of the Chapters

The following is a fairly lengthy summary of the six chapters of this volume. It is meant to orient readers to the overall flow of the material, to touch on and preview many of its details, and to give readers an overall sense of the argument of the volume. However, since much of this summary is repeated, often verbatim, in the respective chapters, readers may prefer to skip the summary offered here and go directly to the chapters.

Part One: SRP → TRP (Scientific Research Program → Theological Research Program)

In chapter 1, I lay out as groundwork for this entire volume Pannenberg's Trinitarian conception of eternity and omnipresence. In section A,

I describe Pannenberg's Trinitarian conception of eternity in relation to the time of creation. Pannenberg views God's eternity both as the source of creaturely time and as overcoming the effect of the present moment in separating and dividing past from future. He discusses the role of duration as forming what Augustine called a psychological "time-bridging" present, now extended by Pannenberg (as I read him) to include duration in nature. I also describe Pannenberg's understanding of the relation between time in creation and time in the eschatological New Creation. In section B, I look at Pannenberg's Trinitarian conception of omnipresence in relation to the space of creation, including the impact of Newtonian science and the shifting understanding of space and time in special and general relativity on this relation. In sections C–E, I look at Pannenberg's additional insights about the relations between space, time, omnipresence, and eternity, the way in which God as Trinity is crucial to these relations, and the role Pannenberg gives to Hegel's concept of infinity in his discussion of the divine attributes. Finally, in section F, I turn again to eschatology and focus on Pannenberg's two concepts of the relation between the eschatological future and the present: the causal priority of the immediate future and the prolepsis of the eschatological future in history at the first Easter.

In chapter 2, I explore a variety of ideas drawn from mathematics, physics, and cosmology in an effort to illuminate two of Pannenberg's key concepts, co-presence and prolepsis, within the theme of time and eternity. The first involves the relation between time as we currently know it and the co-present duration of the divine eternity. The second involves the relation between creaturely time and the eternity of the eschatological New Creation as anticipated proleptically in the resurrection of Jesus. I will focus on co-presence in section B and on prolepsis in section C.

As described above, I coined the term co-presence to characterize Pannenberg's unique understanding of the structure of temporal duration in the divine eternity. To reiterate, we begin with our experience of time in daily life. Here each present moment has a unique past and future which I refer to as the "ppf structure" of time. In our ordinary experience of time, the present constantly changes as what was a set of possible and indeterminate future events becomes a unique, determinate present event, and then immediately becomes an event in the ever-receding and irretrievable past. Such a view of time is often called "flowing time" by physicists and philosophers. Co-presence is meant to signify Pannenberg's concept of the special kind of unity that temporal events in life and nature are given as they are taken up, even now, into the eternity of God. According to Pannenberg, in the divine eternity all events are co-present: each event as a present moment retains its distinct past and future even while all such present events are held together in the differentiated unity of eternity.

To characterize his unique contribution to the second issue I have adopted Pannenberg's key term, prolepsis. Here prolepsis signifies the resurrection of Jesus as the appearance and culmination in history of the eschatological future and its ultimate consummation in the reign of Jesus Christ in the New Creation. It is through both prolepsis and the immediate causality of the future that the eschatological future appears and is active in the present Creation.

But before proceeding to explore these key themes—co-presence and prolepsis—we must recognize that the reality of flowing time is widely, though not universally, disputed by scientists and Anglo-American analytic philosophers even while it is taken for granted in the rich temporal ontology of Continental metaphysics such as Pannenberg deploys. These scholars typically opt for a timeless, static view of nature and a tenseless view of language, claiming they can reduce flowing time entirely to human subjectivity, a view often referred to as the "block universe." Where is Pannenberg in all this? Obviously Pannenberg holds to, even simply presupposes, the reality of flowing time, as do essentially all theologians. But it is not entirely clear to me whether Pannenberg believes that flowing time is merely a subjective phenomenon with no basis in the physical world or whether he believes that flowing time is grounded in nature. Nevertheless I will make the assumption in this work that Pannenberg holds that the physical world is dynamic and characterized by flowing time.

Hence my first task in chapter 2, section A is to put what I take to be Pannenberg's views on "flowing time" in dialogue with the spectrum of views held by Anglo-American philosophers, while recognizing that his views may not fit all that well within this spectrum. I will start with a very brief background to the philosophical debate over the dynamic versus static nature of time. Then, following Pannenberg's commitment to relationality in the Trinitarian being of God, I develop what I take to be, in some ways at least, a new approach to flowing time. The key will be to treat past and future not as properties of events but as relations between events and a given present moment. This will lead me to propose what I will call a relational and inhomogeneous temporal ontology. I also suggest that flowing time entails an underlying "fractal-like" temporal ontology that leads, in turn, to new insights into the extraordinary richness of temporality in nature.

It is important to stress at the outset, however, that one of the main arguments for a timeless view of nature, or the "block universe" view, is based on Einstein's special theory of relativity (SR). Since I wish to reconstruct Pannenberg's discussion of eternity and omnipresence in light of SR (chapter 4) I must at some point respond to relativity's challenge to flowing time. I will do so in chapter 5 where I will suggest that Pannenberg's theology, once reconstructed in light of special relativity, can in fact lead to a new argument for the *philosophical* defense of flowing time in light of relativity. This argument, in turn, will be based on the relational and inhomogeneous temporal ontology articulated in chapter 2. Thus the reconstructive work for theology in light of science is placed in part one of this volume, while the implications of this reconstructed theology for philosophy is placed in part two, following the method of CMI.

Next I offer an extended discussion of Pannenberg's concept of eternal co-presence (section B) and of prolepsis (section C) in light of my approach to flowing time. In chapter 2, section B I lay out an analogy between libraries that have "closed" versus "open" stacks, on the one hand, and creaturely time versus eternal co-presence on the other hand. I then turn briefly to the mathematics of non-Hausdorff manifolds in theoretical physics and cosmology because they can offer a second analogy for the theological concept that time in eternity retains its unique past, present, and future (ppf) structure without events being separable into isolated present moments. I suggest ways in which the phenomenon of entanglement in quantum mechanics, where spatially separate objects still remain fundamentally related, can offer a third analogy from physics for co-presence in eternity. In chapter 2, section C I begin with the idea of the eschatological transformation of the creation into the New Creation based by analogy on the bodily resurrection of Jesus. I then offer a rudimentary diagrammatic approach to illustrate this concept of transformation and the importance of prolepsis to it. Finally, I explore a topological approach to eschatology through such ideas as singularities in spacetime and multiple connections in some forms of inflationary cosmology. I close by thinking imagistically about multiple prolepses between the death of individuals and the eschaton based on the prolepsis of Jesus Christ.

In chapter 3, I explore ways in which the concept of infinity in modern mathematics can offer important insights for reformulating Pannenberg's doctrine of God, particularly in regard to his use of Hegel's concept of infinity in discussing the divine attributes. My specific focus will be on Georg Cantor's groundbreaking mathematical work on infinity in relation to Pannenberg's use of infinity drawn from Hegel. The choice to focus on Cantor's mathematics arises because of Cantor's breakthrough from our traditional way of thinking about infinity in which the infinite is understood in sharp contrast to the finite. This traditional way led theology to follow the *via negativa:* a way of coming to know God through a complete separation between the finite world and God its creator. This way of thinking about infinity dates back to the ancient Greeks, where infinity is defined as the *apeiron*—the unbounded, the unlimited, the formless. Recent developments in mathematics starting in the nineteenth century, though foreshadowed by Galileo Galilei's discoveries in the seventeenth century, have shed provocative new light on infinity. Cantor in particular has given us a new mathematical conception of infinity in which there are an endless variety of infinities, which he called the "transfinites," lying beyond the finite and yet beyond them there is an unreachable *Absolute Infinity*. These revolutionary discoveries in mathematics can lead us to exciting new insights into the concept of infinity in Pannenberg's explication of the divine attributes.

We begin chapter 3, section A with a historical note on the concept of infinity in Greek thought before turning to the modern understanding of infinity in mathematics. Next, I discuss Cantor's fundamental developments in finite set theory, including his concept of a set, the characterization of sets as countable and uncountable, the cardinal and ordinal numbers of a set, and so on. Cantor's breakthrough was to show how to apply these ideas by analogy to an infinite set. He claimed that infinite sets, such as the set of natural numbers, could be thought of not just as potentially infinite but as *actually* infinite or "transfinite." Cantor also showed that there can be transfinite sets that are "bigger" than the set of natural numbers, leading to an unending series of transfinites. This in turn led him to propose what he called Absolute Infinity as that which lies forever beyond the realm of the transfinites in the mind of God. For Cantor, Absolute Infinity is inconceivable, and yet in a subtle way it is conceivable. He showed this by using what is called a "reflection principle" to claim that the properties of Absolute Infinity can be known in that they are shared with those of the transfinites, and yet this sharing leaves Absolute Infinity indistinguishable from the transfinites, and thus unknowable in itself.

But Cantor's set theory generates a series of antinomies: two assertions both apparently true but which are mutually contradictory. Cantor's antinomies are actually indicative of a larger set of crises in the foundations of mathematics, related to the failures of logicism, intuitionism, and formalism and their associated philosophies—realism, constructivism, and nominalism. After briefly discussing these issues I suggest that Cantor's work is still applicable theologically if we are seeking to use it as a conceptual tool to enhance the way Pannenberg explicates the role of infinity in the doctrine of the divine attributes.

I then return in more detail to Cantor's conception of Absolute Infinity and his theological motivation for this concept in order to explore its potential fruitfulness for theology. This will include a brief historical survey of the theological objections to Cantor's work in his own time, where the context was Pope Leo XIII's support of neo-Thomistic philosophy and his 1879 encyclical, *Aeterni Patris*. It was in response to these objections that Cantor proposed his idea of "Absolute Infinity" as lying forever beyond the realm of the transfinites within the mind of God. In addition he distinguished between the eternal, uncreated, Absolute Infinite related to God and the created, transfinite infinity which can be found in nature.

In chapter 3, section B I first summarize Pannenberg's use of Hegel's understanding of "true Infinity" in his discussion of the doctrine of God, where for Pannenberg it serves as the underlying structure of the divine attributes, eternity and omnipresence. I then explore the relation between Cantor's conception of the transfinites and Hegel's concept of infinity. I suggest that Cantor's concept of the transfinites and Absolute Infinity and his use of the reflection principle offer new resources for Pannenberg's conception of the infinite in his theology. For example, Cantor allows us to notice both differences and similarities between the finites and the transfinites, leading me to claim that the finite and the transfinite, while primarily distinguishable, are still in an important sense indistinguishable. Cantor's use of the reflection principle leads to new insights about the incomprehensibility of God who, when thought of metaphorically in terms of Absolute Infinity, is revealed and yet hidden through the properties it shares with the transfinites. I then suggest that the transfinites can be understood as forming a "veil that discloses God," making God conceivable even while it hides God, leaving God as ultimately inconceivable.

In the final portion of chapter 3 I explore Pannenberg's work on time and eternity and the eschatological transformation of the world into the New Creation in light of Cantor's mathematics. Pannenberg's understanding of God as entering immanently into the world to transform it eschatologically while remaining transcendent to it requires that we make two assumptions: 1) that the world as created by God is both finite and transfinite, and 2) that the transfinites stand in relation to Absolute Infinity as depicted by the reflection principle. My point will then be that a "fully finite" world, as creation is traditionally conceived, could not, in principle, be open to God's holiness and divine Spirit the way Pannenberg proposes without leading either to pantheism and the divinizing of the world or to atheism and the secularizing of God. But if the world that God did, in fact, create is both finite and transfinite, it is a world which can be, from the beginning, infused with God's holiness and life-giving Spirit even while remaining creaturely. Finally I suggest that in the New Creation the transfinite character of creation will take on increasing significance compared with the present context in which the world is primarily seen as finite. In sum, Cantor's mathematical language about infinity gives us a way to express something of what Pannenberg tells us about eschatology.

In chapter 4 I turn to Albert Einstein's theory of special relativity (SR) and the task of reformulating Pannenberg's treatment of eternity and omnipresence in light of the "spacetime" interpretation of SR. According to this interpretation given by the mathematician Hermann Minkowski just two years after Einstein published SR, space and time cease to be entirely separate dimensions of the world as they are in ordinary experience and in classical physics. Instead they form a single, four-dimensional geometry called "spacetime." My goal is to offer a reformulation of the theological relation between eternity and omnipresence based on the spacetime interpretation of SR. Once this is in place, we will explore several new theological insights into the interweaving of eternity and omnipresence in relation to spacetime. We will also explore the way a reformulation of divine omnipresence offers a response to the philosophical problem regarding a way to conceive of the unity of the world.

First, however, I stress again that SR poses a striking challenge to theology. This is because the spacetime interpretation of SR is itself subject to an interpretation called the "block universe." According to this interpretation of spacetime, all events in the world—past, present, and future—are equally present and real. There is no objective distinction between what we call past and future. Instead all events in life, history, and the universe are just "there" in the frozen geometry of spacetime, and the flow of time that is so deeply given to our personal experience is an illusion. Such an interpretation of spacetime obviously undercuts the kind of "flowing time" interpretation of time which I will defend in chapter 2 as applying not only to our personal experience but to nature itself. And flowing time is, arguably, essential to most forms of religious experience and their theological systematization. In chapter 2 we will see how this would be true especially for Pannenberg, even as he adds subtle layers of nuance in his interpretation of the relations between events in time and their being taken up into the divine eternity. Hence the challenge of the block universe interpretation of the spacetime interpretation of SR must be met if we are to incorporate the spacetime interpretation, and with it a flowing time interpretation of the spacetime interpretation, into our discussion of the divine attributes. In chapter 4 I will reformulate Pannenberg's understanding of eternity and omnipresence in light of SR, as reflecting the goal of part one of the volume. I will ask the reader to suspend judgment temporarily about the possibility of addressing the challenge raised by the block universe interpretation of SR until chapter 5. There, as appropriate to its being located in part two, I will reverse the direction of the argument and suggest how a reformulated interpretation of the relation between eternity and omnipresence can lead to new insights in the philosophy of time for a "flowing time" interpretation of SR.

I start chapter 4 with an overview of SR (section A). This overview includes Einstein's two postulates, spacetime diagrams, time dilation, the "downfall of the present," the Lorentz transformations and their consequences, and the famous SR "paradoxes," which arise out of the Lorentz transformations. I focus on one paradox in particular, the so-called pole-in-the-barn paradox. First, this paradox embodies many of the results already discussed and it leads to a crucial shift in perspective: we move from viewing SR in terms of separate spacetime diagrams, one for each observer in relative motion (e.g., one respectively for the pole's point of view and the barn's point of view), to viewing SR in terms of a single, "generalized" spacetime diagram that incorporates both points of view seamlessly and transparently. This in turn leads to a profound insight into the non-paradoxical view of nature offered by SR that we can incorporate into theology. Because of its fruitfulness, the pole-in-the-barn paradox will play a crucial role in reformulating both eternity and omnipresence in light of the spacetime interpretation of SR. In chapter 5, this paradox will be pivotal in deploying a flowing time interpretation of spacetime against its competitor interpretation, the block universe. Finally, through the generalized spacetime diagram for the pole-in-the-barn paradox, we will discover a compelling view of the interwoven temporal character of what we take to be our individual narratives of the physical processes in nature.

I have indicated already that I will adopt the spacetime interpretation of SR, while foregoing its usual interpretation, the block universe, and construct a new flowing time interpretation of spacetime in light of Pannenberg's reformulated theology (chapter 5). But my introduction to SR would not be complete without a clear presentation of the reasons why the spacetime interpretation is now almost universally accepted. I present reasons for this before closing chapter 4, section A by exploring how the classical global present is an unneeded anthropocentric illusion. The technical material in that section may be off-putting to readers without a background in physics. Thus, in this section I include short summaries of each topic in SR for those wishing to skip the details. I also cite several online resources that are particularly helpful in explaining SR in non-technical language.

In chapter 4, section B I reflect on Pannenberg's comments on special and general relativity before turning directly to the theological task at hand: a reformulation of Pannenberg's discussion of the divine attributes in light of SR. In essence the spacetime interpretation of SR entails that a strict separation of the temporal and spatial conceptualities underlying eternity is no longer possible—at least in principle. How then are we to treat the divine attributes? My response in section C is to reformulate the discussion of the divine attributes modeled by what I call a "covariant correlation of eternity and omnipresence." In section C.1 I give the basic argument. In C.2 I return to the pole-in-the-barn paradox to suggest how the complex interweaving of the worldviews of observers in relative motion enlarges our understanding of the divine eternity and omnipresence as endlessly interwoven together in their relation to creation. In C.3 I explore the astonishing complexity of the elsewhen region of spacetime associated with every spacetime event. This, in turn, has implications for God's particular omnipresence to each observer with his or her unique view of the world. Finally in C.4, I claim that it is God's particular omnipresence to events in space that gives to these distinct and separate events a differentiated spatial unity, and I relate this to Pannenberg's discussion of the debates over space and omnipresence in the writings of Newton, Leibniz, and Clarke. The chapter closes with section C, where, as stated above, I reformulate the discussion of the two divine attributes—eternity and omnipresence—in light of the spacetime interpretation of SR and modeled by a covariant correlation of eternity and omnipresence. This reformulation is crucial to the goal of part one.

Part Two: TRP → SRP (Theological Research Program → Scientific Research Program)

Chapter 5 reflects the goal of part two: to demonstrate that theology, when reformulated in light of mathematics and science, can offer fruitful directions for research in nontheological disciplines—here the problem of time in the philosophy of science. In this chapter I address the challenge that SR poses to theology, since the spacetime interpretation of SR is itself subject to an interpretation called the block universe, where all events in life, history, and the universe are just "there" in the frozen geometry of spacetime and the flow of time that is so deeply given to our personal experience is merely an illusion. I thus lay out and defend a new flowing time interpretation of special relativity based on my interpretation of Pannenberg's concept of eternity, namely co-presence (chapter 2), and on my reformulation of Pannenberg's understanding of eternity and omnipresence, which I refer to as the covariant correlation of eternity and omnipresence (chapter 4).

In chapter 5, section A I present a brief summary of the debate over the best interpretation of SR: Should it be taken as supporting a philosophy of being (the block universe) or a philosophy of becoming (flowing time)? We have already discussed the more general form of these arguments in chapter 2. Here we focus on the specific form they take in the context of SR. I begin with a standard, and compelling, example of the argument for the block universe by Edwin F. Taylor and John Archibald Wheeler. Then, to illustrate the debate between the block universe and flowing time interpretations of SR, I have chosen early essays by Olivier Costa de Beauregard and Milič Čapek, and a more recent joint essay by Chris Isham and John Polkinghorne. I close this section with an unusual flowing time interpretation of SR frequently referred to as neo-Lorentzian and supported by William Lane Craig and other scholars. Unlike the conventional flowing time interpretation, the neo-Lorentzian framework offers a unique global present but at the cost of treating time dilation and Lorentz contraction as real physical effects and not just aspects of the geometrical concept of spacetime.

In chapter 5, section B I lay out, as promised, a new flowing time interpretation of SR drawing on my theological reconstruction of Pannenberg's work on eternity and omnipresence found in chapter 4. The key move starts with the assumption that a relational and inhomogeneous ontology, such as I develop in chapter 2, is coherent with and supports Pannenberg's theological concepts of eternity and omnipresence when they are reformulated in light of SR. I then propose that we reverse the move and explore whether such an ontology, because of its importance to the task of reformulating Pannenberg's theology, might be preferable, compared to its competitors, when we return to the context of SR. If so, then in this way Pannenberg's theology can be seen to provide fruitful implications for research in the philosophy of physical time, specifically the philosophical interpretation of SR. To accomplish this task I apply the relational and inhomogeneous ontology (chapter 2) to the spacetime interpretation of SR. I then assess the attempt to widen the search for a physical global present by turning from SR to general relativity and big bang cosmology. My conclusion is that the problems raised by SR for a physical global present remain even in this wider context and that they must therefore be addressed directly. Finally, in section C, I spell out what I consider to be the real lessons of SR: that relativity does not lead to epistemic relativism and that the "simultaneity richness" of the elsewhen compared to the "austere paucity" of the classical view of a unique global present leads to a more satisfying understanding of nature, one to be celebrated rather than explained away.

The style of chapter 6 is highly schematic and unapologetically speculative since we are entering truly unexplored territory: the search for research programs in physics and cosmology that in some way reflect the directions one might take if starting from a reformulated theology such as we have explored here based on Pannenberg's work. I try to strike a balance between an exposition of individual scientific topics that is sufficiently general and detailed to make it readable for the nonscientist and a succinct itemization of a diversity of scientific topics that, to the practicing scientist, will convey something of the vastness of the landscape for such research. Let me acknowledge at the outset that some of the scientific research described in this chapter awaits further study and eventual confirmation or disconfirmation by the scientific community. Still, the fact that it is here serves as a "proof of concept" that theology can offer creative suggestions for new research directions in science and additional criteria of theory choice between competing scientific research programs, all the while respecting and endorsing the methodological naturalism that underlines and shapes the natural sciences. For a discussion of the guidelines for this research, see appendix to the introduction, section E.

In this discussion I also lay out what constitutes the overarching conceptual structure of chapter 6: that "resurrection as transformation" means that some of the preconditions for the possibility of the New Creation that God is bringing about starting at Easter and ending in the eschatological future must already be present. Such preconditions constitute one element of continuity between the world as we know it now through the natural sciences and the New Creation into which it is both already being transformed (i.e., "realized eschatology") and into which it will be radically transformed (i.e., "apocalyptic eschatology"). Of course there are considerably more elements of discontinuity between "now and then" than elements of continuity, but the point here is that there must be some form of continuity. Finally, I will make a key move and presuppose that some elements of continuity include those themes that I have drawn and interpreted from Pannenberg and reconstructed in light of science and mathematics: duration, co-presence, and prolepsis.

In this chapter I look at nature as we know it through the lens of mathematics, physics, and cosmology, searching for those features of nature which might be suggestive of duration, co-presence, and prolepsis. After an overview of the chapter (section A), I turn to duration in general in section B, drawing on the metaphysics of Whitehead since it embodies a concept of duration akin to what is found in Pannenberg's writings. I then turn to the ongoing scientific research programs initiated by David Bohm and by Ilya Prigogine since they represent views of duration that in some ways represent elements of Whitehead's philosophy. I also survey current research in string theory which might be interpreted as pointing to a rudimentary form of duration in nature.

In chapter 6, section C I turn to co-presence as the particular structure of duration found in Pannenberg's conception of eternity, and I investigate its possible implications for research physics. Here I explore the idea of co-presence as non-separability in time by way of analogy to the idea of non-separability in space. I then identify a number of current scientific research programs which deal with spatial non-separability both in quantum mechanics and in non-Hausdorff spacetimes. In section D, I explore Pannenberg's concept of prolepsis, including both the causal efficacy of the future on the present and the topological "reaching back" of the eschaton from the New Creation to Easter. Hints of what might be the preconditions for prolepsis as the causality of the future can be found in Fred Hoyle and J. V. Narlikar's work on steady-state cosmology. I describe the roots of their research in the "time symmetric" formulation of electromagnetism as developed by John Wheeler and Richard Feynman. I then point to aspects of their work which continue to appear in the cosmological research of Stephen Hawking and G. F. R. Ellis and in the time symmetric approach to quantum mechanics by Yakir Aharonov, Jeff Tollaksen, Paul Davies, and Brian Greene. Hints of prolepsis as a topological "reaching back" might be found in the areas in physics where non-Hausdorff manifolds apply. I explore its implications as a response to issues such as natural theodicy and I discuss the difference in the purpose of using non-Hausdorff manifolds here and in chapter 2. Finally, in chapter 6, section E I acknowledge the problem raised by the apparent hiddenness of eternal time in ordinary experience: Why does time in nature seem "linear" and its present moments separated into the unavailable future and the unretrievable past? I offer two responses to this problem of hiddenness, one that draws on Luther's "theology of the cross" and the other that utilizes John Hick's argument about "epistemic distance."

D. CONCLUSION

As I note at the outset of this introduction the theological topic of "time and eternity" is central to the relation of God and the world in two contexts. In the first context, eternity is considered in relation to time as we know it now. Here, the concept of eternity involves three competing notions: eternity as timeless, eternity as unending time, or, for many twentieth-century theologians, eternity as the supratemporal source of time. Eternity understood this way includes the concept of duration, and for Pannenberg in particular, duration is internally structured in terms of what I call co-presence. In the second context, the eschatology of New Creation, the theological topic of "time and eternity" is crucial when eschatology is based analogously on the bodily resurrection of Jesus and involves the idea of prolepsis by which the eschatological New Creation is already realized in the Easter miracle.

In my view the topic of "time and eternity" in both of these contexts is most clearly and persuasively articulated in contemporary Christian theology by Wolfhart Pannenberg. The motivation for this book is to address this topic and explore its richness for Christian theology in light of the massive contributions by Pannenberg. The task is twofold: first we must reformulate his theology in light of the natural sciences; then we can explore new insights coming from his thought to access their fruitfulness for research in the natural sciences and the contemporary philosophy of time.

In the process two challenges must be faced, as we saw above. The first is that timeless philosophies of nature based, in particular, on the physics of special relativity threaten to undermine the very cogency of any theological treatment of "time and eternity" with its implicit assumption of "flowing time." The second is that scientific cosmology, from the big bang to quantum cosmology, severely challenges the sheer intelligibility of eschatology when its scope expands endlessly to embrace the universe as a whole as God's creation and therefore as the subject of God's eschatological transformation into the New Creation. Thus in addressing "time and eternity" we must address these two fundamental challenges. My approach here is to seek to turn these challenges into creative opportunities for constructive work both in theology and in the philosophy of science and in physics and cosmology. I do so in particular in chapters 2, 4, 5, and 6.

But there are additional reasons that motivate and challenge this volume. The material in the appendix spells out these motivations and challenges in detail. There I include brief notes on three issues involving creation, cosmology, and evolution that take us directly to the challenge raised to Christian eschatology by scientific cosmology and, within this challenge, the core problem of "time and eternity." Second, because my treatment of Christian eschatology depends crucially on the bodily interpretation of the resurrection of Jesus, I give some details on the New Testament debates over this interpretation (see appendix to the introduction, section B). The scholars surveyed make a convincing case for the bodily Resurrection and an eschatology of cosmic transformation. Nevertheless they by and large ignore the challenges from science. To make clear what the challenges are I give a brief overview of big bang, inflation, and quantum cosmology. As mentioned above, these areas pose a severe challenge not only to Christian eschatology but also to the theology-science dialogue in general. I view this challenge as constituting a crucial "test case" for the feasibility of the dialogue as a whole. In the appendix I also include a survey of a variety of eschatological responses to cosmology, most of which either wall it off by a focus on other concerns or acknowledge it but offer little substantive response to it (appendix to the introduction, section C).

In response to the challenges from special relativity and from cosmology, as well as the additional challenge mentioned in section A above, I have relied in this volume on the method of Creative Mutual Interaction and I have structured the volume accordingly. The result, I hope, will be new insights for reformulating Pannenberg's theology in light of contemporary science and mathematics and, in turn, new suggestions for research in both the philosophy of science and in cutting edge topics in contemporary physics. And this result, in turn, bears directly on what I quoted from Pannenberg in the epigraph: "Without an answer to the question regarding time and eternity, the relation of God to this world remains inconceivable."[10] I hope to have contributed at least part of an answer to this question in this volume. If so, I attribute whatever success might be found here to the radiant theological vision of Wolfhart Pannenberg, to whom this book is dedicated.

Appendix to the Introduction

Background Material

This appendix presents a detailed discussion of the topics in theology and science involving creation, cosmology, evolution, Christology, and eschatology that serve as additional sources of the motivation for, and challenges to, this volume. Readers familiar with this material might skip to chapter 1. Readers less familiar with it might want to look through it briefly and read portions of interest to them in some depth, returning to others later as they work through the chapters that follow, or simply read through it in its entirety before beginning the volume.

The background material is divided into six sections. In section A, I set down brief notes on the key three theological issues mentioned in the introduction, issues involving the sciences of cosmology and evolution: creation *ex nihilo;* theistic evolution; and natural evil. We will see that each one takes us directly to the centerpiece: Christian eschatology and the challenge of cosmology. We will also see that the sequence from the first to the second and to the third also takes us inevitably to the centerpiece. The discussion of Christian eschatology and the challenge of cosmology thus becomes essential to progress in theology and science today. My treatment of Christian eschatology depends crucially on the "bodily" interpretation of the resurrection of Jesus; in section B I will provide an extended account of the current state of the New Testament debates over this interpretation. While scholars who interpret the Resurrection as "bodily" make a convincing case against their competitors, they tend to avoid or ignore the challenges from science. Because I depend on their work in shaping my interpretation of Christian eschatology and because I do so in the explicit context of theology and science, I must find a way forward, and once again the "method of creative mutual interaction" (CMI) will, I hope, prove fruitful.

The New Testament research that supports the bodily resurrection of Jesus reaches out necessarily to include eschatology and, by implication in my view, scientific cosmology. Therefore, in section C I assess current resources in eschatology for engaging physics and cosmology. To make clear what the challenge is from physical cosmology, I give a brief overview of big bang, inflation, and quantum cosmologies. Such topics are especially relevant here because of the severe challenge they bring to the theology-science dialogue. In this sense, physical cosmology constitutes a crucial "test case" for the feasibility of this dialogue as a whole. I include a survey of a variety of responses to cosmology, most of which either wall it off by a focus on other concerns or, when acknowledged, offer little substantive response to it.

In section D, I provide an introduction to Creative Mutual Interaction, or CMI, in theology and science. The requirement that cosmology be taken seriously by scholars working in eschatology is particularly acute for those laboring in theology and science. Surprisingly, few of my colleagues here have given sustained attention to this task. This is particularly disappointing, given that the field has been built on a very specific methodology for taking the theories in science with the utmost seriousness while still supporting the claim that theological concepts cannot be reduced to those of science. To move us forward I have developed the method of CMI between theology and science. Such a method draws on the concept of epistemic emergence and on the claim for an analogy between methods of reasoning in science and in theology. CMI lays out the ways in which theology is not only open to influence by science but the ways in which theology, in turn, can offer creative suggestions for scientific research. It is my hope that this methodology will move the conversation between eschatology and cosmology beyond its current impasse to more fruitful grounds.

In section E, I offer a series of guidelines that operate at a metalevel for determining which new directions might be worthy of further research and which ones should be set aside. Throughout this volume, I apply these guidelines as criteria for the assessment of a variety of possible research options, including suggestions for new research in science and the phi-losophy of science. Finally in section F I provide a special note on Wolf-hart Pannenberg's 1960s discussion of the New Testament debates that, while dated, still has an enormous influence on my construction of these guidelines, especially the first one. Pannenberg argues that the historicity of the resurrection of Jesus is something which secular scholars cannot rule out without succumbing to what I would call metaphysical reduc-tionism. On the other hand Pannenberg is more willing than I am to chal-lenge secular historians to be open to the possibility of Resurrection.[1] I believe that the resurrection of Jesus is a theological category that cannot be ruled out by secular history (à la Pannenberg), but it is also one that cannot be placed within secular history (and here I believe I differ from Pannenberg). Instead the Resurrection as historical is a theological claim that embraces the historical elements entailed, such as the empty tomb and the experiences of the disciples of a postmortem Jesus. In this vol-ume I go on to explore the additional elements entailed for nature in terms of duration, co-presence, and prolepsis.

A. THREE ISSUES INVOLVING CREATION, COSMOLOGY, AND EVOLUTION

1. Creation *Ex Nihilo* and Contemporary Cosmology: Consonance and Dissonance around t = 0 and the Anthropic Principle

Many Christian scholars start with the basic contents of creation *ex ni-hilo:* the Triune God creates the universe as a whole and all its parts for all time, giving the universe contingent rationality through the divine Logos in the power of God's Spirit. But the tradition has often affirmed that this universe has an absolute beginning including the beginning of time. The discovery of the initial singularity, or beginning of time, "t = 0" in big bang cosmology circa the 1950s and 1960s naturally inspired some Christian theologians to see the singularity as directly relevant to the doc-trine of creation *ex nihilo.* Others, though, found it irrelevant. Some, in-cluding myself, viewed it as *indirectly* relevant with the philosophical category of contingency, playing a mediating role between a finite past, marked by t = 0, as an empirical form of contingency and creation *ex nihilo* as a theological form of contingency.[2] The temporal beginning of the universe in big bang cosmology was thus considered "consonant" with creation theology, to use the term coined by Ernan McMullin.[3] But the same cosmology offered two prospects for the cosmic future: endless expansion and exponentially decreasing temperature ("freeze") or inevitable recontraction and exponentially increasing temperature ("fry"). And both prospects are clearly "dissonant" (as I termed them[4]) with a Christian eschatology that takes the fate of the universe on board—and as stressed by the discussion of the challenges of cosmology to eschatology in this volume.

By the 1970s and 1980s, all of this began to change rapidly with the onset of inflationary big bang cosmology, a variety of approaches to quantum cosmology and, recently, superstring theory and the multiverse. The empirical and theoretical justification for claiming that our universe has an absolute beginning at t = 0 is challenged by these more recent cosmologies. At the same time, the discovery of the "fine-tuning" of the physical laws and constants of nature has led to a new round of cosmological design arguments over a century since Darwin had effectively silenced Paley's claims—and now in the context of physical cosmology! Why should the actual physical laws have the exact form, and these constants the precise value to within six orders of magnitude, that are required for the possibility for life to evolve in our universe? Is the fine-tuning evidence of God, as the anthropic principle suggests? Or is it the result of our universe being only one of a multitude of universes, either as different domains of a single inflationary big bang universe, as part of a vast sea of interconnected universes generated by "eternal inflation," or as part of a practically limitless multiverse as string-cosmology suggests?

Some scholars support the design interpretation; others opt for one or another many-universe explanation. I have argued instead that there are elements of design and many worlds (or necessity and contingency) in both of these positions. I have also suggested that the fine-tuning of *our* universe, which is empirically undeniable, underscores the claim that biology and physics are much more closely interrelated than we might have guessed. This interrelatedness in turn leads to consonance between the phenomena of life in our universe, and with it on earth at least sentient, self-conscious creatures capable of culture, religion, ethics, art, and science, and the sheer physical structures of our universe. In this sense at least we are truly "at home" in our universe and not an irrelevant and meaningless happenstance in a dysteleological universe, a vast domain of unfeeling materiality as writers from Monod to Weinberg proclaim.[5]

Moreover, regardless of the radical changes in physical cosmology, the sheer *fact* of the existence of the "universe"—whatever science currently means by this concept—will always be grounds for the cosmological argument. Meanwhile theological discussions over t = 0 and the anthropic principle can be allowed to recede as having been useful and illuminating, for they mark a high point in the constructive conversations between theology and science. What remains is a deep sense of the grounding of life in the very specific physics of our universe.

But what about the destiny of our universe: Is it going to be "freeze or fry" as scientific cosmology predicts (and, most likely, "freeze," given the recent discoveries about the acceleration of the universe's rate of expansion)? Or are there credible grounds for hope in an eschatological future? In this way, the issue of creation and cosmology drives us to the topic of this book.

2. Continuous Creation and Biological Evolution: Theistic Evolution

Since the 1980s, theological discussion of biological evolution has increasingly flourished, particularly as the centerpiece, the spectrum of views widely described as "theistic evolution," has matured and developed its fundamental theme:[6] the Triune God who creates the universe in its sheer being as a whole and in all its parts for its entire history and future also acts as the divine Spirit within nature, in, through, and under (to use Arthur Peacocke's enduring phrase) the processes of physics and biology, processes that issue into the evolution of life and the diversity of species, as *creatio continua* affirms. But what does "in, through, and under" actually mean? Granted, what nature does by its own efficient causality is also the result of God's activity in nature, what one can call "general providence." But does God's action make a difference in what happens in particular events in nature, events that would not have happened without God's "special providence" and that natural processes by themselves cannot achieve?

I have argued that if there are one or more levels of complexity in nature in which some events are not the result of a sufficient efficient natural cause, then we can think of God acting objectively to bring about such differences, yet without violating or suspending the processes of nature, that is, without "intervention." I call this "non-interventionist objective divine action" (NIODA). It requires ontological indeterminism at one or more levels of complexity in nature, an ontological openness to objective divine action (fig. A.1). The theological premise is that God created the universe *ex nihilo* with ontological indeterminism in one or more levels of complexity in order that God's action in nature will not disrupt the ordinary natural processes but instead bring them gently to their intended purposes. The philosophical question is whether any theory in the natural sciences can be interpreted as pointing to ontological indeterminism in nature.[7]

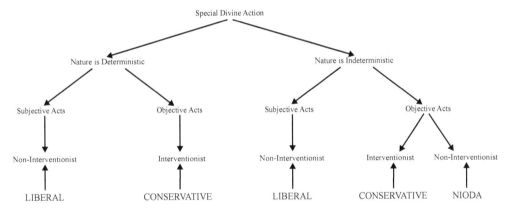

Figure A.1. Options for Divine Action Given Ontological Determinism or Indeterminism in Nature

Several scholars, including George Ellis, Nancey Murphy, Tom Tracy, and myself, argue in detail that quantum mechanics (QM) provides one such theory. Here the Heisenberg uncertainty principle can be interpreted philosophically as pointing to ontological indeterminism at the subatomic level in nature.[8] If valid, this is of particular importance to evolutionary biology because quantum mechanics is relevant to many of the molecular processes involved in the production of genetic mutations, and these in turn play some role (essential? minimal?) in driving the evolution of species. Thus a QM-based NIODA would deliver a robust version of the-istic evolution: God really makes a difference in the biological history of life on earth.[9]

What, then, is the relation between God's objective special provi-dence via NIODA and the eschatological transformation of the universe into the New Creation? Is NIODA-type divine action enough to bring about the eschaton? I have argued that it is not, that much more is in-volved in God's radical new action in raising Jesus from the dead, as I will suggest below. Nevertheless the possibility of a robust form of the-istic evolution through a NIODA-type approach to divine action does offer a starting point for considering the meaning of Christian escha-tology in light of physics, biology, and cosmology. We will return to this possibility below.

3. Creation, Natural Evil, and Natural Theodicy

With the success of the paradigm, theistic evolution, particularly in its robust form using QM-NIODA, the phenomena of "natural evil"[10] throughout the evolution of life raises a crucial challenge: what is God's relation to natural evil? The challenge issues in what is often called "natu-ral theodicy." Of course, one way to try to avoid the challenge—or at least deflect it—is to pull back on the robust form by arguing that God does not really act in nature beyond holding it in existence. In fact, God really had no choice: if God were to create life through natural means working themselves out on their own, then natural selection would gen-erate these evils.[11] But such a deistic view of God is a high price to pay for seeming to lessen the intensity of the "bite" of predation in nature.

I have explored two main approaches to theodicy, starting with tradi-tional and current responses to the problem of moral evil and extending them to the problem of natural evil.[12] The first draws on the Augustinian free will defense as retrieved by Reinhold Niebuhr. In the first volume of his Gifford Lectures[13] Niebuhr claims that the inner logic of the Augus-tinian defense is that sin is unnecessary (i.e., it is not in our nature to sin, the anti-Manichean argument that defends God as Creator from the fact of sin) and inevitable (i.e., no amount of spiritual exercise can, without God's grace, heal us from sinfulness, the anti-Pelagian argument that puts the lie to such current fads as the self-help movement). However, unlike Augustine, we have no recourse to a contingent historical event, the Fall, to ground the origins of human sinfulness. Sin is not a consequence of a singular event, even of cosmic proportions (such as the "Fall of the angels"). Instead we must face the slow development of various forms of natural evil as constitutive of the evolution of life on earth.[14]

This in turn led me to deploy a transcendental argument for the gradual development of the necessary preconditions for the possibility of human evil in the underlying biology and physics of nature. My claim was that without something like entropy in thermodynamics, including the increase in disorder in closed systems, the physical consequences of sinful behavior could not play themselves out. At the same time entropy in open systems contributes to the increase in complexity and beauty, making possible much of what is good and beautiful in human behavior. So the evolution of life on earth is an unfolding of a series of increasingly complex conditions for the possibility of what eventually become both human virtue and vice (and with prefigurings in the higher primates at least). This approach avoids the deistic view since God's activity of redemption can be seen as working in and through the evolutionary processes even as natural evil expands its complexity and domain. But it still leaves a fundamental question open: Why did God create *this* universe with these laws of physics and natural constants? Couldn't there have been another kind of universe in which life would evolve but without, or with radically less, natural evil? This ultimate formulation of the problem I call "cosmic theodicy" and I claim it is essentially unanswerable.[15] Scientifically we simply don't know enough about the kind of universes that would have arisen with different physical laws and constants to know whether they could be "anthropic." Philosophically we do not know how to "measure" the lesser of two evils or whether there would not always be a "worst evil."[16] Theologically it is a Leibnizian-type theodicy with all the attendant problems so well criticized by Karl Barth.[17] But it also offers an insight on eschatology: just as Augustine wrote that on earth we cannot not sin but in heaven we cannot sin, so we could suggest that on earth, we cannot not experience entropy, but in the New Creation we cannot experience entropy—at least in the ways it contributes to dissipation, destruction, and death but not in the ways it contributes to organization, complexification, and life.

The other response I have developed draws on John Hick's reformulation of the theodicy of Irenaeus and of Schleiermacher into what he calls the "soul-making" theodicy.[18] I find value in this approach for its explanation of why there cannot be a one-to-one correspondence between moral behavior and divine blessings: that is, why, instead, bad things happen to good people. But I do not endorse the kind of "means-end" reasoning that seems to underlie Hick's approach, particularly if it suggested—which he does not—that humanity was somehow justification for the history of suffering in the evolution of life. Instead, every form of life must be its own "end" and have its intrinsic value both as creation and as participant of the New Creation.[19] Moreover, like Hick, I recognize that the real challenge comes from unmitigated, overwhelming evil—"horrendous evil," to use Marilyn Adams' expression,[20] which defies any possible justification vis-à-vis spiritual growth. However, I part company with many of my colleagues who start with Hick and extend his argument by claiming that God withdraws divine power in order that nature have the freedom to be itself—what John Polkinghorne has called the "free process defense."[21] It has been developed further through a theology of kenosis that starts with Christ's Incarnation and his suffering with humanity on the cross and extends it to all of life on earth. I appreciate the intention here and find it much more helpful than the deistic response. However I view God as giving freedom to creatures by acting in and with them, not by withdrawing divine power from them, and I find it a stretch, to say the least, to use kenosis, whose roots are in Christology, to support a view of divine power as withdrawn from nature in order that nature be free. Instead while I believe the extension of Christ's suffering beyond humanity to all of life is correct I also believe we must relocate the problem of natural evil from the locus of creation theology and into the locus of redemption theology, the natural location for the Christian response to moral evil.[22] In short, the full goodness of creation does not lie in a lost Eden but in an eschatological future, as Ted Peters stresses.[23]

Redemption theology, however, requires we do not stop with the cross but move instead directly to the Easter event, the bodily resurrection of Jesus of Nazareth. Here is where the central claims of Christian faith lie, and here too the gravest challenges from science to that faith, particularly to eschatology. Once again this leads us to the topic of the present book, time *in* eternity. As part of my response I have also sought to address the challenge of theodicy by inverting it into a set of creative guidelines for assessing any viable eschatology (see guideline 7 below):

Following Irenaeus and Schleiermacher, any such eschatology must not be a "means-end" argument, but instead value each and every individual life in nature. Following Augustine's insight it must not involve a New Creation where the biological and physical conditions required for natural and moral evil are still at play; instead such features as thermodynamics must be "wiped away"—at least to the extent that they contribute to suffering in nature. We shall see that these criteria serve a very creative purpose in helping form guidelines for evaluating the promise of types of eschatologies yet to be developed.

B. NEW TESTAMENT ISSUES CONCERNING THE BODILY RESURRECTION

We now turn to the first of several substantive and detailed expositions of the themes that form the background and motivation for this present book: New Testament issues around the bodily resurrection of Jesus and the varieties of available eschatologies. I have published widely on the previous issues and thus have only summarized them here. The issues regarding the bodily resurrection of Jesus and Christian eschatology require a much more detailed explication to demonstrate their essential importance and motivation for the topic of this book, the relation of time to eternity in light of both a theology of Christian eschatology and the challenges from the natural sciences.

My intention here is not to enter in detail into the technical debate between New Testament (NT) scholars about the meaning of the resurrection narratives. Instead, I will adopt as a working hypothesis the view that the resurrection of Jesus is "bodily": it is neither a *resuscitation* (like the resurrection of Lazarus[24]), nor a *docetic* escape from the body left behind in the grave, nor is it *entirely reducible* to the experience of renewed hope in the post-crucifixion lives of the disciples, though it clearly involves their experience. It follows that in this hypothesis the Empty Tomb (ET) as an historical fact is essential in holding us accountable to take "bodilyness"—that is, whatever we mean by such terms as "materiality" and "physicality"—seriously in interpreting the resurrection of Jesus. Although it is not the basis of faith in the Resurrection, the ET gives the cognitive content of that faith an irreducibly *worldly* dimension. Below I discuss the profound implications the bodily resurrection poses for the constructive dialogue between theology and science. It is important at the outset, however, to show that this view—that Jesus rose bodily from the dead—is well defended by NT scholars and Christian theologians, so that the challenges raised to it by science are, to say the least, not easily avoidable.

1. Bodily Resurrection of Jesus: Objective and Subjective Interpretations

I use the terms "objective interpretation" and "subjective interpretation" of the resurrection of Jesus in the following way, drawing from their frequent usage in the NT literature.

Objective interpretation. The claims that the disciples made about the resurrection of Jesus of Nazareth cannot be reduced entirely to the subjective experiences of the disciples as recorded in the appearances and the empty tomb traditions. Instead, something actually happened objectively to Jesus after his crucifixion, death, and burial such that he is the living Risen Christ, present throughout history and present to us and to our church communities. In short, God raised Jesus from the dead. N. T. Wright gives us a compelling argument in support of what I am calling the "objective interpretation" of the bodily resurrection of Jesus: "The *only* possible reason why early Christianity began and took the shape it did is that the tomb really was empty and that people really did meet Jesus, alive again, . . . and that, though admitting it involves accepting a challenge at the level of worldview itself, the best historical explanation for all these phenomena is that Jesus was indeed bodily raised from the dead."[25]

Subjective interpretation. The reports found in the appearances and empty tomb traditions are entirely reducible to the subjective experiences of the disciples following Jesus' death. While the disciples described these subjective experiences as though they had actually happened objectively to Jesus after his crucifixion and death, they were in fact entirely about the subjective experiences of the disciples. They were expressed as though they actually happened in the world but they were not truly about en-counters with a postmortem Jesus. A classic example of the "subjective interpretation" comes from Rudolf Bultmann: "This hope of Jesus and of the early Christian community was not fulfilled. The same world still exists and history continues. The course of history has refuted mythology. . . . Modern science does not believe that the course of nature can be interrupted or, so to speak, perforated, by supernatural powers."[26]

What is at issue here is "what/why" reductionism: in the *subjective* interpretation, "what happened to Jesus" (i.e., language about Jesus' transformation to eternal life with God the Father) is reduced without remainder to "why the disciples believed it" (i.e., their experience of Jesus after his death).[27] For scholars who adopt a subjective interpretation, the "resurrection of Jesus" is only and entirely a way of speaking about the experiences of the disciples. It is not about purported events in the new life given to Jesus by God after his death and burial. These scholars include Rudolf Bultmann, John Dominic Crossan, John Hick, Gordon Kaufman, Hans Küng, Willi Marxsen, Sallie McFague, Norman Perrin, and Maurice Wiles. I will refer to this as the Bultmannian school.

O'Collins cites John Hick as an example. Hick conflates "(i) the experiences that caused the first disciples to know and believe something new after Jesus' crucifixion (= why they believed) with (ii) what they claimed had happened to Jesus himself in 'the original resurrection event' (= what they believed)."[28] For Norman Perrin the empty tomb (ET) is "an interpretation of the event—a way of saying 'Jesus is risen!'—rather than a description of an aspect of the event itself."[29] According to Marxsen, "the evangelists want to show that the activity of Jesus goes on. . . . They express this in pictorial terms. But what they mean to say is simply: 'We have come to believe.'"[30] McFague follows Perrin in interpreting the resurrection as "a way of speaking about an awareness that the presence of God in Jesus is a permanent presence in our midst."[31] Küng writes that "historical criticism has made the ET a dubious fact . . . the stories of the tomb are legendary elaborations of the message of the resurrection."[32] As Stephen Davis quips, these scholars depict the NT writers as either "obtuse communicators" or "deceptive communicators."[33]

In the *objective* interpretation, the contents of the claims that the disciples made about the resurrection of Jesus as something which happened to Jesus of Nazareth after his death and burial cannot be reduced entirely to the experiences of the disciples as reported in the appearances and the ET traditions. Scholars who support an objective interpretation include William Alston, Karl Barth,[34] Raymond Brown, Gerald O'Collins,

William Lane Craig, Stephen Davis, R. H. Fuller, Wolfhart Pannenberg, Pheme Perkins, Sandra Schneiders, Janet Martin Soskice, Richard Swinburne, and N. T. Wright.[35] I will call this the "Barthian school." According to Brown, "our generation must be obedient . . . to what *God* has chosen to do in Jesus . . . we cannot impose on that picture what we think God should have done."[36] Davis argues that "the New Testament writers should be interpreted as saying that the resurrection is essentially and primarily something that happened to Jesus and not to the disciples."[37]

I will reject the subjective interpretation here and work entirely with the objective interpretation. There are, in turn, crucial differences within the objective interpretation which can be seen clearly in discussing the appearances and the ET traditions, to which we now turn.

First, for convenience, the references for the Empty Tomb accounts are Mark 16:1–8; Matthew 28:1–10 (11–15); Luke 24:1–11; John 20:1, 11–18, and those of the Appearances[38] are Matthew 28:9f, 16–20; Luke 24:13–35, 36–49 (–53); John 20:14–18, 19–23, 24–29; John 21:1–14, 15–17. A more detailed exposition and discussion of the texts relating not only to the Empty Tomb but also to the appearances and other elements of the New Testament witness to the Risen Lord would require an historical-critical exegesis beyond the limited scope of this volume. These texts of course include not only 1 Corinthians 15 (which is discussed below) but also those such as Colossians 15:1–20. I leave this to future research.[39]

Objective Interpretation of the Resurrection of Jesus:
Bodily or Personal?

The objective interpretation of the resurrection of Jesus includes a variety of views, but for expediency they can be grouped roughly into two versions, which I will call "the bodily resurrection of Jesus" and "the personal resurrection of Jesus." Both approaches include elements of continuity and discontinuity between Jesus of Nazareth and the risen Jesus. Both versions seek to hold these elements in tension by such phrases as "identity-in-transformation."[40] But while both versions acknowledge that elements of *discontinuity* are to be found in *all* aspects of the person of Jesus (i.e., what could be referred to as the physical/material, the mental/psychological, the spiritual, etc.), they differ sharply over the elements of *continuity*.

According to the bodily version of the objective interpretation there are elements of continuity between Jesus of Nazareth and the risen

Jesus in *all* aspects of the person of Jesus, including what can be called the physical/material, the mental/psychological, and the spiritual. In essence, the bodily version of the objective interpretation of the resurrection of Jesus is *incompatible* with the claim that his body remained in the tomb and suffered the same processes of decay that ours will after our death.

According to the personal version of the objective interpretation there are elements of mental/psychological and the spiritual continuity between Jesus of Nazareth and the risen Jesus, but *no* elements of physical/material continuity. In essence, the personal version of the objective interpretation of the resurrection of Jesus is compatible with the claim that his body remained in the tomb and suffered the same processes of decay that ours will after death. Thus while the appearances as well as the ET tradition are crucial for the bodily version of the resurrection, the appearances alone are crucial for the personal version of the resurrection but one can dismiss the ET.

It is noteworthy that two of the most prominent scholars in theology and science—John Polkinghorne and Arthur Peacocke—are divided over these issues. While they both support the objective interpretation of the resurrection, Polkinghorne defends what I am calling the bodily version of the objective resurrection of Jesus. Peacocke can be read as open to both bodily and personal versions.

According to Peacocke, we can support the resurrection of Jesus either by (1) affirming that Jesus' body undergoes a "fundamental transformation" into his risen state such that the tomb is truly empty or by (2) remaining agnostic and leaving open the question of bodily continuity (i.e., the tomb might actually be closed and its contents subject to ordinary decay).[41] Both views, in Peacocke's opinion, retain the "essential core" of Christian belief: that the whole person of Jesus was glorified, his identity was preserved, he now exists united with God, and he appeared to the disciples. Still Peacocke seems to favor option 2, which includes the possibility that Jesus' body decayed just like ours will. This is indicated by his reliance on Pheme Perkins who, in turn, is paraphrasing Hans Küng's comments on Jesus' resurrection.[42] It is also indicated by his claim that option 1 makes the relation of Jesus' death to ours problematic. Unlike the constituents of Jesus' resurrected body, our bodies will decay and their constituents will disperse about the globe where they will contribute to other organisms. "So within a few years, there are no physical remains that could, logically, possibly be the vehicle of any continuity for our particular identity. This would constitute an insuperable, logical gulf—even for God—between what could happen to us and what happened to Jesus."[43] Finally it is suggested by his discussion of eschatology in which God could provide us with a new form of "embodiment" and where the ultimate destiny of humanity is expressed in terms of *theosis* (the "deification" of the human person) and the "beatific vision" of Dante's *Paradiso*.[44]

Polkinghorne has argued extensively for the historicity of the empty tomb. He disagrees with the claim that Jesus must share our lot vis-à-vis bodily corruption in the grave. Instead the empty tomb means that "matter has a destiny," though a transformed one. Christian faith does not deny the reality of death, but it denies it as the ultimate reality. "We do not proclaim a message of survival but a gospel of death (real death) and resurrection (as God's real re-creative act of a whole man [sic], not a disembodied spirit)." The meaning of "corporeality" in the appearance accounts should not be given an "exclusively spiritual" interpretation, nor be equated with a "mere resuscitation." And we see the crucial importance of the empty tomb in its eschatological significance: Just as Jesus' body was transformed into the risen and glorified body, so the "matter" of this new environment must come from "the transformed matter of this world": "The new creation is not a second attempt by God at what he [sic] had first tried to do in the old creation. . . . The first creation was *ex nihilo* while the new creation will be *ex vetere* . . . the new creation is the divine redemption of the old. . . . [This idea] does not imply the abolition of the old but rather its transformation." Again, we will return to this below (section E, Guidelines for New Research in Eschatology and in Scientific Cosmology).[45]

I focus on scholars who support the bodily version, since here the detailed arguments against the subjectivist interpretation of the resurrection of Jesus are most clearly deployed. But more important, this approach is most clearly challenged by science and thus it poses the hardest case for retaining a creative dialogue between theology and science. Conversely, if we can create a convincing response that keeps the dialogue going then we will be able to support the bodily resurrection of Jesus *and* discover extraordinary new insights for theology from science—and perhaps vice versa (as CMI proposes).

Support for the Bodily Version of the Objective Interpretation
of the Resurrection of Jesus

Because the texts of the appearances and the ET traditions are among the
primary sources for the bodily version of the objective interpretation of
the resurrection of Jesus, it is crucial to first demonstrate their historical
reliability. The case for their historical reliability is considerably strength-
ened if it can be shown that these traditions are of independent origin.

In brief, scholars who argue for independent origination tend to take
one of two approaches to the way the appearances and ET traditions are
understood and defended. Scholars in both approaches *agree* on the
following claims: 1) The ET by itself does not lead inevitably to faith
in, and certainly does not prove, the resurrection, a point already found
in the NT itself (John 20:2, 13, 15). 2) However, when placed within
the context of the appearances to the disciples, the ET tradition greatly
strengthens the objective interpretation. 3) Conversely, had Jesus' corpse
been available, it is hard to see how the objective interpretation could
have arisen. Nevertheless, these same scholars *differ* in the relation they
see between the appearances and the ET traditions. In the first approach,
the appearances are understood to be the primary source of the disciples'
faith in the resurrection of Jesus, while the ET tradition plays a secondary
role. This is supported in part by the idea of "the flight of the disciples" as
an hypothesis concerning where the appearances first occurred.[46] According to this hypothesis, the disciples fled Jerusalem during the crucifixion
and returned to Galilee. It was there in Galilee that the risen Lord appeared
to them. After the appearances, they returned to Jerusalem and were told
about the ET by the women who had remained there and discovered it.
Differences between the ET accounts are minimized by scholars in this
approach but differences between the appearances and the ET traditions
are stressed to support their claim that they are of independent origin.
In the second approach, the appearances and the ET traditions are considered to be of comparable importance. Some scholars support the hypothesis of the "flight of the disciples" while others reject it, but for most
the Gospel accounts of the discovery of the ET by the women and the
disciples are taken as relatively certain. Independent origination of the
traditions is argued by stressing the differences within the ET accounts
themselves.

In the current debate many of the arguments by scholars such as N. T. Wright against others such as John Dominic Crossan can be traced back to Wolfhart Pannenberg's early writings on Christology in his now classic 1968 text, *Jesus—God and Man*.[47] In section F, I offer a brief summary of Pannenberg's discussion of debates on such related issues as the reliability of the appearances and the ET traditions, the historical meaning of the resurrection of Jesus, and the challenge of natural science. However, one point requires us to note it explicitly here, as it will play a crucial role in helping us undercut the apparent challenge posed by scientific cosmology to Christian eschatology: the question of the historicity of the resurrection of Jesus.

How then should the historian attempt to reconstruct the events triggering the emergence of primitive Christianity? Pannenberg's response to this question may be his most important contribution to the debate, as it challenges the Enlightenment assumptions about the uniformity of history. Instead, according to Pannenberg, the possibilities to be admitted will depend upon the understanding of reality that is brought to the task by the historian. If the historian assumes unquestionably that "the dead do not rise" then it is a foregone conclusion that Jesus did not rise. On the other hand, if the apocalyptic expectation of resurrection is admitted as a possibility, then this must be considered in reconstructing these events—even if it entails using metaphorical language such as the disciples used.[48] With this in mind, the resurrection of Jesus would be designated as "a historical event" in the following sense: "If the emergence of primitive Christianity . . . can be understood in spite of all critical examination of the tradition only if one examines it in the light of the eschatological hope for a resurrection from the dead, then that which is so designated is a historical event, even if we do not know anything more in particular about it. Then an event that is expressible only in the language of the eschatological expectation is to be asserted as a historical occurrence."[49]

2. The Bodily Resurrection of Jesus in Its Eschatological Context: Resurrection as Transformation from Creation to New Creation and the World/Cosmos Conflation

Why can we not just side with those who argue for the bodily resurrection of Jesus and view it as a "once off" event, a solitary divine miracle that is insulated from any implications for the rest of the natural world and thus from the challenge from science? The reason, simple logically but profound in its implications, is that scholars who argue for the bodily resurrection of Jesus use Paul (and other NT texts) to connect his resurrection with the general resurrection "at the end of time" and, in turn, to the "new creation" consisting of a "new heaven and earth" as the context of the general resurrection.

A pivotal text for this connection is Paul's writings in 1 Corinthians 15:12–19:

> 12 But if it is preached that Christ has been raised from the dead, how can some of you say that there is no resurrection of the dead? 13 *If there is no resurrection of the dead, then not even Christ has been raised.* 14 And if Christ has not been raised, our preaching is useless and so is your faith. 15 More than that, we are then found to be false witnesses about God, for *we have testified about God that he raised Christ from the dead. But he did not raise him if in fact the dead are not raised.* 16 *For if the dead are not raised, then Christ has not been raised either.* 17 And if Christ has not been raised, your faith is futile; you are still in your sins. 18 Then those also who have fallen asleep in Christ are lost. 19 If only for this life we have hope in Christ, we are to be pitied more than all men. (NIV, my italics)

For many NT scholars, the main point of this and related texts is that the context of the general resurrection is the New Creation viewed as a transformation of the world as a whole, and not just a transformation of individual human life. This transformation is not, on the one hand, a dualistic denial of the value of, and abandonment of, the present creation, a docetic/Gnostic "snatch and grab" in which only the "soul" of Jesus is taken from this transitory material world to another, eternal one. Nor is the New Creation a second, entirely new creation *ex nihilo* but part of the single, universal result of God's Trinitarian act of creation. Even more so, it is not just the natural product of an evolutionary universe, one arising smoothly within the ordinary processes of the world as we now know them (i.e., what some call "evolutionary eschatology"). Instead it is a *return* of the risen Christ to *this* world in order that this world might be transformed into an eternal world without death, decay, sorrow, and the irrevocable passage of time, that is, the New Creation. Four scholars typify this position in various ways, Raymond Brown, Janet Martin Soskice, Gerald O'Collins, and N. T. Wright.

Raymond Brown provides two starkly different eschatological possibilities: "the model of eventual destruction and new creation, or the model of transformation . . . into the city of God." These models are directly linked to our view of the bodily resurrection of Jesus: "If Jesus' body corrupted in the tomb so that his victory over death did not involve bodily resurrection, then the model of destruction and new creation is indicated. If Jesus rose bodily from the dead, then the Christian model should be one of transformation." These models in turn determine our attitudes and values towards the world in profound ways: "What will be destroyed can have only a passing value; what is to be transformed retains its importance. Is the body a shell that one sheds, or is it an intrinsic part of the personality that will forever identify a man [sic]?"[50]

As a self-acknowledged "unrepentant empty tomber," Janet Martin Soskice is concerned with what she calls an "etiolated orthodoxy": a minimalist account of resurrection faith in which Jesus was raised (including the empty tomb) and we shall be too. For Soskice, these claims, though true, offer an inadequate account of the "full-blooded" resurrection faith that led Galilean peasants to follow Jesus, let alone that led some early Christians to accept martyrdom. What is needed is an understanding of the resurrection of Jesus as "the beginning of the restoration which will bring a new heaven and a new earth . . . the resurrection of Jesus inaugurated a 'new creation' in which all our relationships with one another and with the world around us are changed."[51] Because an etiolated orthodoxy focuses too exclusively on hope for human life in another world, it leaves "no hope of the triumph of God's justice on earth, no point in praying that God's kingdom will come and will be done *on earth* as it is in heaven, and no salvation for non-human creations. . . . Trees and valleys, and even all the animals, are indeed no more than the great 'stage set' on which the drama of the saving of souls is acted out."[52] Thus our concern with the resurrection of the body must lead us to understand our "social embodiment" with all believers, all peoples, and with the environment.

According to Gerald O'Collins, "the material world *will* share in the glorious destiny which Christ's resurrection promises to all men and women." Indeed resurrected life without a new "material environment" would seem more like an "immortal existence of souls" rather than a bodily resurrection. And if there is to be a bodily resurrection for humanity, the entire material creation will participate in the "transformation of our cosmos" culminating in the "new creation." O'Collins goes even further in a bold move: "The resurrection has *already* changed the material world. . . . Can one point to any differences that Christ's resurrection has made" in it? For O'Collins, the purview of eschatology is truly cosmic. "At the end the universe will share in the transforming power of the Lord's resurrection. The image of the 'new heaven and new earth' expresses the last state to which our whole material environment will be called home, (revealing) the purpose of what God did in creating 'heaven and earth.'"[53] "The first Easter began the work of finally bringing our universe home to its ultimate destiny . . . . [It] is God's radical sign that redemption is not an escape to a better world but a wonderful transformation of our world."[54] Finally, O'Collins returns to the key question for this book: "To what extent can modern science illuminate . . . the nature of a transfigured world to come?"[55]

For N. T. Wright, the New Creation "is the *transformation*, not merely the replacement, of the old [creation]." It will be "in important senses continuous with the present one." It is this New Creation, made by God out of the old, which provides the context for the general resurrection of the dead. In this New Creation and at Christ's *parousia* we will be raised from the dead for judgment and everlasting life. Between our individual death and the general resurrection we wait patiently in the "intermediate state," resting in "the blissful garden, the parkland of rest and tranquility."

Wright summarizes the exegesis we have briefly touched on here with a now famous phrase: "life *after* life after death." According to Wright, "resurrection . . . wasn't a way of talking about life after death. It was a way of talking about a new bodily life *after* whatever state of existence one might enter immediately upon death." What then will our resurrected bodies be like? Wright argues that in the resurrected life we will experience a "new mode of physicality which stands in relation to our present body as our present body does to a ghost. It will be as much more real, more firmed up, more *bodily*, than our present body as our present body is more substantial, more touchable, than a disembodied spirit."[56] With these powerful images Wright underscores the irreducible importance of the "materiality" of the "new creation" both in continuity and in discontinuity with the present age.[57]

World/Cosmos Conflation in Discussions of Eschatology

While a variety of complex, interrelated, and crucial issues arise in the eschatological discussions such as the above, underlying all of them and posing the hardest challenge to cosmic eschatology is the far future predicted by contemporary scientific cosmology.[58] Surprisingly, this challenge is hardly mentioned in the scholarly literature on NT exegesis such as we have surveyed above. Instead, we find what I call the world/cosmos conflation: in discussing the "new creation" as above the term "cosmos" is often used interchangeably with the term "world," thereby undermining the conceptual challenge "cosmos" should convey in pointing to the physical universe of big bang and related cosmologies.

This may be a mere linguistic ambiguity, but I fear it often covers over a deep disconnect. To be direct, when we refer to the transformation of the "world" into the "new creation," do we really have the actual universe explicitly in mind, or are we implicitly limiting the scope of eschatology to "planet Earth"? I am confident that the scholars I have cited would claim they mean the actual universe and would readily acknowledge the challenge raised by physics and cosmology, but with the verbal slippage between the term "cosmos" and "world," it is all too easy to let the challenge go unmentioned. The conflation may be due simply to the fact that the Greek word κόσμος can be interpreted as either "world" or "cosmos." I suspect, however, that this slippage continues to occur, in part, because it plays a somewhat seductive role by allowing one to implicitly maintain contact with the cosmology of the ancient Near East and not the cosmology of today. The problem is that if we really think about eschatology in terms of the transformation of the *universe as understood by science,* is eschatology still intelligible, let alone credible?

3. Summary: The Resurrection of Jesus and the Challenge from Science

I have suggested that there are two widely populated, and widely varying, schools of interpretation of the resurrection of Jesus. These optional interpretations are sketched in figure A.2 below. For brevity I cite them as rooted in the subjective theology of Bultmann and the objective theology of Barth.

According to the Bultmannian school, the resurrection of Jesus can be reduced without remainder to the subjective faith experience of the disciples. For the Barthian school, the resurrection of Jesus is an objective event in the postmortem life of the Risen Christ even while it is one that is received through the subjective faith of the disciples. Hence for both interpretations the appearances are crucial, but for different reasons. For the Bultmannian school, the appearances are all we have; without them there would never have been Christianity. But the ET experiences are irrelevant; they are just a pre-scientific myth with no objective consequences for the world which science studies. Thus the Bultmannian school experiences no conflicts with science since, in effect, it was designed to recast the meaning of Jesus' resurrection to avoid such conflicts. Theirs is a classic example of the "two worlds" approach.

For the Barthian school, the appearances and the ET are interpreted together as leading to an objective theology of the resurrection of Jesus, one in which language about his resurrection refers to a postmortem event in the continuing life of Jesus. But the Barthian school splits into two options. One is the "personal" version of the objective interpretation of the resurrection of Jesus as supported by Peacocke. The other is the "bodily" version of the objective interpretation of the resurrection of Jesus as supported by Polkinghorne and by most Christian theologians. For the former, the ET is once again irrelevant; for the latter, the ET is crucial.

Now for the latter "bodily" version, there are two additional options: The first is to interpret the Resurrection as a miraculous resuscitation— an extraordinary event regarding Jesus, but one that leaves his surrounding environment and, indeed, the world as a whole and with it the laws of physics and biology, unchanged. In this interpretation, Jesus, like Lazarus, will eventually die a natural death. The second is to view the Resurrection as an event that I call, for lack of a better terminology, "more than a miracle." This view leads to the idea of a change in the entire environment of the ET and indeed to a change, eventually, in the world as a whole. This, then, is a change that ultimately extends to the universe and the foundational laws of physics.[59] In such a view, the Resurrection entails the beginning of a radical and ongoing transformation of the universe by God into the New Creation and, with it, the ending of suffering, disease, death, and extinction (i.e., natural evil). I refer to this as FINLONC: the "first instantiation of a new law of the New Creation" (see section E.1, guideline 6b, below).

Thus the bodily version of the objective interpretation of the resurrection of Jesus, which I support, runs directly into the seemingly overwhelming challenge of science. But this, I think, is its virtue, for if we can make some headway, no matter how small, in addressing this challenge, we will, in turn, have offered new strength to the Christian tradition, broadly defined, and to Pannenberg's theological agenda in specific. And that is, after all, our task here. And equally we will have a theology robust enough to "push back" to science to ask it to address new kinds of questions and search for otherwise hidden factors in its massive explanatory paradigm of the natural world. And this, too, is our task here, especially recalling Pannenberg's essay published some three decades ago and entitled "Theological Questions to Scientists."[60]

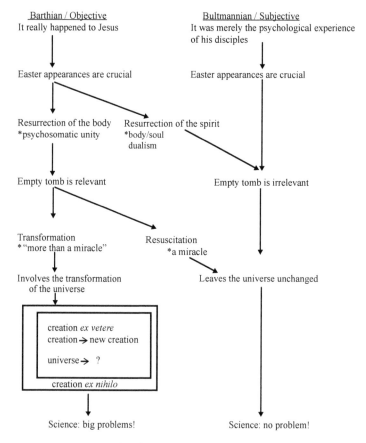

Figure A.2. The Resurrection of Jesus:
Competing Interpretations and the Challenge from Science

C. ESCHATOLOGY AND COSMOLOGY

1. Introduction

Whether the topic of eschatology arises in the context of a scholarly text in systematic theology or in a nontechnical introduction to Christian theology, few authors acknowledge, let alone seek to respond to, the challenge raised to it by scientific cosmology.[61] In this section, I briefly comment on the centrality of eschatology to Christian theology and then give a short overview of scientific cosmology and its predictions of "freeze or fry" for the far cosmic future, including the recent discovery of the accelerating expansion of the universe and its very strong evidence for the "freeze" scenario. It is these predictions that explicitly challenge those versions of Christian eschatology that entail God's transformation of the universe as a whole into the New Creation. I lay out a typology of eschatologies configured around the responses, or lack of responses, to the challenge of scientific cosmology. I conclude with an appreciative assessment of several recent approaches to eschatology that do in fact seek to address this challenge and that serve as a starting point for the proposals in this volume.[62]

2. The Centrality of Eschatology to Christian Theology

Historical theology routinely treated eschatology in terms of the "last things": general resurrection, last judgment, heaven and hell, the end of the world, and so on. Beginning with the eighteenth-century Enlightenment, however, eschatology was frequently reinterpreted in ethical, social, political, and economic categories with little attention to traditional issues. This reinterpretation was challenged on historical-critical grounds in 1906 with Albert Schweitzer's publication, *The Quest of the Historical Jesus*.[63] Here, Schweitzer argued that Jesus' understanding of his mission as well as early Christian faith and praxis was grounded in Jewish apocalyptic eschatology. He concluded that since Jesus' eschatology had been proven wrong by history, his "interim-ethics" could not be applied directly to contemporary society. Still Jesus' personality, with its world-negating perspective, was acclaimed by Schweitzer to be of timeless relevance to society.

The twentieth century saw a variety of responses to the theological crisis caused by Schweitzer's work. Because of what he saw as its conflict with modern science, Rudolf Bultmann viewed eschatology as mythology and reinterpreted it in existentialist categories. Charles H. Dodd spoke of "realized eschatology" to emphasize the idea that the kingdom of God has already come about in the life of Jesus. More recently, theologies of liberation and orthopraxis have emphasized the world-transforming power of eschatology to challenge racism, sexism, political and economic oppression, the abuse of the environment, etc. Scholars in the Jesus Seminar offer what I would call an "entirely non-eschatological interpretation." They claim that the apocalyptic texts in the New Testament do not originate with the historical Jesus.

A more promising contemporary form of eschatology combines future hope for a universal transformation of the world ("apocalyptic eschatology") with the present partial realization of that hope in the world ("realized eschatology"), and then makes this combination central to Christian theology as a whole. It is this type of eschatology that will be taken up here because of its critical dependence on, and its creative pushback to, natural science.

Karl Barth took a decisive step in the direction of making eschatology central to theology (although he did not emphasize its relation to science) by his claim that "Christianity that is not entirely and altogether eschatology has entirely and altogether nothing to do with Christ."[64] Paul Tillich saw the eschaton both in terms of our present experience of the eternal and as the aim and end of history in its "elevation" into the eternal. For Tillich this "aim and end" included the universe itself. Although he did not develop this insight extensively, Tillich did write that "without the consideration of the end of history and *of the universe,* even the problem of the eternal destiny of the individual cannot be answered."[65] A central document of the Second Vatican Council, *Lumen Gentium,* points to the immediacy and the futurity of the Kingdom of God in proclaiming that "the human race as well as the entire world . . . will be perfectly reestablished in Christ. . . . The final age of the world has already come upon us."[66] Although primarily emphasizing the present-day implications of eschatology for political, social, and economic liberation, Jürgen Moltmann has also stressed the importance of nature to Christian eschatology and thus to the coming of the universal New

Creation.[67] And just as some scholars interpret our own resurrection not only as immediately following our death but also as a coming, future eschatological event at the end of the age,[68] by analogy others suggest that the transformation of the world happens not only synchronically at the end of time but also diachronically throughout the entire course of its history.[69]

Still, it is Wolfhart Pannenberg who, in my opinion, has given eschatology its most creative, novel, and all-encompassing formulation in contemporary systematic theology. Indeed the role Pannenberg gives to eschatology is that of determining the entire content of his systematics. We will explore his work on eschatology in more detail in chapter 1 as well as throughout this volume. Still it will be helpful to give a brief summary here in the appendix to the introduction.

According to Pannenberg, God as Trinity acts from the eschatological future through the proleptic event of the resurrection of Jesus to transform history into its eschatological goal. Pannenberg argues that the very deity of God depends, in one sense, on the eschatological consummation of the world when the Son hands back lordship to the Father.[70] An eschatology such as Pannenberg articulates views the New Creation not as a "replacement" of the present creation—that is, not as a second *ex nihilo*—nor as the mere working out of the present natural processes and potentialities of the world—a kind of "evolutionary" or "physical" eschatology. Instead, eschatology involves the complete *transformation* of the world by a radically new act of God beginning at Easter and continuing into the future of the world. The Easter event thus becomes a proleptic manifestation in time of what is the not-yet/still-future eschatological-apocalyptic destiny for all the world. In the Easter event the New Creation, having been transformed by God out of the original creation, reaches back over and into, and is manifested in, the world. Then through the Easter prolepsis, all of history, prior to it and following it, is filled with the promise of New Creation.

Pannenberg's eschatology also involves an emphasis on realized eschatology through his unique concept of "anticipation as the arrival of the future" and his idea of the causal priority of the immediate future. In essence, the present is the result not only of the efficient causal efficacy of the past but also the causal priority of the immediate future. Pannenberg combined these two concepts—prolepsis and the causal priority of the immediate future—to achieve a truly novel synthesis of eschatology as realized and eschatology as future.

It should now be self-evident that the eschatologies of the transformation of the universe, advocated by Pannenberg and some New Testament scholars, face the severest challenge from contemporary science, particularly from cosmology. When we turn the domain of eschatology from an anthropological (personal, socio-political) and even a terrestrial (environmental, ecological) context to a *cosmological* horizon we encounter a grim scientific prediction: all life in the universe will inevitably be extinguished as stars collapse into black holes, and following this the cosmic future will be one of either unimaginable heat in the eventual recollapse of the universe or, what is much more likely, endless cold in the infinite expansion of the universe.

What response can we give to this challenge? Before addressing this question we must review its basis in scientific cosmology in order to grasp the depth of the challenge.

3. The Challenge of Scientific Cosmology

Scientific cosmology has undergone stunning developments this century.[71] For convenience, I group these developments into three models of cosmology: big bang, inflation, and quantum cosmologies.

Einstein's General Theory of Relativity (GR)/Big Bang Cosmology. In 1905, Albert Einstein proposed the special theory of relativity (SR), which was quickly given a geometrical interpretation by Hermann Minkowski: space and time, independent in Newtonian physics, are united as a four-dimensional "spacetime" geometry.[72] A decade later, Einstein proposed the general theory of relativity (GR) that treats the force of gravity in a way that is consistent with SR. According to GR, the non-Euclidean curvature of spacetime due to the sun accounts for earth's orbit around the sun. Contrast this with Newton's theory in which the sun exerts a gravitational force on the earth that deflects it from uniform motion in Euclidean space, resulting in its orbit of the sun. Einstein formalized this idea in the field equations, $R_{\mu\nu} - \frac{1}{2}Rg_{\mu\nu} = 8\pi T_{\mu\nu}$. In an apt description: "spacetime tells mass how to move; mass tells spacetime how to curve."[73]

It soon became apparent that the field equations of GR could not describe the kind of universe Einstein presupposed ours to be, namely one that is eternal, finite in size and static in time. Consequently, Einstein modified his equations with the introduction of the "cosmological constant" Λ. The new equations, $\Lambda g_{\mu\nu} + R_{\mu\nu} - \frac{1}{2} R g_{\mu\nu} = 8\pi T_{\mu\nu}$, allow an Einsteinian static universe. Meanwhile, however, Edwin Hubble and other astronomers in the 1920s reported that light from distant galaxies is redshifted and that the redshift is proportional to their distance d from us. Within a decade most scientists came to see the redshift as resulting from the galaxies moving away from us with a velocity v proportional to their distance as given by Hubble's law: $v = H \times d$ (H is Hubble's constant).[74] But were galaxies merely receding from us in a static spacetime or was space itself "expanding" in time? During this same period theoretical evidence mounted in favor of an expanding universe. The Russian scientist Alexander Friedmann and the Belgian priest Georges Lemaître produced a variety of theoretical models of an expanding universe. When the expansion is traced back in time we approach the "absolute initial singularity": an event marking the beginning of time, labeled "t = 0," in which the density and the temperature of the universe go to infinity as its size approaches zero. The combination of these theoretical and observational arguments resulted in a general consensus among scientists by the 1950s that we do indeed live in an expanding universe with the event "t = 0" labeled the "big bang." Evidence today suggests that the origin of the universe at t = 0 is some 13.7 billion years in our past.[75]

Einstein's field equations actually allow for three kinds of big bang cosmologies that differ radically in their depiction of the far future. In the first two models (open universes with negative and flat curvatures) the universe is always actually infinite in size and it is destined to expand endlessly as its temperature falls exponentially towards absolute zero. The third model (closed universe with positive curvature) is finite in size and will eventually stop expanding to begin to recontract toward a future singularity much like t = 0, one in which the temperature of the universe soars to infinity.[76] The far future in the first two models is aptly termed "freeze," in the third model, "fry."[77] See chapter 5, section B.3 and figure 5.5.

Inflationary (Hot) Big Bang Models. The early big bang models were beset by a number of important technical problems,[78] the most significant of which is the absolute singularity, t = 0. Inflationary models were originally developed in the 1980s by Alan Guth to address these problems. They depict the very early universe (circa the so-called Planck time,

10^{-43} seconds) as undergoing an exponentially rapid expansion before settling down to expansion rates predicted by the original big bang scenarios. As a result of this initial rapid expansion, the universe is thought to be much larger than what the original big bang models suggested (fig. A.3).

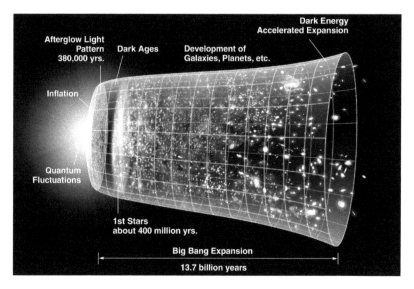

Figure A.3. Diagram of Inflationary Big Bang Cosmology
Source: NASA/WMAP Science Team, "Time Line of the Universe," *Wikipedia,*
October 26, 2010, http://en.wikipedia.org/wiki/File:CMB_Timeline300_no_WMAP.jpg.

In inflationary cosmology the universe consists of countless domains or regions, one of which is our entire visible universe. The value of the physical constants, such as the speed of light, c, or Planck's constant, h, may vary from domain to domain. Inflation solved several of big bang's technical problems but it left the status of t = 0 unsettled: in these models it may be impossible in principle to decide whether or not the universe emerged from an absolute singularity, t = 0.[79]

An example of a different inflationary cosmology is Andrei Linde's "eternal Chaotic inflation."[80] Here our universe emerges from a prior "superspace" of universes that are multiply connected like a string or web of expanding universes. Some are like ours, others might be radically different. These universes endlessly replicate as new bubble or daughter universes form, creating an overall structure that continues forever. Which inflationary model best describes the universe, and our domain in it, is a hotly contested issue in cosmology today.

Quantum Gravity/Quantum Cosmology. A crucial goal in theoretical physics over the past several decades has been to unite general relativity with quantum mechanics to produce a theory of quantum gravity. The link between quantum mechanics and gravity becomes crucially important to cosmology because as we think back toward $t = 0$, the size of the universe shrinks endlessly and the universe must ultimately be treated as a quantum mechanical "object." Quantum gravity could lead us beyond inflationary big bang models to what is called "quantum cosmology." Early approaches to quantum cosmology include the Hartle/Hawking model[81] and that of Alex Vilenkin.[82] More recent work includes the Turok/Hawking instanton, pre–big bang scenarios, and multiverse or brane cosmology drawing on superstring theory and including as many as 10^{500} "universes," etc. Though these scenarios differ strikingly, in most cases they result in a standard big bang universe following an initial, inflationary epoch (we will discuss string theory briefly in chapter 6, section B).

Quantum cosmology, however, is a highly speculative field because quantum gravity is notoriously hard to test. Moreover, our basic understanding of what we mean by "the universe" is further complicated by the complex philosophical issues still associated with quantum mechanics: what is the meaning of quantum indeterminism, the observer/the measurement problem, nonlocality, and quantum correlations when applied to the universe as a whole.[83]

An Open, Accelerating Universe

There is now growing evidence that the matter density in the visible universe is far below the critical density required for a closed universe. For this and other reasons most scientists believe that the universe is marginally *open* (approximately flat) and thus will expand forever. Moreover, recent theoretical and observational evidence indicates that its expansion rate is not slowing, as it would in the standard flat or open big bang models; instead it is actually *speeding up*.[84] Saul Perlmutter, Adam Riess, and Brian Schmidt shared the 2011 Nobel Prize in physics for their roles in the discovery of the accelerated expansion. Such an acceleration might be accounted for by the existence of a nonzero cosmological constant, Λ, in Einstein's field equations, or what is now commonly referred to as "dark energy." So the evidence for an open universe seems increasingly strong—although still not conclusive.[85]

The "Bottom Line" Regarding the Cosmological Far Future and the Possibility of Life in It

Where does this leave us regarding the scientific prognosis for biological life in the cosmic far future of the universe, be it closed or open (and, in particular, accelerating in its expansion)?[86] And what implications does this prognosis raise for Christian eschatology?[87]

According to most accounts, the differences in the cosmic future between a closed and an open universe do not really matter for the long-term prospects for biological life: in either case it is doomed inevitably to universal extinction. This in turn means that neither cosmic future offers any serious grounds for a Christian eschatology such as I am exploring here. Frank Tipler and John Barrow draw out the details of this grim prognosis for biological life in both closed and open universes:[88]

- In 5 billion years, the sun will become a red giant, engulfing the orbit of the earth and Mars, and eventually becoming a white dwarf.
- In 40–50 billion years, star formation will have ended in our galaxy.
- In 10^{11}–10^{12} years, all massive stars will have become neutron stars or black holes.
- If the universe is closed, then in 10^{12} years the universe will reach its maximum size and then recollapse back to a singularity like the original hot big bang.
- In 10^{19} years, dead stars near the galactic edge will drift off into intergalactic space; stars near the center will collapse together forming a massive black hole.
- In 10^{20} years, orbits of planets will decay via gravitational radiation.
- In 10^{31} years, protons and neutrons will decay into positrons, electrons, neutrinos, and photons.
- In 10^{34} years, dead planets, black dwarfs, and neutron stars will disappear, their mass completely converted into energy, leaving only

black holes, electron-positron plasma, and radiation. All carbon-based life-forms will inevitably become extinct. Beyond this, solar mass, galactic mass, and finally supercluster mass black holes will evaporate by Hawking radiation.

- *If the universe is open, it will continue to cool and expand forever. All traces of its early structure, from galaxies to living organisms to dust, will vanish without a trace, never to recur again.*

The upshot is clear: "Proton decay spells ultimate doom for life based on protons and neutrons, like *Homo sapiens* and all forms of life constructed of atoms."[89] So, according to science, the cosmic future, whether open/freeze or closed/fry, is one in which all life in the universe will inevitably vanish forever. If closed, a structureless sea of fundamental particles will heat without restriction to infinite temperatures. If open, this sea of fundamental particles will expand and cool forever into a darkening future whose time never ends. In *neither* case does the cosmic future include *any* record of our ever having been here—and the "our" here means all forms of life, indeed of physical complexity beyond a random plasma of fundamental particles, on earth and throughout the universe. Finally, recent evidence that the universe is not only open but accelerating in its expansion seems to make this prognosis conclusive. If the predictions of scientific cosmology do indeed come to pass in the future, ours will be a barren universe devoid of any trace that life had ever existed, one which is in no way compatible with the New Testament belief in the general resurrection in the New Creation.

4. Eschatology and Cosmology: A Variety of Minimalist Responses

Eschatology as Falsified by Science ("Outright Conflict")

Not surprisingly, many scientists have given pessimistic, "dysteleological" readings of cosmology. In 1903, even before the predictions of big bang cosmology were on the horizon, Bertrand Russell wrote in regard to the sun's eventual supernova that "all the labors of the ages, all the devotion, all the inspiration, all the noon-day brightness of human genius are destined to extinction in the vast death of the solar system."[90] Some seventy years later, Nobel laureate Steven Weinberg anguished over the fact that "[our visible universe] is just a tiny part of an over-whelmingly hostile universe. . . . [It] has evolved from an unspeakably unfamiliar condition, and faces a future extinction of endless cold or in-tolerable heat. The more the universe seems comprehensible, the more it also seems pointless."[91]

Clearly these statements by scientists threaten to falsify Christian eschatology at least as "universal transformation." Interestingly, similar positions emphasizing the potential falsifiability of scientific cosmology to Christian eschatology can be found among those few theologians who have seriously considered the meaning of eschatology in light of cos-mology. In 1966 John Macquarrie wrote, "If it were shown that the uni-verse is indeed headed for an all-enveloping death, then this might . . . falsify Christian faith and abolish Christian hope."[92] Three decades later, Ted Peters also wrote: "Should the final future as forecasted by the com-bination of big bang cosmology and the second law of thermodynamics come to pass . . . we would have proof that our faith has been in vain. It would turn out to be that there is no God, at least not the God in whom followers of Jesus have put their faith."[93] As early as 1979 in his pio-neering study on the doctrine of creation in relation to contemporary sci-ence, Arthur Peacocke acknowledged that "science raises questions about the ultimate significance of human life in a universe that will eventually surely obliterate it."[94] In an article written in 1981—groundbreaking for its structure as "theological questions to scientists"—Pannenberg listed the conflict between eschatology and cosmology as one of five key ques-tions in the theology/science dialogue. He also offered wise, if rather cau-tious, advice: living with the conflict may well be better than seeking an "easy solution."[95]

Eschatology as Reducible to Science
("Acquiescence to Conflict: Eliminative Reductionism")

Perhaps whatever cosmic future science predicts is simply all there is for Christian eschatology to affirm and we should acquiesce to its elimi-native reductionism. Yet such reductionism might not be the "end of the line" if we are willing to countenance a highly truncated eschatology. It is here that the work of Freeman Dyson and colleagues becomes impor-tant in offering a glimmer of hope against the radically dysteleological predictions we saw above.

In 1979 Dyson advanced an unprecedented argument about "life" in an open universe: If life can be reduced to the physics of information processing and thus freed from its biological basis, then life in a modest form can continue into the *infinite* future, processing new experiences, storing them through new forms of *non*-biologically based memory, and ultimately remolding the universe to its own purposes. "An open universe need not evolve into a state of permanent quiescence. . . . So far as we can imagine into the future, things continue to happen. In the open cosmology, history has no end."[96] A decade later, Frank Tipler and John Barrow took up Dyson's arguments and focused them on the closed universe with its "fry" scenario for the far future.[97] Like Dyson, they too defined life reductively as mere "information processing." Their crucial insight was that if the rate of information processing could increase exponentially in time, an infinite amount of information could be processed even in the finite time remaining before the end of the closed universes, and this would constitute what they saw as "eternal life."

The scientific details of these scenarios are fascinating, and to his immense credit Dyson did challenge Weinberg's dysteleological read even while accepting the reductionist assumption according to which scientific cosmology is the only framework within which hope for life in the cosmic future can arise.[98] But what are the theological consequences of this move? On the one hand, I appreciate the fact that Dyson, Tipler, and Barrow all mentioned the potential connections between physical cosmology and Christian eschatology in the spirit of Pierre Teilhard de Chardin. Recently Tipler has even claimed to provide a scientific basis for God, resurrection, and immortality via his Omega Point theory.[99] On the other hand, while these moves should not go unacknowledged they have received very mixed reviews. Willem B. Drees pointed to theological strengths and weaknesses in their work.[100] John Polkinghorne, Ian Barbour, Peacocke, Philip Clayton, and Mark Worthing criticized them on both theological and philosophical grounds.[101] Tipler and Pannenberg have engaged in an extensive interaction to which Drees, Russell, and others have replied, and Tipler's scientific claims have been attacked aggressively by George Ellis and William R. Stoeger.[102]

It is clear that "physical eschatology" does *not* hold out genuine promise for an eschatology of "new creation." Nevertheless if we set aside the reductionist assumptions, theological oversimplifications, and scientific controversies that habitually accompany both Weinberg's conclusions and the Dyson-Barrow-Tipler response to it, we may yet discover some vital clues for our attempt to relate scientific cosmology to Christian eschatology.

Eschatology and the Separation from Science ("Two Worlds")

Before turning in a constructive direction, we must address a beguiling "solution" to the challenge: perhaps cosmology is simply irrelevant to eschatology; if so, the conflict would be over. There are several forms this approach can take.

One approach is offered by Peacocke, whose views we touched on above in section B.1. While contributing immensely to the constructive theology/science dialogue in general, Peacocke provides an intriguing case for a "two worlds" view when it comes to eschatology and cosmology. First, he asserts that scientific cosmology cannot provide a basis for eschatology, but he actually goes further. Although he disagrees with scholars who reduce the resurrection of Jesus to the subjective experience of his disciples, he does not insist that the resurrection of Jesus is "bodily" and thus connect it with God's eschatological transformation of the universe. Instead, eschatology refers to "our movement towards and into God beginning *in the present* . . . it transcends any literal sense of 'the future.'" It is "beyond space and time within the very being of God." Resurrection in the "bodily" sense and, in turn, the redemption of nature through its transformation into the New Creation, are mostly set aside.[103] In taking this position Peacocke avoids a conflict with cosmology, but at an enormous cost: the physical universe has no eschatological destiny.

Another approach is to appeal to the provisionality of science. Clearly scientific theories change in time and it is not inconceivable that a future cosmology will be less incompatible with eschatology than contemporary cosmology is. In light of this possibility perhaps we need not be too concerned about the present conflict between eschatology and scientific cosmology; instead we should "wait and see." There is some wisdom to this view, given that physical theories, and their attendant cosmologies, are eventually replaced by, or incorporated into, new ones. But there is also reason to be cautious about relying too heavily on the provisionality of science. It is one thing to acknowledge it as a temporary response to conflict. It is quite another thing to drift from provisionality into a rigid and totalizing "two worlds" position that the dialogue between theology and science is meant to avoid.[104]

A similar problem arises when process theology reduces eschatological hope to a form of immortality that leaves out bodily resurrection. The starting point is Whitehead's bipolar theism with its joint concepts of the primordial nature of God, the source of eternal possibilities for the world and the initial aim of every actual occasion, and the consequent nature of God by which God's life is continually enriched as God prehends every actual occasion in the world.[105] The prehension of the world by God is the basis for process eschatology. Whitehead used the term "objective immortality" for the claim that all actual occasions are objectively present in God's prehension of the world and contribute to the enrichment of God's consequent nature. In this way the "perpetual perishing" of the world in the concrescence of every actual occasion is taken up into God's everlasting enjoyment of the world through God's unending prehension of the world. But are we included personally in God's consequent nature as "living" or are we just remembered as "having lived"?

In attempting to bring Whitehead's views closer to traditional Christian eschatology some process theologians, such as Ian Barbour and Marjorie Suchocki, argue for "subjective immortality" in which a person continues after death to be "a center of experience" within God.[106] Notably, while John Cobb and David Griffin acknowledge that Whitehead's philosophical system was open to this view, they insist that Whitehead himself remained neutral about it.[107] Nevertheless, in a particularly moving treatment of the problem, Suchocki describes our resurrection as something more than mere objective immortality. Instead, based on the resurrection of Jesus, it is a transformation from this world into the "infinite context" of God. She also affirms the resurrection as bodily because she argues that when God prehends every actual occasion in the whole universe, the result is to transform the physical universe into a new heaven and a new earth. Thus Suchocki says "there is a home in God, a home for the whole universe."[108]

Still, do Barbour and Suchocki grapple seriously with the New Testament accounts of the bodily resurrection of Jesus, particularly the empty tomb traditions? And do they take on board and respond to the challenge science poses about the far future to Christian eschatology? Here again, as with Peacocke, the future of the universe seems unimportant to process eschatology, but now for a philosophical, rather than a theological, reason: Whiteheadian metaphysics provides process theology with a way to speak of at least a truncated form of eschatology while ignoring the scientific predictions for the far future of our universe. Thus for process eschatology the scientific predictions about the far future are irrelevant.[109]

A problem related to that arising in process theology, but one which takes a more serious turn, is found in the writings of John Haught, where his view of the relation between eschatology and cosmology seems at the very least deeply ambiguous and perhaps even incoherent. Haught inherits the problems already noted in process theology, but the difficulties are exacerbated because he attempts to combine process perspectives with Catholic theology drawn from Teilhard de Chardin and Karl Rahner. The result is that Haught proffers three different and mutually inconsistent perspectives on the relationship between eschatology and cosmology.[110] 1) In the first perspective Haught adopts a process approach. Every moment of time will be redeemed by God since it will be held in God's memory (i.e., "objective immortality"). Haught writes that "everything whatsoever that occurs in evolution . . . is 'saved' by being taken eternally into God's own feeling of the world."[111] This allows him to ignore the scientific scenarios for the cosmic far future. "The so-called 'heat death' that may be awaiting the universe is not inconsistent with the notion that each moment of the entire cosmic process is taken perpetually into . . . God."[112] 2) The second perspective is an entirely naturalistic one, relying strictly on the inherent power of nature, not on God's redeeming action in Christ. According to this perspective, "the fifteen billion years of cosmic evolution now appear, in the perspective of faith, to have always been seeded with promise . . . [and] bursting with potential for surprising future outcomes."[113] To compound matters, Haught ignores the fact that whatever "seeds of promise" might have flowered until now in a reductive, naturalistic account of the universe will inevitably be cauterized by the death of the sun, the devolution of galaxies, and the decay of protons—long before the endless freeze of the open universe reduces all matter to dust. As Peacocke and others have tried to warn us, there is no eschatological hope within naturalism. 3) The third perspective is grounded specifically on a Catholic view of bodily resurrection now extended to the cosmos in the mode of Teilhard de Chardin and Rahner. According to Haught "the whole physical universe . . . somehow shares in our destiny." Here, though, where the scientific scenarios about the future of the cosmos would seem highly relevant—and apparently challenging to his Catholic roots—Haught simply ignores them.[114]

5. Eschatology and Cosmology: Initial Constructive Proposals

Eschatology as necessarily including not only the resurrection of the body but also the transformation of the universe into the new creation by a new act of the Trinitarian God is a foundational commitment to a variety of theologians today. In this section I briefly survey several of the most promising directions for developing this commitment in light of science.

Jürgen Moltmann offers several crucial reasons for requiring that eschatology be cosmic in scope. One reason is given in his doctrine of the Trinity in which the Redeemer is the Creator: "Cosmic eschatology is . . . necessary for God's sake. There are not two Gods, a Creator God and a Redeemer God. There is one God. It is for his sake that the unity of redemption and creation has to be thought." A second reason is because Moltmann's theological anthropology requires that the redemption of humanity include the redemption of the universe out of which we evolved:

> Because there is . . . no humanity detached from nature . . . there is no redemption for human beings either without the redemption of nature. . . . Consequently it is impossible to conceive of any salvation for men and women without "a new heaven and a new earth." There can be no eternal life for human beings without the change in the cosmic conditions of life.[115]

Finally, cosmic eschatology is required if we are to avoid a gnostic reading of redemption: "Christian eschatology must be broadened out into cosmic eschatology, for otherwise it becomes a gnostic doctrine of redemption, and is bound to teach, no longer the redemption of the world but a redemption from the world, no longer the redemption of the body but a deliverance of the soul from the body."[116]

Denis Edwards constructs eschatology in relation to the writings of Teilhard de Chardin, Moltmann, and Rahner, one that reflects his commitment to a relational view of the Trinity.[117] He too bases eschatology on the bodily resurrection of Jesus and the transformation of the universe into the New Creation. In doing so he raises a crucial question:

will "every sparrow that falls" be redeemed (Matt. 10:29, Lk. 12:6)? While Moltmann answers yes, Edwards leaves this question open.

Ted Peters develops the theme of prolepsis found in Pannenberg in terms of what he calls "temporal holism."[118] For Peters the cosmos is a unity of time and space which is both created *and* redeemed proleptically from the future by the Trinitarian God revealed in Jesus Christ.[119] Prolepsis ties together *futurum,* the ordinary sense of future resulting from present causes, and *adventus,* the appearance of something absolutely new, namely the kingdom of God, the renewal of creation.[120] The creation will be consummated and transformed into the eschatological future which lies beyond, but which will include, this creation. Having said this, Peters is ruthlessly honest about the challenge from science: If the cosmic future is truly "freeze or fry," our faith in the Christian God is in vain.

I have already touched on Pannenberg's way of combining eschatology as already-realized and eschatology as future-apocalyptic, namely through the idea of the "causality of the future" and of Easter as the "prolepsis" of the transformed New Creation within history. Pannenberg grounds his eschatology squarely on his formulation of the Trinitarian doctrine of God, as we shall see in chapter 1. In addition his eschatology is dependent directly on his detailed discussion of the resurrection of Jesus of Nazareth, as I surveyed in section 2 (above), with its insistence on the objectivity of the bodily resurrection and the historicity of the Easter event. But does Pannenberg succeed in connecting these theological moves with the predictions of scientific cosmology and its challenge to eschatology?

During the period between the publication of the second and third volumes of *Systematic Theology,* Pannenberg engaged in a sustained interaction with the arguments of Dyson, Barrow, and Tipler that we discussed above.[121] In his treatment of the doctrine of creation (volume 2) Pannenberg suggests that some of the underlying ideas in the proposal of Barrow and Tipler might be used theologically without adopting their model whole-cloth. These ideas include the permanent role of intelligent life in the universe; the divine reality as both emerging at the final Omega Point and as present throughout the history of the universe; and the constitutive role eschatology plays for the whole universe.[122] In volume 3, however, where the doctrine of eschatology is discussed in detail, little more is added about science. Here, Pannenberg states that Christian faith in the end of the world can neither be supported by, nor be in contradiction to, scientific knowledge. He notes that the closed model, which is finite in space and time, is "undoubtedly more compatible with the biblical view of the world than that of a world that is infinite and imperishable."[123] Still he sharply distinguishes between the biblical expectation of an imminent end of the world and the scientific view of a "remote" end, and he concludes that they may not even "relate to the same event." With this move, Pannenberg seems to be drawing back from his discussions with science into a "two worlds" position, but this is certainly better, as he has said, than relying on "easy solutions." This, in turn, leaves the playing field open to the current project: to take up Pannenberg's profound theological creativity about eschatology and begin to make clearer connections between it and contemporary science.

It seems clear that the initial proposals surveyed above, although recognizing the enormous conceptual challenge posed by physical cosmology to Christian eschatology, do not take us very far in responding to these challenges. An important exception, however, is the brilliant contributions by John Polkinghorne, touched on previously in our discussion of the bodily resurrection and now extended to eschatology and cosmology.

In his Gifford Lectures, Polkinghorne turns to the relation between the resurrection of Jesus and an eschatology of the New Creation.[124] Here we see the crucial importance of the empty tomb: Just as Jesus' body was transformed into the risen and glorified body, so the "matter" of the new creation must come from "the transformed matter of this world." This leads Polkinghorne to develop the idea of "creation *ex vetere*":

> The new creation is not a second attempt by God at what he [sic] had first tried to do in the old creation. . . . The first creation was *ex nihilo* while the new creation will be *ex vetere* . . . the new creation is the divine redemption of the old . . . [This idea] does not imply the abolition of the old but rather its transformation.[125]

There are clues as to what this new creation, this "new heaven and new earth," will be like. The resurrection of Christ within history is an anticipation of the eschaton beyond history. The incarnation is a bridge between Creator and creature. Themes of continuity and discontinuity in the gospel accounts of the resurrection of Jesus indicate something of what the transformed, eschatological "matter" will be like, as does the Eucharist for us today. The New Creation will be temporal and everlasting in some sense. Panentheism will hold for the new creation, though not for the present one. The new creation will be a "totally sacramental world, suffused with the divine presence [and] free from suffering." Still, being created as an *ex vetere* act of God means that "the present created order [has] . . . a profound significance, for it is the raw material from which the new will come." It underscores our concern for this world, even this planetary environment, and it offers the only viable response to the problem of suffering and thus theodicy. Thus "eschatology is indispensable to theology."[126]

These themes surface again in Polkinghorne's more recent writings. "God [did] not straightaway create a world free from death and suffering . . . the new creation is not due to God's wiping the cosmic slate clean and starting again. Instead, what is brought about is the divine redemptive transformation of the old creation. The new is not a second creation *ex nihilo,* but it is a resurrected world created *ex vetere.*" Since there is both continuity and discontinuity in the transformation of the present universe, then science "may have something to contribute" to our understanding of this transformation. In particular, the continuities might lie within the province of science, or more precisely, what Polkinghorne calls "metascience . . . the distillation of certain general ideas from the scientific exploration of many particulars." These include the significance of relationality and holism; the concept of information of a "pattern-forming kind" in addition to the familiar ideas of matter and energy; mathematics; and a dynamic view of reality as "open becoming."[127]

In this volume I build on the combined work of these scholars to take us further in the fruitful interaction between eschatology and scientific cosmology. As mentioned above, I do so by using a method that I have developed over the past decade, Creative Mutual Interaction (CMI), for relating theology and science.

D. CMI: THE METHOD OF CREATIVE MUTUAL INTERACTION WITH EIGHT PATHS BETWEEN THEOLOGY AND SCIENCE

The construction of CMI draws on several crucial achievements in theology and science: the defense of epistemic emergence (holism) against epistemic reductionism and the articulation of a methodological analogy between the ways theories are constructed and tested in science and in theology (where they are called "doctrines").[128]

1. Epistemic Emergence

The method of CMI defends the humanities in general, and theology in particular, from eliminative epistemic reductionism through an articulation and defense of "epistemic emergence" championed by Ian Barbour, Francisco Ayala, Arthur Peacocke, and others as early as the 1970s.[129] Here "emergence" includes two distinct claims: 1) Against reductionism the disciplines of the sciences and the humanities, including theology, can be placed in a series of epistemic levels that reflect the increasing complexity of the phenomena they study. In this "epistemic hierarchy," the disciplines that study phenomena at higher levels of complexity include processes and properties that cannot be reduced entirely to the description of the processes and properties offered by disciplines that study phenomena at lower levels of complexity. 2) Against a "two worlds" approach it is argued that the lower-level disciplines place crucial constraints on the higher-level ones. For example, while physics, as the bottom level, places constraints on biology, the processes, properties, and laws of biology cannot be reduced without remainder to those of physics. Because physical cosmology, including big bang, inflation, and quantum cosmology, is part of physics, it constrains all of the other disciplines including theology and thus eschatology. (We saw this concern above in the challenge raised by scientific predictions about the far future of the universe.) Therefore the predictions of "freeze or fry" must place constraints on and challenge what theology can claim eschatologically. No appeal to contingency, quantum physics, chaos theory, Whiteheadian novelty, emergence, the potency of nature, or metaphysics alone will be sufficient to solve this problem.[130]

2. Methodological Analogy

The methodology of CMI also draws on Ian G. Barbour's groundbreaking insight dating back to the 1960s and fully articulated in his 1990 Gifford Lectures.[131] According to Barbour, the disciplines of the sciences and the humanities, particularly theology, involve a pivotal analogy between their respective methods of theory construction and testing. Thus theological method can be seen, to a large extent, as analogous to scientific method though with several important differences. This claim is both a description of the way many theologians actually work and a prescription for progress in theological research. Like scientific theories, Barbour views theological doctrines as working hypotheses held fallibly (à la Popper), constructed through metaphors that refer even if partially (à la Ricoeur) and then through models that are sustained sets of metaphors. Doctrines, in turn, are tested in light of the data of theology, now including the results of the sciences as well as such traditional sources as scripture, tradition, reason, and experience.[132]

3. Creative Mutual Interaction (CMI)

How then are we to proceed? To address this question my proposal is to combine Peacocke's epistemic emergence with Barbour's analogy between scientific and theological rationality. The result is to place theology and physics into a single framework, as indicated by figure A.4.[133]

I call this method "creative mutual interaction" because CMI is truly interactive: it not only includes five distinct paths from science to theology but in a rare move in the theology-science dialogue it also identifies three different paths back from theology to science. The first five paths may be taken together as representing what is generally called a "theology of nature," where theology is reconstructed in light of science. The problem is that this term in itself does not differentiate between these five paths to show each of their unique characteristics. CMI is markedly different from what is frequently called a "systematic synthesis" of theology and science based on a given metaphysical system such as Ian Barbour offers under the rubric of "integration" in his famous fourfold typology.[134]

By identifying the distinctiveness of each of these eight paths and clarifying their differences, we can gain a much clearer sense of how to compare the intentions and results of specific research projects undertaken by scholars in the field. To see this in more detail, I will set out very brief examples of each of these paths.[135]

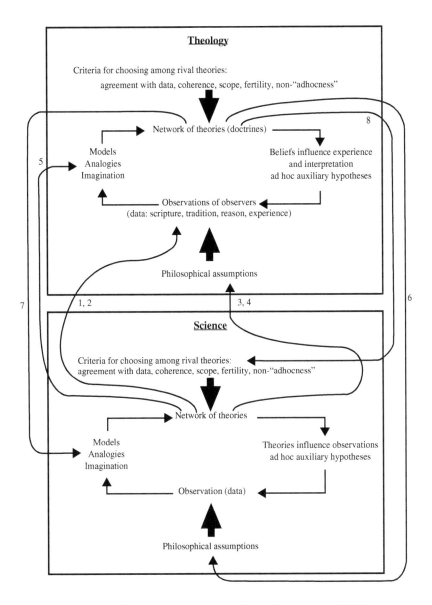

Figure A.4. The Method of Creative Mutual Interaction (CMI)

Five Routine Paths from Science to Theology: SRP → TRP

I use the symbol SRP → TRP (Scientific Research Program → Theological Research Program) suggested to me by George F. R. Ellis to indicate that the following five paths represent ways in which scientific research programs (SRPs) can influence theological research programs (TRPs).

Path 1. Theories in physics can act directly as data that places constraints on theology. So, for example, a theological theory about divine action should not violate special relativity.

Path 2. Theories can act directly as data either to be "explained" by theology or as the basis for a theological constructive argument. Thus t = 0 in standard big bang cosmology was often explained theologically via *creation ex nihilo*.

Path 3. Theories in physics, after philosophical analysis, can act indirectly as data for theology. For example, an indeterministic interpretation of quantum mechanics can function within philosophical theology as making intelligible the idea of non-interventionist objective divine action.

Path 4. Theories in physics can also act indirectly as the data for theology when they are incorporated into a fully articulated philosophy of nature (e.g., that of Alfred North Whitehead).

Path 5. Theories in physics can function heuristically in the theological context of discovery, by providing conceptual or aesthetic inspiration, etc. So biological evolution may inspire a sense of God's immanence in nature.

Three New Paths from Theology to Science: TRP → SRP

I use the symbol TRP → SRP (Theological Research Program → Scientific Research Program), also suggested by George F. R. Ellis, to indicate that the following three paths represent ways in which theological research programs can influence research in the sciences and its underlying philosophical assumptions.

Path 6. It is now abundantly clear that theological ideas based in the doctrine of creation *ex nihilo* provided some of the philosophical assumptions such as contingency and rationality that underlay and fed into the birth of scientific methodology.

Path 7. Theological theories can act as sources of inspiration in the scientific "context of discovery," that is, in the construction of new scientific theories. An interesting example is the subtle influence of atheism on Hoyle's search for a "steady state" cosmology.

Path 8. Theological theories can lead to "selection rules" within the criteria of theory choice in physics. For example, if one considers a theological theory as true, then one can delineate what conditions must obtain within physics for the possibility of its being true. These conditions in turn can serve as motivations for an individual research scientist or group of colleagues to choose to pursue a particular scientific theory.

The *asymmetry* in the interaction between theology and science should now be quite apparent: theological theories do not act as data for science, placing constraints on which theories can be constructed in the way that scientific theories do for theology. This, again, reflects the prior assumption of epistemological emergence, namely that the natural sciences are structured in an epistemic hierarchy of constraints and irreducibility, and that this hierarchy then extends to the neurosciences and cognitive sciences, the social sciences, the psychological sciences, etc., and finally to the humanities including art and music, economics, jurisprudence, political science, and so on, and, finally, theology. It also safeguards science from any normative claims by theology or the other humanities. Nevertheless it celebrates the creative role theology has played historically in the rise of modern science and the role it is playing today, even if mutely, within the diversity of scientific research projects in such areas as theoretical physics and cosmology. Together these eight paths portray science and theology in a much more interactive mode. Through this method of CMI, we can begin to delineate the conditions needed for real progress in "theology and science" through a series of guidelines for evaluating which directions to take for research progress.

E. GUIDELINES FOR NEW RESEARCH IN ESCHATOLOGY AND IN SCIENTIFIC COSMOLOGY

Given CMI we are prepared to engage in a twofold project: By following paths 1–5 in part one of this volume, we will construct a more nuanced understanding of Pannenberg's eschatology in light of physics and cosmology, as indicated by the symbol SRP → TRP. Then, in part two, by following paths 6–8 we will search for a fresh interpretation of the philosophy of physics (specifically the philosophy of time) as well as potentially fruitful new research directions in science as indicated by the symbol TRP → SRP. If such a project is at all successful, it might eventually be possible to bring these two trajectories together at least in a very preliminary way to give a more coherent overall view than is now possible of the history and destiny of the universe in light of the resurrection of Jesus and its eschatological completion in the parousia.

This book is the first full-length instantiation of the CMI project. Beyond this volume, the project will clearly be a long-term undertaking requiring the participation of scholars from a variety of fields in the sciences, philosophy, and theology working together in joint, collaborative interdisciplinary research. But for now, how do we proceed? My sense is that we first need some broad guidelines or criteria of theory choice for deciding between new directions in theology and in science, criteria that will point us in a fruitful direction. Using them, we can begin to explore specific ways to enter into the research. A metaphor I like to use to illustrate the need for such guidelines is that of searching for a needle in a haystack—when one is confronted by an immense field of haystacks. Instead of just starting with the one closest at hand we first need some way to decide which haystacks are more likely to have a needle and which ones less likely or even unlikely to have one. These guidelines are meant to help us at this metalevel in which we are first searching for candidate haystacks and leaving to the future more detailed searches of individual haystacks which pass the test of the guidelines.

On a more personal note, the motivation and encouragement for this CMI project came to me in large measure in the spring of 2001, through a simple, yet life-changing, insight:

> If it is impossible, it cannot be true.
> But if it is true, it cannot be impossible.

In essence if an eschatology based on the resurrection of Jesus is impossible because of science, then its report by the New Testament cannot be true and must instead be considered mythological. But if the New Testament witness is in some sense true, then the resurrection of Jesus cannot be impossible and our task is to find hints of the ways in which its possibility not only points toward the true eschatological future but to elements already present in the universe as we know it through science.

Over the past decade I have also been encouraged by Dante's moving insight that if our reasoning undercuts our faith, then the truth of faith will be hidden from us.[136] This has motivated me to prioritize faith but to insist on the crucial role of reason in relation to faith, perhaps never more wonderfully put than by Augustine: *fides quaerens intellectum* (roughly translated as "faith seeking understanding"). It is my hope that CMI and the following guidelines, which are meant to "flesh out" these ideas in a bit more detail, will reflect Augustine's dictum and launch us into the project of this current volume.

1. SRP → TRP: Guidelines for Revising Eschatology in Light of Contemporary Science

Guideline 1. Rejection of two philosophical assumptions about science: the argument from analogy and its representation as nomological universality. The first guideline deals with the fundamental challenge physical cosmology poses to the kind of eschatology we are considering here: namely, one based on our choice of the "hardest case"—the bodily resurrection of Jesus. In bare form, the challenge from *science to theology* is stark: If the predictions of contemporary scientific cosmology come to pass then the general resurrection will never happen. But the logic of Paul in 1 Corinthians 15 is inexorable: if there will never be a general resurrection, then Christ has not been raised from the dead, and our hope is in vain.[137] Thus the challenge from science is *first* to the general resurrection and *then* to the resurrection of Jesus. In this sense I am taking on board the "hardest case" of the challenge from science, one aimed at the objective interpretation of the bodily resurrection of Jesus, or what I am calling the Barthian tradition, and not at other accounts of his resurrection that are subjective interpretations in the so-called Bultmannian school.

The challenge can also be seen as coming from *theology to science:* If it is in fact true that Jesus rose bodily from the dead, then the general resurrection cannot be impossible. This must in turn mean that the future of the universe will not be what scientific cosmology predicts.

We seem to be at loggerheads. How are we to resolve this fundamental impasse? My response is that the impasse is not technically from science per se but from a philosophical assumption that we routinely bring to science, namely that the predictions of well-winnowed theories hold without qualification. It is quite possible, however, to accept a very different philosophical assumption about the future predictions of science while accepting all that science describes and explains about the past history and present state of the universe. The careful reader will note my dependence here on Pannenberg's arguments for the historicity of the bodily resurrection of Jesus discussed above (section B).

The first step forward is deciding whether the laws of nature are descriptive or prescriptive. Science, alone, cannot settle the matter but a strong case can be made on philosophical grounds that the laws of nature are descriptive.[138] The second step is to claim on theological grounds that the processes of nature described by science in terms of the laws of physics, biology, and so on, are actually the result of God's ongoing action as Creator within nature and not of nature acting entirely on its own. The regularity of natural processes is ultimately the result of God's faithfulness, even if God bequeaths a significant degree of causal autonomy to nature. Finally, if God is free to act in radically new ways (which of course God is!) not only in human history but in the ongoing natural history of the universe, then the future of the cosmos will not be what science predicts. Instead, the cosmic far future will be based on a radically new kind of divine action that began with the resurrection of Jesus, and this new act of God cannot be reduced to, or explained by, the current laws of nature, that is, by God's action in the past history of the universe.

In short, we could say that the "freeze" or "fry" predictions for the cosmological future might have applied had God *not* acted at Easter and if God were *not* to continue to act to bring forth the ongoing eschatological transformation of the universe. But because God *did* act at Easter to raise Jesus from the dead, and because God promises to complete this action in the now-and-coming eschatological New Creation, the "freeze" or "fry" predictions based on science will *not* come to pass.

Guideline 2. Eschatology should embrace methodological naturalism regarding the cosmic past and present. Any eschatology which we might construct must be consistent with science in its description of the past history of the universe. More precisely, it must be constrained by methodological naturalism in its description of the past: it should not invoke God in its scientific explanation of the causes, processes, and properties of nature. Indeed there is a strong theological reason for supporting methodological naturalism as the basis for the natural sciences, one to which I alluded earlier in pointing to the role of the doctrine of creation *ex nihilo* in the historical rise of modern science. This guideline separates my proposal as sharply as possible from movements such as "intelligent design" which criticize current theories in the physical and biological sciences for not including (divine) agency in science.

Guideline 3. "Relativistically correct eschatology": constructing eschatology in light of contemporary physics (paths 1, 2). We must be prepared to reconstruct current work in eschatology and related areas in theology in light of what contemporary physics tells us about the universe, following paths 1 and 2. In chapter 4 I refer to this project as the attempt to construct a Lorentz-covariant interpretation of eternity and omnipresence.

Guideline 4. Big bang and inflationary cosmology as an epistemic constraint on any revised eschatology (path 1): the material argument. Guideline 4 follows path 1 by stating that standard and inflationary big bang cosmologies, or any other current scientific cosmology, place an epistemic constraint on any possible eschatology. All we know about the development of the universe from the past up to the present, and all we know about the evolution of life in it, will be data for, and will put constraints on, eschatology.

Guideline 5. Metaphysical options: limited but not forced (paths 3, 4). In revising contemporary eschatology there are various metaphysical options from which we may choose; they are not determined exclusively by science. These options include physicalism, emergent monism, dual-aspect monism, ontological emergence, and panexperientialism (Whiteheadian metaphysics).

We must now begin the enormous job of reconstructing eschatology in a way that takes up and incorporates our theological requirements (i.e., the resurrection of Jesus as bodily transformation and new creation, and the problems of theodicy raised by evolution) as well as all the findings of science, and particularly scientific cosmology, without falling back into the philosophical problem guideline 1 is meant to address.

Guideline 6. Eschatology in light of the "resurrection of the body."

Guideline 6a. "Transformability" and the formal conditions for its possibility (the "such that" or "transcendental" argument). Our starting point, based on the bodily resurrection of Jesus, is that the new creation is not a replacement of the old creation, or a second and separate creation *ex nihilo.* Instead, God will transform God's creation, the universe, into the new creation *ex vetere,* to use Polkinghorne's phrase. It follows that God must have created the universe such that it is transformable, that is, that it can be transformed, by God's action. In particular, God must have created it with precisely those conditions and characteristics which it needs as preconditions in order to be transformable by God's new act. Moreover, if it is to be transformed and not replaced, God must have created it with precisely those conditions and characteristics which will be part of the new creation. Since science offers a profound understanding of the past and present history of the universe (guidelines 2–4), then science can be of immense help to the theological task of understanding something about that transformation if we can find a way to identify, with at least some likelihood, these needed conditions, characteristics, and preconditions. I will refer to these as "elements of continuity." Science might also shed light on which conditions and characteristics of the present creation we do not expect to be continued into the new creation; these can be called "elements of discontinuity" between creation and new creation. There must also be conditions and characteristics of the New Creation which, as additional "elements of discontinuity," are not present in the creation as we know it. It is less clear whether science can shed light on these conditions and characteristics.

Guideline 6b. Continuity within discontinuity: inverting the relationship. Closely related to the previous guideline is a second formal argument about the relative importance of the elements of continuity and discontinuity. So far in theology and science, discontinuity has played a secondary role within the underlying theme of continuity in nature as suggested by the term "emergence." Accordingly, irreducibly new processes and properties (i.e., discontinuity) arise within the overall, pervasive and sustained background of nature (i.e., continuity). Thus biological phenomena evolve out of the nexus of the physical world, the organism is built from its underlying structure of cells and organs, mind arises in the context of neurophysiology, and so on. Now, however, when we come to the resurrection and eschatology, I propose we "invert" the relation: the elements of "continuity" will be present, but within a more radical and underlying "discontinuity" as is denoted by the transformation of the universe by a new act of God *ex vetere.* With this inversion, discontinuity as fundamental signals the break with naturalistic and reductionistic views such as "physical eschatology," while continuity, even if secondary, eliminates a "two worlds" eschatology. It also eliminates non-interventionist objective special divine action (NIODA) as a candidate, since NIODA does not involve a transformation of the whole of nature but instead assumes that the present way nature works is compatible with God's acting without intervening.

Together, guidelines 6a and 6b can be thought of as representing what I previously called "first instantiation contingency" or FINLON, the "first instance of a new law of nature." I developed this idea while analyzing Pannenberg's writings on contingency, particularly as found in *Jesus—God and Man*. In a "weak" sense, FINLON refers to first instances of phenomena within creation, such as the emergence of life on earth. Here, however, in the context of the resurrection of Jesus, FINLON refers to the first instance of a radically new phenomenon, one which might better be termed "the first instance of a new law of the new creation" (FINLONC).[139]

Guideline 7. Eschatology in light of the challenge of evolutionary theodicy. In previous writings I have studied the problem of theodicy raised by evolution and with it the four-billion-year history of natural evils on planet earth. In response I have explored two approaches. The first is based on the Augustinian tradition as developed in the twentieth century by Reinhold Niebuhr. The second is based on John Hick's appropriation of the theodicy of Friedrich Schleiermacher (which Hick associates with the theology of Irenaeus). Both lead to insights and problems. These in turn were used to generate criteria which any acceptable eschatology must address. I thus changed them from problems to criteria of theory choice in theology.

Guideline 7a. Drawing on the Augustinian/Niebuhrian theodicy. The severe problems arising in the Augustinian/Niebuhrian theodicy in light of the challenge of evolutionary biology, physics, and cosmology brings with it an exceptional gift: It gives us a crucial insight based on the idea that in the new creation it will be impossible to commit moral evil (Augustine: *non posse peccare*). Then by analogy the new creation will not include natural evil either, and this in turn has implications for the role of physics such as thermodynamics not only in creation but in the new creation. For example it might mean that the new creation will not include thermodynamics to the extent that it contributes to natural evil through the law of increasing entropy but that it would include it to the extent that it contributes to natural good through the way the processes of open thermodynamic systems help make possible biological evolution and living systems.[140]

Guideline 7b. Drawing on the Schleiermacher/Hick theodicy. We saw that the gravest challenge raised by the Schleiermacher-Hick trajectory in theodicy is the combination of excessive suffering in the world and the attempt to justify it by a "means-end" argument. Hick's response was that only eschatology can provide a context for addressing the challenge of theodicy. The overall goodness of creation—its truly "greater good"—must lie not in the present, as with Augustine, but in the eschatological future.[141]

As before, this implies that the eschatological "new creation" must include not only humanity, but all the species and individual creatures in the history of life on earth—and not merely as included through the human experience of redemption. In particular, every moment in the history of the evolution of life, and not just the "end" of historical time, must be taken up and transformed eschatologically by God into eternal life. One might say that, rather than a means/end eschatology it must be such that every means is also an end in itself. The involuntary suffering of all of nature—each species and each individual creature—must be taken up into the voluntary suffering of Christ on the cross (theopassianism) and through it the voluntary suffering of God the Father (patripassianism).[142]

2. TRP → SRP: Guidelines for Constructive Work in Science in Light of a Revised Eschatology

Our project also involves the question of whether such revisions in theology might be of any interest to contemporary science—at least for individual theorists who share eschatological concerns such as developed here and are interested in whether they might stimulate a creative insight into research science.

Guideline 8. Theological reconceptualization of nature leading to philosophical and scientific revisions (path 6). Here we move along path 6 in discovering whether a richer theological conception of nature both as creation and as new creation can generate important revisions in the philosophy of nature that currently underlies the natural sciences, the philosophy of space, time, matter, and causality in contemporary physics and cosmology.[143]

Guideline 9. Theology as suggesting criteria of theory choice between existing theories (path 7). We can also move along path 7 to explore philosophical differences in current options in theoretical physics and cosmology. The theological views of research scientists might play a role in selecting which theoretical programs to pursue among those already "on the table" (for example, the variety of approaches to quantum gravity).

Guideline 10. Theology as suggesting new scientific research programs (path 8). Finally we can move along path 8 and suggest the construction of new scientific research programs whose motivation stems, in part, from theological interests.

F. SPECIAL NOTE: WOLFHART PANNENBERG AND THE NEW TESTAMENT DEBATES

Pannenberg's early writings on the resurrection of Jesus, particularly the issue of its historicity, deserve special attention here. I include it in this appendix because, even though it is not entirely central to the project of this volume, it is in many ways essential background reading. It is divided into issues relating to the appearances and the empty tomb.

Before proceeding, please note that on the one hand, I agree with Pannenberg that the secular historical sciences should not rule out of bounds a priori the possibility of Jesus' resurrection, since secular history's categories are not sufficient to contain all that happens in history (and in this way I rule out an epistemic reductionism of theology to history). On the other hand I do not agree with Pannenberg, if he says this, that secular historical science should be adjusted to contain the conceptual tools needed in order to be open to the possibility of Jesus' resurrection. Theology's categories transcend those of secular history (and in this way I support an epistemic emergence of the knowledge available to theology from that for history), but they do not alter those of secular history.

1. The Appearances

Are the appearances historical? In Pannenberg's view the gospel accounts "have such a strongly legendary character that one can scarcely find a historical kernel of their own in them." Still, these legendary elements are meant to emphasize the "corporeality of the appearances." But the key source for the historical question of the appearances of the risen Lord is 1 Corinthians 15:1–11. Here Paul's intention is "clearly to give proof by means of witnesses for the facticity of Jesus' resurrection." Pannenberg clarifies what he means by "historical proof." It is not "a historical proof in the modern sense" if that means "without personal engagement," since Paul clearly argues from "an inner involvement." Instead, the "vital interest of the historian already lies at the basis of all historical investigation, even though such an interest certainly cannot be permitted to prejudice the results of the inquiry." In particular, we know that historical inquiry involves "preunderstanding" (*Vorverständnis*), and if that is the case, Paul is certainly offering a historical proof by his standards and by ours (JGM, 89).

Were the appearances really experienced by the disciples? Pannenberg claims that Paul's account in 1 Corinthians was probably written in the spring of 56 or 57 in Ephesus, and it includes earlier knowledge of the resurrection that he received. Now Galatians 1:18 places Paul in Jerusalem three years after his conversion, and the conversion was probably in 33 or 35. The mention of the appearance to James may date back to Paul's visit to Jerusalem. If Jesus died in 30, then Paul was in Jerusalem between six and eight years afterward. So the statements in 1 Corinthians 15 are most likely within a few years of the actual appearances. The formula itself probably originated in the first five years after Jesus' death. "In view of the age of the formulated traditions used by Paul and of the proximity of Paul to the events, the assumption that appearances of the resurrected Lord were really experienced by a number of members of the primitive Christian community and not perhaps freely invented in the course of later legendary development has good historical foundation." Given this, the attempt to explain the origin of Christianity by making parallels with the histories of other religions is "an idle venture." But even if we conclude that the disciples were convinced that the risen Lord appeared to them, we cannot on this basis alone determine what sort of experiences these may have been. Here "the greatest difficulties begin to arise" (JGM, 91).

What sort of experiences were these? For the probable nature of the appearances, Pannenberg moves to the accounts in Acts (i.e., Acts

9:1–22; 22:3–21; 26:1–23), which he claims also agree with Paul's statements in Galatians 1:12, 16f. From these he draws five conclusions: 1) It is Jesus who appears to Paul. 2) Paul saw a spiritual body, not a person with an earthly body. 3) The encounter takes place from "heaven"; it is not an encounter on earth. This corresponds to the oldest New Testament tradition in which the Resurrection and Ascension coincide (Phil. 2:9; Acts 2:36; 5:30f; Mk. 14:62). 4) The Damascus appearance happened as a light phenomenon (Acts 9:3f), though the terminology of light here is "a general stylistic element." 5) The appearance was also connected to an audition, conveying to Paul the gospel of freedom from the law (Gal. 1:12).

These elements can be presupposed for the other appearances, with the possible exception of the fourth. If we view "the completely alien reality" of these appearances as an encounter with someone raised from the dead, we can only explain it "from the presupposition of a particular form of the apocalyptic expectation of the resurrection of the dead" (JGM, 93).

What are we to make of the term "vision" regarding the appearances? Pannenberg offers a lengthy discussion of attempts since D. F. Strauss to explain the appearances in epistemological, psychological, or psychiatric terms. He grants that the appearances may have involved "an extraordinary vision, not an event that was visible to everyone." But this does not mean we can presuppose that the subjective aspect of "vision" means that the NT appearances lacked an "extrasubjective reality." Pannenberg claims that the "subjective vision hypothesis" has failed for several reasons: First, "the Easter appearances are not to be explained from the Easter faith of the disciples; rather, conversely, the Easter faith of the disciples is to be explained from the appearances." Second, it cannot account for the number and temporal distribution of the appearances, which arose in three stages: 1) the appearance to Peter in Galilee; 2) the appearance to James, who only later joined the community in Jerusalem; and 3) the appearance to Paul. Moreover the appearances to the Twelve, the apostles, and the five hundred come within these three stages and may also be historical, particularly the five hundred since Paul claims that some are still alive and can be checked (JGM, 96).

How then should the historian attempt to reconstruct the events triggering the emergence of primitive Christianity? Pannenberg's response to this question may be his most important contribution to the debate, as it challenges the Enlightenment assumptions about the uniformity of history. Instead, according to Pannenberg, the possibilities to be admitted will depend upon the understanding of reality that is brought to the task by the historian. If the historian assumes unquestionably that "the dead do not rise" (1 Cor. 15:16) then it is a foregone conclusion that Jesus did not rise. On the other hand, if the apocalyptic expectation of resurrection is admitted as a possibility, then this must be considered in reconstructing these events—even if it entails using metaphorical language such as the disciples used (JGM, 74). With this in mind, the resurrection of Jesus would be designated as "a historical event" in the following sense: "If the emergence of primitive Christianity . . . can be understood in spite of all critical examination of the tradition only if one examines it in the light of the eschatological hope for a resurrection from the dead, then that which is so designated is a historical event, even if we do not know anything more in particular about it. Then an event that is expressible only in the language of the eschatological expectation is to be asserted as a historical occurrence" (JGM, 98)

What about conflicts with science? Some have argued that the resurrection conflicts with science and violates the laws of nature. Pannenberg challenges this argument with two insights: 1) only some of these laws are ever known; 2) the world as a whole is a singular, irreversible process in which individual events are never completely determined by natural law. Instead he asserts that both law and contingency are operative, and in one sense everything that happens is contingent—even the laws of nature. Thus natural science can only give the probability of an event's occurrence, but not rule it out as impossible. "The judgment about whether an event, however unfamiliar, has happened or not is in the final analysis a matter for the historian and cannot be prejudged by the knowledge of natural science."

Can historical research help decide the historicity of the Resurrection? Some theologians have argued against the Resurrection as historical, since it entails the beginning of a new aeos which cannot be perceived with the eyes of the old aeon and since historians are restricted to the standards of the old aeon. Pannenberg admits that "there is something quite correct about this argument." The resurrection of Jesus involves the new creation, so that the Lord is "not perceptible as one object among others in this world." Instead the experiences of the resurrected

Lord must be called extraordinary experiences or visions, describable only in metaphorical language. At the same time, these visions occurred in the midst of our reality at definite times to particular people. Thus these events, whether affirmed or denied, are historical events. "There is no justification for affirming Jesus' resurrection as an event that really happened, if it is not to be affirmed as a historical event as such." It is not made certain by faith but by historical research, the only method for deciding on the historicity of past events (JGM, 99).

2. The Empty Tomb Traditions

Moving next to issues surrounding the Empty Tomb (ET) traditions, Pannenberg first discusses Paul's silence. Paul nowhere mentions the ET, but this does not necessarily undermine the trustworthiness of the ET tradition. One reason is that "the ET does not affect the parallel between the Christ event and the destiny of the believers, which Paul explains again and again in his letters." Another is that the ET is unique to Jesus: after a very short time he was "transformed" to another life. Because of this the ET was not of interest to Paul, who may not even have been aware of the Jerusalem tradition. The situation was completely different for the primitive community in Jerusalem: they could not have affirmed the resurrection of Jesus if the grave contained his body (JGM, 100).

Is the ET tradition trustworthy? Pannenberg argues that it is, citing the polemic of the Jews against the Christian message which agreed that the tomb was empty but explained it differently. Moreover, the trustworthiness of ET does not depend primarily on Mark 16. "Even if the account of the discovery of Jesus' grave which has been reserved for us should be shown to be a late legend conceived in the Hellenistic community, the weight of the arguments presented here would remain. Only when one restricts oneself one-sidedly to the analysis of the textual tradition for the basis of the historical judgment . . . can one really come to a negative result in the question of Jesus' ET." If we start with historical considerations, the existing tradition confirms what is historically probable: the ET was known in Jerusalem. "Only if the existing text virtually forced one to make the opposite judgment could the weight of the historical argument from the relation between the resurrection proclamation in Jerusalem and Jesus' ET, which this proclamation presupposes, be countered at all" (JGM, 101–2).

How should we interpret Mark 16:7, 8? Pannenberg turns to Mark 16 since the tradition "in its most original form" is preserved there. The two most difficult questions posed by the text are: Is the original conclusion of the Gospel of Mark chapter 16, verse 8? What is the relation between verses 7 and 8? After a detailed discussion, Pannenberg concludes that "even if the traditions carry a strongly legendary overgrowth, they point in the opposite direction, which had to be assumed historically from the very beginning as the presupposition for the resurrection kerygma in the Jerusalem community: the Jews there, as well as the Christians, were familiar with the fact of the ET." The burial is tied to Joseph of Arimathaea. "This can hardly have been invented secondarily, since the entire tradition about Jesus' burial hangs on this name." The claim that "the Jews" buried him (Acts 13:27–29) has no historical value because the removal from the cross was the responsibility of the Roman authorities. Moreover, the term "the Jews" points to the Lucan schema of the speeches in Acts. Instead, the account of the burial and the discovery of the grave has a firm place in the pre-Marcan Jerusalem passion tradition. The Pauline passage in 1 Corinthians 15:4 indicates the age of the grave tradition (JGM, 103).

CONCLUSION

What is the relation between the ET and the appearances? According to Pannenberg, most scholars argue that the appearances took place in Galilee, and of course the tomb was in Jerusalem. The relation between them hinges on the idea of the "disciples' flight" to Galilee: did the disciples flee to Galilee immediately after Jesus was taken prisoner or remain in Jerusalem at first? If the former, then they would not have been part of the discovery of the ET, and the Easter appearances would have been an independent tradition. According to von Campenhausen, the "disciples' flight" is rejected as a "legend of the critics," but Pannenberg argues against him. His conclusion: the disciples fled to Galilee, where the appearances took place, and came to know about the grave only after their return later to Jerusalem. Thus the ET and the appearances traditions came into existence independently of each other. This in turn supports the historicity of the Resurrection. "If the appearance tradition and the grave tradition came into existence independently, then by their mutually complementing each other they let the assertion of the reality of Jesus' resurrection, in the sense explained above, appear as historically very probable, and that always means in historical inquiry that it is to be presupposed until contrary evidence appears" (JGM, 105). Pannenberg has made a very persuasive case, even if it is now some four decades old. Clearly responses would need to be taken into account for any current assessment of what Pannenberg wrote here. Nevertheless his clarity and foresight are self-evident and praiseworthy.[144]

PART ONE

SRP → TRP

Revising Pannenberg's Trinitarian Conception of Eternity and Omnipresence and Its Role in Eschatology in Light of Mathematics, Physics, and Cosmology

The Trinitarian Conception of Eternity and Omnipresence in the Theology of Wolfhart Pannenberg

*The relation between time and eternity is the crucial prob-
lem in eschatology, and its solution has implications for all
parts of Christian doctrine.*

—Wolfhart Pannenberg, *Systematic Theology*

In this first chapter I provide in detail what can be broadly called Wolfhart Pannenberg's Trinitarian conception of eternity and omnipresence.
This chapter serves as a basis for the developments of the rest of this
volume, all of which follow and exemplify my interdisciplinary method,
Creative Mutual Interaction, or CMI. Given its culminating role in his
lifetime of theological research, Wolfhart Pannenberg's massive three-
volume *Systematic Theology* is my primary focus here.[1] However, I also
use material from Pannenberg's *Metaphysics,* from his early work *Theology and the Kingdom of God,* from a chapter in his recently published
The Historicity of Nature, and articles from various journal sources.[2]
These sources are in no way meant to be exhaustive of Pannenberg's writings on the topic of eternity and omnipresence, but I do see them as representing his main position and as offering fertile grounds for new directions of inquiry. Hopefully the clear limitations on the scope of this
present volume can be superseded in future research, using this work as

a point of departure. Again, my approach is an effort to give a careful, detailed, and "friendly read" of Pannenberg's work in order to discover the profound insights that can be gleaned from this "pearl of great price" (Matt. 13:45–46) and to place them within and reformulate them in light of the ongoing interdisciplinary conversations between theology and natural science.

A. PANNENBERG'S TRINITARIAN CONCEPTION OF ETERNITY IN RELATION TO THE TIME OF CREATION

1. God's Eternity as the Source of Time

According to Pannenberg, the Old Testament depicts the eternity of God through a series of four claims.[3] First, God endures forever, "from everlasting to everlasting." Second, as opposed to all created things, God is incorruptible and unchangeable. Third, God is "the source of all life and thus has unrestricted life in himself." Fourth, God is eternal in that all of time, past and future, is present to God. That which fades into the past for us, or that which lies in our remote future, is present in the limitless duration of God's eternity.[4] In the New Testament, God's eternity includes all of creaturely time. In the Gospel of Mark, Jesus affirms God's eternity as including all of the past and the future. The book of Revelation asserts that Jesus Christ shares in the eternal life of the Father.[5] I summarize Pannenberg's understanding of the biblical concept of the eternity of God as follows: the *eternity* of God is duration as unlimited and unending time, and the eternal *God* is unchangeably the same God throughout the unlimited and unending duration of eternity.

Turning to early Christian theology, Pannenberg first describes how Greek Platonism offered crucial conceptual resources for interpreting the biblical conception of eternity. "Platonic teaching about the eternity of the ideas and the deity . . . seemed to be closely akin to Christian beliefs."[6] While Plato derived time in part from the motion of the planets and described time as the "moving image of eternity," his basic view was that eternity is timeless and unchanging, the antithesis of all that is temporal and changeable.[7] The result was consistency between the changelessness of Platonic eternity and the biblical idea of the changelessness of God, but inconsistency between Platonic eternity as timeless and the biblical insistence that all moments in time are present to God.

Plotinus and Boethius are even more pivotal sources for Pannenberg's view of time and eternity.[8] Plotinus understood eternity not as timeless but as "the source of time" and "the presence of the totality of life."[9] Eternity is not Plato's antithesis of time; instead, eternity is "the whole of life" and "the presupposition of understanding it."[10] Because eternity includes temporality, the moments of our life, which we experience as separate and transitory, can be related to each other and become part of the whole of time "if we refer them to the totality of eternity." Unlike our experience of time, in which the momentary present divides time into a vanished past and a not-yet future, the eternal present is a present that "comprehends all time, that has no future outside itself. . . . A present can be eternal only if it is not separate from the future and if nothing sinks from it into the past."[11] It is a present that includes all that is past and all that is future, and it is this temporal unity which provides the linkage for what is separated in time. In short, the sequence of events we experience in time proceeds from eternity and is "constantly comprehended by it."[12]

What, then, caused the difference between the temporal unity of eternity and the temporal fragmentation of ordinary experience? According to Pannenberg's reading of Plotinus, the World-Soul could, in principle, mediate the temporal unity of eternity to us.[13] However, because of the fall of the World-Soul from eternity, the unity and totality of life is dissolved into the separate moments (*diastasis*), which we experience as the passage of time.[14] As a consequence, the totality of time is only a future goal of life and "the path to this goal is time."[15] As Pannenberg writes, "the future thus became constitutive of the nature of time because only in terms of the future could the totality be given to time which makes possible the unity and continuity of time's processes."[16] This, in turn, leads to one of Pannenberg's trademark theses: "when the theory of time is oriented toward the eternal totality, the consequence is a primacy of the future for the understanding of time."[17]

Pannenberg then points out the influence of these philosophical views of time on the theologies of Augustine and Boethius. In developing the doctrine of creation, Augustine rejected Plotinus's idea of the World-Soul, its role in Christian Gnosticism, the fall of the World-Soul, and the basis the fall offered for the distinction between moments of time. Instead, "if God positively willed the world and all its creatures, the same applies to the temporal form of their existence."[18] Augustine then incorporated the Platonic antithesis between time and eternity into his doctrine of creation. Hence time is created along with the finite world out of God's timeless eternity. "For Augustine, there was no time before bodily movement in the world of creatures. There was thus no time in God's eternity."[19]

Boethius, however, worked with Plotinus's concept of eternity as "the simultaneous and perfect presence of unlimited life."[20] It is Boethius's view of eternity that has been of enormous consequence for some sectors in twentieth-century theology. Pannenberg tells us that there is now "widespread agreement . . . that eternity does not mean timelessness or the endlessness of time." Following Boethius's view, the divine eternity must instead be such that all created things are present to God "at one and the same time" and in a way that preserves their intrinsic temporal differences. He then makes a crucial claim: "This is possible only if the reality of God is not understood as undifferentiated identity but as intrinsically differentiated unity. But this demands the doctrine of the Trinity."[21] Pannenberg particularly applauds Karl Barth for supporting Boethius's understanding of eternity not as the antithesis of time but instead "as authentic duration and therefore as the source, epitome, and basis of time."[22] Barth "bewailed" the way Boethius's view was overlooked by earlier theologians. He criticized Schleiermacher in particular for viewing eternity as completely timeless in order to free God from every aspect of temporality. Instead Barth argued for an "order and succession" within the divine life and with it a "before" and an "after."[23] Pannenberg then breaks into his commentary on Barth to make a second crucial point: The claim that there is order and succession, or before and after, within the divine eternity "can only be made with reference to the manifestation of the Trinity in the economy of salvation. It corresponds to the realization that the immanent Trinity is identical with the economic Trinity."[24]

Returning to Barth, Pannenberg states that God's eternity can then be understood to include the entire span of creaturely time from creation to eschatological consummation, and he coins the terms "pre-, super-, and post-temporality" to refer to the differentiated temporality of God's eternity in relation to creation. Our creaturely temporality is based on

God's eternal temporality, and creaturely temporality moves toward the future which is God, the "source, epitome, and basis" of all time.[25] Barth even says that our time is "embedded" in God's eternal present.[26] Because of this, the past does not dissolve away but remains in the eternal present of God.[27] Pannenberg claims that underlying Barth's use of Boethius is Plotinus's understanding of eternity as the "simultaneous presence of the whole of time." Our experience of time is that of a succession of moments. Because of this, "we can understand the nature of time only in relation to [Plotinus's view of] eternity, since otherwise transitions from one moment of time to another make no sense."[28]

As I read him, Pannenberg makes the case, drawing primarily on his reading of Plotinus and Boethius, that eternity is the source of the temporality of creation while also overcoming the loss of the present into the past and the unavailability of the future to the present that characterizes the temporality of creation. Pannenberg begins with the straightforward assumption that every present moment has a unique past and a unique future, a past and future that we could say define, to a large extent, the distinctive meaning of *this* present moment and distinguish it from all others. But on a deeper level, the present not only distinguishes but radically separates and divides time into a set of past events that once were present but *will never again be experienced as present* and a set of future events that *one day may be present but that are never experienced in this present as present.* The key point is this: it is this latter separation and division of time *by the present* into past moments and future moments that is overcome in eternity, but not the fact that each present moment has associated with it a distinct past and future.[29] In eternity, the present, along with its distinctive past and future, is brought into a differentiated temporal unity with all other present moments, each with their own unique pasts and futures, and this differentiated temporal unity provides the reconnection for what is separated and divided in our ordinary experience of time. Finally, it is the Trinitarian character of God that provides the basis for the intrinsically differentiated unity of eternity (more about this below). In short, the passing and irreversible sequence of separated events that we experience in time is constantly comprehended by eternity as an attribute of the differentiated unity that is Trinity.

In closing this section I should note that these key concepts—the present as having a distinctive past and future, the present as separating time into a past of once present moments and a future of one day to be present moments, and eternity as overcoming this temporal separation while retaining the unique past and future for each moment in time—will be central to what I will call Pannenberg's concept of eternity as "co-present" in chapter 2.

2. The Role of Duration in Augustine's "Time-Bridging Present"

According to Pannenberg, Augustine breaks with Plotinus on several points. First, instead of Plotinus's fallen World-Soul, Augustine asserts that each person has an individual soul created by God. Next, the soul participates in eternity not in terms of the future whole of time, as Plotinus believed, but in terms of the "time-bridging present" within the soul itself. It is here that Pannenberg sees Augustine moving beyond Plotinus to "an independent analysis of the *experience* of time, one that has remained definitive for all later treatments of the human consciousness of time."[30]

Specifically, Augustine claims that the experience of the present is not limited to the point-like "now, which divorces past from future and is itself already past at the moment we notice it." Instead the present moment is experienced as "time-bridging" because of an "extension" (*distentio animi*) of the soul beyond the instantaneous now, an extension that includes our memory *(memoria)* of the past and our expectation *(expectatio)* of the future.[31] Moreover, the unity of the extended, time-bridging present arises by means of "attention" (*attentio*), which is directed toward the past and the future, that is, toward memory and expectation. "To the extent that attention can pull together that which is separated within time, and which advances moment by moment, into the unity of *one particular* present, we experience duration, the *spatium temporis*." Thus, it is duration that is the actual image of eternity within the human soul by which the soul senses and participates in eternity, even if in a very limited way.[32]

Pannenberg adds a point of clarification about his concept of duration by distinguishing between "the span of time that we can experience as an actual present"—a span of perhaps two to four seconds—and the much longer span of time that we bring into the duration of the actual present via memory and expectation. The crucial point here is that the longer span of time is experienced within the brief duration of the actual present.[33] Augustine's key examples of the time-bridging present are taken from our human experience of spoken discourse and of listening to music, in which we grasp the unity of the sequence of moments through anticipation and memory.[34] And while duration occurs in every ordered series in nature, and not just in the realm of human experience,[35] time is only experienced as duration within the human soul.[36]

Pannenberg writes that duration, though limited, is "decisive for the independent existence of creatures."[37] Duration is an "overarching present, by which [creatures] are simultaneous to one another and relate to one another" spatially.[38] Duration is both the basis for our distinction from one another and from God, and it is the way in which we sense the "indefinite totality" of our life. This "sensed totality" grows in richness through our recollection of the past and, more pivotally, through expectation of "the future that completes it."[39] Nevertheless, our duration is limited, both by our span of life and by our fragmented abilities of recollection and expectation, while God's duration is limitless. "The eternal Today of God . . . has no need of recollection and expectation. God's day lasts." Our attempt to base the totality of life on "the Now of the I's present . . . is bound to fail" as each momentary present is replaced by another. Only eternity provides "the basis on which to hold all the moments of time together in their sequence."[40]

The "Now" along with objective duration as the basis of recollection and expectation and its movement through time are all part of God's gift of creaturely existence. This "multiplicity of times . . . and events . . . is an essential prerequisite of the variety and wealth of creaturely reality [and also] of the development of creatures to mature existence and . . . creaturely independence. [They are] constituent parts of the good creation of God."[41] And such creaturely independence, resulting from the reintegration of the past into the present, is a "partial participation in the divine eternity." All creatures yearn for this eternity as it finds expression in duration, a totality of life lying in a future that will integrate all of the past moments into a unity. But since death breaks off life, the future we yearn for, and through it the totality of life, lies beyond death as a future that "will manifest the identity of our existence in full correspondence with the will of God as Creator by unbroken participation in the eternal life of God." "The finitude of creaturely being undoubtedly rules out unlimited existence but not the presence of the whole of this limited existence in the form of duration as full participation in eternity."[42] Only participation in God's eternity "can overcome the disintegration of human life" by the flow of time.[43]

In sum, the independence of creaturely existence requires temporality as the form of creaturely existence, and this includes both duration and the differentiation of the future toward which we act and the past in terms of which the self is fashioned.[44] "The fruits of an independently led life can still persist in eternity insofar as we view temporal existence in the simultaneity of the eternal present." However, "without inclusion of the difference between time and eternity we would have no conception of the process of fashioning an independent and finite being that has its own center."[45]

3. Time, Duration, and Eschatology

Commenting on time, duration, and eschatology, Pannenberg writes that "the future of consummation is the entry of eternity into time." Here again eternity is the totality of life as an undifferentiated unity of what we know in life as disintegrated time, with its fragmented past, present, and future.[46] The eschatological future "is the basis for the lasting essence of each creature" as manifested during its life but fully achieved only in the eschatological future.[47] We are both becoming ourselves and yet we already are the person we shall become.[48] Such eschatological statements, in turn, require an "ontology of the present reality of being as this is constituted by the eschatological future."[49] This reversal, in which the future is hidden in the present, provides an answer to the question of the identity of the present with the future consummation. The future is not totally different from the present; instead the present is a manifestation of the future. "The relation of the essential reality of things to their present appearance is mediated by the relation between eternity and time."[50] This essential reality is simultaneity "purged of all the heterogeneous admixtures, perversions, and woundings of their earthly existence, not of traces of the cross, but of traces and consequences of evil" that arise in the ways creatures seek to achieve independence from God. This means that the question of personal identity between our death and our resurrected life cannot be understood by analogy with our present continuity in the course of time. Instead this identity is based on the eternity of God for whom "all things that were are always present," and thus on our existence in God's eternal presence. It is the consequence of an act by the Spirit of God that also includes the renewal of all creation. Thus, all creatures go to eternity directly at death, but they only receive the totality of their existence at the end of the ages to live in common with all others in eternity.

This relation between time and eternity also applies to the kingdom of God as the consummation of society. "The end of time, like the death of the individual, is to be seen as the event of the dissolving of time in eternity." Even the different moments of time are "no longer seen apart," although their differences are "not erased." In eternity the "antagonisms" between individuals and society are overcome and the existence of all creatures is simultaneous. "Only in the sphere of eternity can there be an unrestricted actualizing of the unity of our destiny as individuals with that of humanity as a species."[51]

Moreover, Christian eschatology hopes for "the end of this aeon [as] more than an epochal turning point in the flux of time." Here, time as we know it will end because God will overcome the separation of past, present, and future. While the distinctions between past, present, and future will remain, "the separation will cease when creation participates in the eternity of God." This in turn means that "the separation in the sequence of time cannot be one of the conditions of finitude as such."[52] In this way, Christian eschatology can take up and complete Plotinus's argument where "the search for the future totality remained an empty illusion in the endlessness of the march of time." From a Trinitarian perspective, creation and history are "embraced by the economy of God for which world history is the path that leads to the future of God's glory."[53]

B. PANNENBERG'S TRINITARIAN CONCEPTION OF OMNIPRESENCE IN RELATION TO THE SPACE OF CREATION

1. Omnipresence and the Space of Creation

According to Pannenberg, God's eternity "implies his omnipresence." But while eternity means that all things are present to God, the concept of omnipresence means that God is present "to all things at the place of their existence."[54] But how is God omnipresent to them? God's omnipresence does not mean a "localized presence" nor a presence that is "extended across the whole world" for that would imply that God exists as a body.[55] Moreover, God's presence "does not exclude the simultaneous presence of other things in the same place." Instead omnipresence is "a presence that fills all things." It "comprehends all things . . . but is not comprehended by any."[56] As with the divine eternity, omnipresence includes both divine transcendence and divine immanence as one would expect of "the criterion of the true Infinite." Nevertheless, if the element of transcendence in God's omnipresence is to be more than the abstract distinction between the finite and the infinite, it requires the concept of space. This concept is crucial because, as Pannenberg reminds us, God's relation to creatures is not only an "expression of the freedom of his essence," but it is grounded in God's essence and it affects God's essence. Since, for Pannenberg, essence is "a relational concept,"[57] the idea of space provides exactly what is needed to carry "the distinction between that which is in relation and the relation itself."[58]

Scripture speaks of God as dwelling in heaven, where heaven is a spatial image expressing the differentiation between God and creation. Through this image, Scripture can claim that God gives "his earthly creatures room to live their own lives in their own present but alongside him." The concept of space as presupposed by this image rests on the "simultaneity of the eternal presence of God." In sum, earth is the space of creation in which creatures live alongside one another and not just in succession to one another, while heaven is the presence of God to creatures at the place of their existence.[59]

2. Omnipresence and Space in the Rise of Modern Science

Pannenberg enters into the eighteenth-century debate between Isaac Newton, Samuel Clarke, and Gottfried Wilhelm Leibniz on the relation between divine omnipresence and physical space. In reading his argument it is crucial to keep in mind that Pannenberg is differentiating between two well-known views of space often called "receptacle space" and "relational space."[60] In the receptacle view, space is a "container of things" while in the relational view, space is "the epitome of the relations of bodies."[61] Although Newton and Leibniz independently invented the calculus that made classical mechanics possible, they radically disagreed over the concept of space. Newton thought of space as a receptacle of matter while Leibniz believed it represented the set of causal relations between matter. Thus, for Newton, the idea of empty space is readily conceivable while for Leibniz it is inconceivable.

We should also recall that, in order to make classical mechanics coherent, Newton needed a way to distinguish between inertial and non-inertial (accelerating) motion, and this led him to stipulate the existence of what he called "absolute space" and "absolute time." Here, absolute space is an infinite receptacle space ontologically prior to physical space, which, together with absolute time, determines which motions are inertial. Finally, Clarke appealed to a third concept of space, one that is infinite, unitary, and undivided, in his defense of Newton against Leibniz. Clarke's notion is somewhat problematic because, although Newton's absolute space is infinite and unaffected by matter, it is not unitary and indivisible. Instead, as Pannenberg points out, it is composite and divisible, as evidenced from the fact that it includes the Euclidean metric by which inertial motion in physical space can be distinguished from accelerated motion.[62] In contrast, Clarke's infinite and undivided space, as Pannenberg notes, "is prior to all division or to every relation of what is divided," and Clarke associates it with the space of God's omnipresence.

Pannenberg writes that "Newton viewed physical space as the form of God's omnipresence with his creatures . . . calling space the *sensorium Dei*."[63] Leibniz objected to this idea, claiming that if physical space is the *sensorium* or *organon* of God's omnipresence, it leads to pantheism since God's omnipresence would then be composite and divisible.[64] Clarke defended Newton by claiming that God creates physical space as a means for giving each creature its own place. Division and divisibility are thus a part of finite creation. "Absolute space [in contrast] is undivided and indivisible. As such it is identical with the divine immensity."[65] Being identical with absolute space, the divine omnipresence is "the effect of God's infinity in his relationship with the world of his creatures." Moreover, Clarke stressed that "*sensorium* denotes the place, not the organ, of perception."[66] However, this makes Clarke's interpretation of Newton's concept of absolute space different from Newton's own concept of it. "Clarke's insistence that the space of God's omnipresence was not only infinite but also undivided made it difficult to identify this space with Newton's own concept of absolute space because that absolute space had to have a metrical structure" to distinguish between the inertial and non-inertial motion.[67]

Pannenberg also argues that Leibniz's understanding of space as relational reflects the Augustinian tradition that space, like time, comes into being with God's act of creating the world.[68] He points out the similarity between this conception of divine omnipresence and Plotinus's argument that eternity is the condition for making intelligible the transition from one moment of time to the next.[69] Pannenberg concludes that "Newton was unable satisfactorily to explain the union of transcendence and presence in God's relation to his creature because he did not develop his thought in terms of trinitarian theology."[70]

3. Space and Time in the Special Theory of Relativity

In *Systematics,* volume 2, Pannenberg briefly comments on the relation of space and time in Einstein's special theory of relativity. He begins by claiming that time is more basic than space: the concept of time is "constitutive" for the concept of space because "the simultaneity of what is different constitutes space." Theologically, the concept of the presence of God in space presupposes the "reduction of space to time." He notes that this constitutive relation lends "philosophical plausibility" to special relativity's idea of spacetime "as a multidimensional continuum." While acknowledging that the concept of *absolute* simultaneity is challenged by special relativity, Pannenberg's response is that relativity does not eliminate simultaneity altogether. Instead it defines simultaneity "relative to the standpoint of the observer," and in doing so it relativizes spatial measurements as well. He then returns to his discussion of time in Augustine's thought: "In the case of our sense of time [relative simultaneity] is made possible by the phenomenon of the present that bridges time as Augustine first described it."[71]

In recent work Pannenberg articulates these themes in a slightly different form. He defines space as "the order of togetherness of simultaneous phenomena" and time as "the order of their sequence." Because of the central role of the concept of simultaneity in the definition of space, Pannenberg reaffirms his claim that time is more fundamental than space.

Because of this, the theory of relativity, with its challenge to the concept of simultaneity, had "incisive consequences for the concept of space." The distinction between space and time is now seen as an "approxima-tion" of what is better described as spacetime—although, in human ex-perience the distinction remains.[72]

4. Space and Time in the General Theory of Relativity

Pannenberg claims that Einstein, in his general theory of relativity (GR), followed Leibniz in opposing Newton's presupposition of "absolute space as the receptacle of things," opting instead for a relational view of space. Einstein formulated GR by using the mathematics of Riemann, who had considered the possibility of relating the geometrical structure of space and the distribution of matter in space. Einstein utilized Riemann's idea in his "geometrical interpretation of gravity."[73] Still, although Einstein's theory is based on a relational understanding of space, it too must pre-suppose the unity of space as stressed by Clarke. The crucial point is that a geometric space is by definition divisible and "a condition of all divi-sion and all relation of spaces is the unity of space." The unity of space is then found in Clarke's view of space as "the divine immensity, which is indivisible as such" and which is presupposed in all geometric views of space, stemming from our intuition of "the Infinite as the supreme condi-tion of all human knowledge."[74]

Pannenberg asserts that these two views of space—space as rela-tional/geometrical and space as undivided—though seeming to form a dichotomy, can actually be brought at least partially together if relational space is used to refer to created space and if undivided space refers to God's omnipresence. Still, maintaining their difference is crucial if we are to avoid the pantheism of Spinoza.[75] Pannenberg also argues that geometric space might be potentially infinite in the sense of being un-limited, but it is not actually infinity. Instead, God's infinity is God's om-nipresence to all creation and a presupposition of our human view of creatures in distinct spatial relations.

Again, Pannenberg returns to these themes in recent writings.[76] Be-cause spacetime is a relational, geometrical concept, it is distinct from Clarke's ontologically prior concept of an undivided space.[77] And while a pantheistic interpretation of spacetime is possible—Einstein, after all,

was sympathetic to the pantheism of Spinoza—Pannenberg is adamant
that "space-time is not eternity." Pannenberg admits that the geometric
description of spacetime may suggest a similarity between spacetime
and eternity insofar as in eternity all events are simultaneous, but this
similarity only holds if one presuppose that spacetime entails a spatialization of time in which "the differences of tense—the distinctions between present, past and future are removed." In contrast, by eternity Pannenberg means a form of simultaneity which preserves these distinctions
"in the eternal possession of the whole of life." This is why "the priority
of undivided infinite space [and time] with regard to any specific conceptions of spatial and temporal units and as a condition of their possibility is so important." This is also why divine omnipresence is "in-
explicable" without a concept of space in which "the space of God's
omnipresence is not a container space." In this way the distinction between God and world is upheld. God is present to the world but never a
component in it (except for the Incarnation).[78]

C. SPACE, TIME, OMNIPRESENCE, AND ETERNITY:
RECENT INSIGHTS AND SUMMARY

In recent writings, Pannenberg develops the relation between eternity
and time further. He argues that "the concept of eternity is certainly opposed to the transience in the temporal succession of events." However,
this does not mean a complete opposition between eternity as a never-
changing present (*nunc stans*) and temporality as a continuously changing present (*nunc fluens*). Augustine opted for this opposition because of
his commitment to divine immutability.[79] Nor does it mean "life everlasting" as merely "time without end." Instead it refers to the "totality of
life as presently experienced." Here Pannenberg returns to the concept of
eternity in Plotinus and Boethius; "what in our experience is separated
by the course of time . . . is present all at once in eternity." This concept
of eternity includes the idea of an unchanging identity through life's sequence of events. This sequence is integrated into a wholeness of life
that is "enjoyed as present in its wholeness and therefore not subject to
change." He reiterates Plotinus's concept of the wholeness of eternity as
a presupposition for understanding the transition from moment to mo-

ment in our life, and Kant's insight that "even the single moment of time would not be conceivable except for an awareness of time as a whole." Finally, Pannenberg, citing Augustine, tells us that the unity of time as an infinite whole is present in the flow of time as the experience of duration which includes memory and anticipation.[80]

Pannenberg calls this idea of eternity "omnitemporal." It includes the infinite unity of time as presupposed in the experience of individual events, but it is not eternity as "empty time." Instead eternity includes "the differentiated fullness of life as simultaneously present." The omnitemporal eternity of God includes "the atemporal existence of God prior to the creation of the world" since, following Augustine, time itself is created in God's (causal but not temporal) act of creating the world. Still, as the "simultaneous possession of the fullness of life," the omnitemporal eternity of God also includes "the participation of the eternal God in the history of his creation, the divine economy which is finally to be consummated in the eschatological participation of creation in God's own eternal life." Since creaturely existence is related to their future participation in the divine life, and since God acts in history to draw creatures into the future participation, "the future of God, which is identical with his eternal present when it becomes the destiny of his creatures, is already the creative source of their existence." Finally, the fact that this eternal God as Trinitarian allows for both differentiation and self-differentiation means that the world of creatures can be both different from God and each other and yet can exist in the divine omnipresence. Thus, the world of creatures is destined to participate in the divine eternity while possessing their finite nature and individual identity. Our acceptance of our difference from God then becomes "the condition of having communion with that God."[81]

There is a similarity, as Pannenberg points out, between God eternally perceiving the temporal process as a whole and a scientist's view of the temporal process as a whole after it has been spatialized in terms of the geometry of spacetime. But the analogy breaks down because within God's eternal knowledge of the temporal process, the difference between temporal events and the specific past and future for each event is preserved. This view of God's eternity reflects the fact that God created time in order for creatures to be independent of each other and of God. Time allows organic creatures the ability to organize their own life and preserve themselves in relation to the environment. Similar points apply for the creature's need for space. "While in God's eternity, simultaneity, the principle of space, and everlasting continuity are united, in the world of creatures, they get separated into space and time as conditions of their finite existence."[82]

In sum, space and time within creation are not independent realities as suggested by the container view of space that traces back to Aristotle. While this view was rejected by Nicene theology[83] it was adopted in Western medieval thought and by Newton in his combination of absolute space and Euclidean geometry. Clarke offered the idea of an undivided, infinite space, a concept which is more appropriate for God's dynamic omnipresence with the world. Pannenberg claims that Newton's concept of an absolute, geometric, and empty-container concept of space "has been destroyed by the relativity theory." Instead, according to this theory, "there is an interdependence between physical objects and the spatial and temporal dimensions of their existence. There is no measurable time and space without creatures. . . . This means that God created time and space when he created the world of finite entities as dimensions of their existence."[84] However, is the undivided, infinite whole of space and time, prior to geometrical space and time, a property of the world as such, or is it, following Clarke, the omnipresence of the eternal God with the world? According to Pannenberg, measurable, geometric space and time belong to the world, but undivided infinite space and time are prior to the world and belong to eternity and omnipresence. "Their finite existence involves their setting in measurable time and space, which is created with them but takes place within some more comprehensive, infinite, and undivided space and time . . . within the orbit of God's eternity and omnipresence."[85]

D. THE TRINITY AS CRUCIAL TO THE RELATIONS BETWEEN TIME AND ETERNITY AND BETWEEN SPACE AND OMNIPRESENCE

As we saw above, following Boethius's view Pannenberg insists that the divine eternity must not be reduced to either mere timelessness or unending time. Instead, it must be such that all created things are present to God "at one and the same time" and in a way that preserves their intrinsic temporal differences. Pannenberg then makes a crucial move: such a view of eternity requires a Trinitarian doctrine of God. In specific, the presence of all things to God with their intrinsic temporality preserved is only possible "if the reality of God is not understood as undifferentiated identity but as intrinsically differentiated unity. But this demands the doctrine of the Trinity."[86] Thus, "to rightly conceive of God's eternity and properly understand eternity's relation to time, we must interpret it as the eternity of the Trinitarian God."[87]

Pannenberg next introduces the Trinity to relate God's transcendence of, and immanence to, creation. The transcendent God is present to creatures first of all as God's Spirit who creates and upholds them. More specifically, the Trinity, through the perichoresis and consubstantiality of the three persons, links the transcendence of the Father in heaven with God's presence in believers on earth in terms of the Son and Spirit. "Thus the trinitarian life of God in his economy of salvation proves to be the true infinity of his omnipresence."[88] Plurality in God as Trinity is the presupposition for the plurality of spaces that arise with God's making of creatures. But this divine plurality does not consist in spatial distinctions or "fixed divisions." Instead, "in the act of self-distinction each of the persons is one with the other from whom it is distinguishing itself." Still the bringing forth of creatures is only indirectly related to the self-distinction between Father and Son. It is "an expression of the overflowing love" of the Father for the Son. Created space, then, is constituted by the relations of finitude and limitations that characterize creatures. "From this standpoint space is the epitome of relations between divided spaces, between points of space."[89]

Finally, as we saw above, the fact that this eternal God as Trinitarian allows for both differentiation and self-differentiation means that the world of creatures can be both different from God and each other and yet can exist in the divine omnipresence. Thus the world of creatures is destined to participate in the divine eternity while possessing their finite nature and individual identity.[90]

But what is the basis for the unity of the Trinity, that is, the relation between the doctrine of the Trinity and the assertion that Christian theology is monotheistic, not tritheistic? Pannenberg is critical of the Eastern tradition of subordination in which the unity of the Trinity is found in the generation of the Son and the Spirit from God the Father.[91] He is also unsatisfied with the argument by the Cappadocians that the unity of the divine persons is grounded in the unity of their activity in the world.[92] Instead he takes up Athanasius's idea that the distinction between the divine persons is constituted by their relations. Given this idea, the divinity of the Father, which was never challenged, could be attributed by these constitutive relations to the Son and to the Spirit, thus establishing their divinity. The constitutive relations and the divinity of the three persons together establish, at least in part,[93] the argument for the Trinity as monotheistic.

> The Father cannot be thought of as Father without the Son. This was [Athanasius's] decisive argument for the full deity of the Son. He transferred the same argument to the relation between the Father and the Spirit. . . . Athanasius . . . argued forcibly against the Arians that the Father would not be the Father without the Son. Does that not mean that in some way the deity of the Father has to be dependent on the relation to the Son, although not in the same way as that of the Son is on the relation to the Father?[94]

E. THE ROLE OF INFINITY IN PANNENBERG'S DOCTRINE OF GOD

The concept of infinity plays a crucial role in Pannenberg's interpretation of the doctrine of God in volume 1 of *Systematic Theology,* particularly in his treatment of the divine attributes of eternity and omnipresence, attributes that are pivotal for the subject at hand.[95] Pannenberg stresses his indebtedness of his understanding of infinity to that of Hegel, and he contrasts it with its meaning in the traditional Greek thought, where infinity as *apeiron* is seen merely as the antithesis of the finite. Here I will briefly summarize the role of the concept of infinity in Pannenberg's doctrine of God, particularly his use of Hegel in understanding this concept.

1. The Unity and Attributes of God

In his extensive discussion of "The Unity and Attributes of the Divine Essence" (ST1, chap. 6), Pannenberg offers some initial comments on the role on infinity in the doctrine of God.[96] Although Scripture stresses "the inconceivable majesty of God which transcends all our concepts," we can still know something of this incomprehensible God when we turn to God's revelation in Jesus. Here "the hidden God himself is manifest." Following Luther, Pannenberg maintains that the Trinitarian distinctions of Father, Son, and Spirit are disclosed in the event of revelation. It is their unity that is hidden in the process of history, and our knowledge of their unity will only be complete in the eschaton. For now the unity of the incomprehensible God who works in history and the God revealed in Christ is hidden in the contradictions of historical experience.[97]

Traditional theology, of course, assumed that the existence and essence of God are "accessible to rational knowledge through the works of creation, while knowledge of God as Trinity is confined to special revelation." Pannenberg notes that Gregory of Nazianzus and Gregory of Nyssa related the incomprehensibility of God to the divine essence, with the latter paying particular attention to the role of the infinity of God in this relationship. He tells us that Gregory of Nyssa argued that "if God is infinite . . . it follows that we cannot ultimately define his essence, for it is indescribable (*adiexitēton*). The concept of infinity is also . . . the basis of the incomprehensibility of the unity of God in relation to the doctrine of the Trinity."[98] Pannenberg then discusses the way Thomas Aquinas based the distinction between God and creatures on the concept of God as first cause of all that exists, and recounts how Aquinas used the doctrine of analogy in order to attribute "positive descriptions of God to the divine essence." Next Pannenberg turns to Duns Scotus, who opposed Aquinas.[99] Scotus used general, univocal concepts, like that of being, to describe both creatures and the divine in order to distinguish between "the infinite being of God and everything finite" and to stress "the remoteness of human knowledge of God from the infinite God." Occam later built on Scotus by combining univocal terms with distinguishing terms to underscore that "God and creatures are infinitely distinct." Finally, Protestant dogmatics began with God's being and then ascribed God's attributes to this being. They thought of being in terms of "spiritual essence" and added to being "as a distinctive concept that of the infinite." Pannenberg concludes this section by arguing that "the main thesis of Aquinas still stood": namely the undivided simplicity of the being of God "which results from its infinity" and the multiplicity of the divine attributes which are ascribed to God "in the mode of undivided unity." Thus, "God's incomprehensibility has to do with his infinity and with the infinite unity of his essence."[100]

2. The Distinction between God's Essence and Existence

Pannenberg continues his analysis of the role of infinity in the doctrine of God by turning from the unity and attributes of God to the distinction between God's essence and existence.[101] John of Damascus believed that knowledge of the existence and perfection of God is intrinsic to our nature even if it is obscured by sin. He derived the infinity of God from the perfection of the divine being, and from God's infinity he obtained the unity of God. Aquinas, however, began with God as the first cause of the world and from there he derived God's simplicity, then, in turn, God's perfection, goodness, infinity, eternity, and unity.[102] Although Occam denied that God's unity, infinity, and omnipotence could be derived from God as first cause, Gregory of Nyssa and, later, Duns Scotus both argued that infinity "is not just one divine attribute among others but has basic significance for the whole concept of God."[103] Descartes continued this trajectory by emphasizing the importance of infinity for the idea of God, particularly because he saw everything finite as limited by the infinite. "We attain a sense of the infinite only by negation of the limit of the finite. It does not precede perception of the finite."[104] Thus, Pannenberg sees Descartes as having reversed the move by John of Damascus to derive God's infinity from God's perfection and then equate the idea of infinity with God. "The idea of God as infinite and perfect being comes first and existence follows from it."[105]

3. The Infinity of God in Relation to the Divine Attributes

But it is with Pannenberg's move in ST1, chapter 6, section 6, "The Infinity of God: His Holiness, Eternity, Omnipotence, and Omnipresence," and continuing into section 7, "The Love of God," that we more clearly find the central role of the concept of infinity in the doctrine of God.[106] In closing section 5, Pannenberg has already pointed out that there are two kinds of divine attributes: 1) Those attributes that define the God who acts (i.e., what "God" means): God as infinite, omnipresent, omniscient, eternal, omnipotent, and holy. 2) Those attributes that describe God's actions (i.e., what God does): God as kind, merciful, faithful, righteous, patient, good, gracious, and wise.[107] "Even if the Infinite is not the essential concept of God from which all the qualities of his essence are to be derived, it is still to be viewed as the initial concept of the divine essence to which all other statements about God's qualities relate as concrete expressions of the divine nature."[108] Pannenberg then tells us that he will first discuss the pivotal role of the concept of the Infinite in the definitional attributes, such as eternity, in section 6, before turning to the second set of attributes in section 7, attributes which, while they are "structurally" related to the concept of the Infinite found in the first set, are most clearly connected with the Johannine proclamation that the essence of God is love (1 John 4:8).[109] This leads Pannenberg into a detailed discussion of the theological meaning of infinity.

Pannenberg's initial definition of infinity is a qualitative one, "that which stands opposed to the finite, to what is defined by something else."[110] Pannenberg refers to this as Hegel's "first simple definition." Infinity only secondarily means something quantitative and mathematical, that is, freedom from limitation or that which is endless. In mathematics, freedom from limitation can be found in the unending generation of an endless, though always finite, series. "The infinite series—[meaning] the indefinite sequence of finite magnitudes in space and time—actualizes the antithesis of the infinite and the finite only in a one-sided way, namely, by an unrestricted addition of finite steps." In its primary, qualitative sense, however, the infinity of God means that God is radically distinct from everything finite, limited, and transitory. When it is linked to the holiness of God, the concept of the infinite separates God from all that is profane.[111]

But the biblical witness is also that the holiness of God brings salvation, election, and hope, which, in the Old Testament postexilic period and in the New Testament, are extended to "all secular reality." This leads Pannenberg to modify the previous sharp distinction between the infinite and the finite. Now God's holiness not only opposes the profane world but "embraces it, bringing it into fellowship with the holy God." In this context Pannenberg sees a "structural affinity" between God's holiness and what he calls "the true Infinite." Here he specifically refers to Hegel's complex understanding of infinity, this time placing his reference to Hegel in the full text:

The Infinite that is merely a negation of the finite is not yet truly seen as the Infinite (as Hegel showed), for it is defined by delimitation from something else, i.e., the finite. . . . The Infinite is truly infinite only when it transcends its own antithesis to the finite. In this sense the holiness of God is truly infinite, for it is opposed to the profane, yet it also enters the profane world, penetrates it, and makes it holy. In the renewed world that is the target of eschatological hope the difference between God and creatures will remain, but that between the holy and the profane will be totally abolished (Zech. 14:20–21).[112]

4. Pannenberg's Assessment of Hegel's Understanding of the Infinite

Pannenberg then makes a crucial claim, namely that there is a *logical* paradox in Hegel's complex concept of the Infinite. This claim surfaces most clearly in sections 6 and 7 of chapter 6. Pannenberg's precise wording invites a careful reading.

In section 6 Pannenberg writes that Hegel's abstract concept of the infinite

> contains a paradox which it does not itself resolve but which it formulates only as a task and a challenge for thought. It tells us that we have to think of the Infinite as negation, as the opposite of the finite, but also that it comprehends this antithesis in itself. But the abstract concept of the true Infinite does not show us how we can do this.[113]

In section 7 he returns to the problem of what he sees as a logical paradox in Hegel's concept of the infinite. According to Pannenberg,

> there appears to be no way of showing how we can combine the unity of the infinite and the finite in a single thought without expunging the difference between them. We cannot solve this problem, as Hegel thought, by the logic of concept and conclusion. The perfect unity of concept and reality in the idea is itself no more than a mere postulate of metaphysical logic. The dynamic that in the process has to be ascribed to the idea leaps over the frontiers of logic.

In Pannenberg's first account (section 6), the paradox is that the abstract concept of the infinite does not show us how it can "comprehend" its antithesis to the finite. In the second account (section 7), the paradox is that the unity of the infinite and the finite would "expunge" their difference. If we tend carefully to these two claims, we will find that they are, in fact, a single claim: logic alone cannot conceive the infinite as negating the finite while containing the finite without being reduced to the finite.

Pannenberg then sets out two *theological* responses which he believes resolve the logical paradox. They too are found in sections 6 and 7. In section 6, Pannenberg's response involves two related themes: God's holiness and the Spirit of God, both of which carry the structure of true Infinity. God's holiness is truly infinite because, while opposing the profane world, it enters into it and makes it holy. This is actually an eschatological process that culminates in the renewed world to come, where God and creature remain different but the distinction between the holy and the profane finally vanishes. Similarly God's Spirit is truly infinite. Although the Spirit opposes the world, it works in creation, gives life to all creatures, and sanctifies them by bringing them into fellowship with God. Thus, in both cases, theology can overcome the paradox by asserting that

> the transcendent God himself is characterized by a vital movement which causes him to invade what is different from himself and to give it a share in his own life. . . . God gives existence to the finite as that which is different from himself, so that his holiness does not mean the abolition of the distinction between the finite and the infinite.[114]

We can gain further insight into how Pannenberg conceives of the paradox he finds in Hegel's thought by a reading of his understanding of the divine attributes. We have discussed the divine attributes of eternity and omnipresence in some detail above. Here we focus specifically on Pannenberg's claims that their proper meaning must finally be grounded on "the structure of the true Infinite."

As the first divine attribute treated by Pannenberg, eternity offers a "paradigmatic illustration and actualization of the structure of the true Infinite which is not just opposed to the finite but also embraces the antithesis." In comparison, Pannenberg rejects eternity as timeless since it would presuppose an "improper" concept of the infinite as defined simply by its opposition to the finite, and in doing so, eternity would be merely finite.[115] In discussing omnipresence, Pannenberg combines divine transcendence and divine immanence via the structure of the true Infinite. Thus God's mode of transcendence makes possible God's presence to everything without God's immanence undercutting God's transcendence of everything.[116] He then grounds this claim explicitly in the doctrine of the Trinity by which the transcendence of the Father in heaven is linked to the presence of the Son and Spirit to believers. "Thus the trinitarian life of God in his economy of salvation proves to be the true Infinity of his omnipresence."[117]

In section 7 Pannenberg fills in the meaning of the dynamic leap he called for in discussing the logical paradox of infinity by drawing on key resources in the doctrine of the Trinity.

> [The dynamic] may be found only in . . . the dynamic of the Spirit . . . in the OT sense, not in that of its fusion with thought . . . [and in] the thought of the divine love. Love, of course, is infinite only as divine love. As infinite love it is divine love only in the trinitarian riches of its living fulfillment. Divine love in its trinitarian concreteness . . . embraces the tension of the infinite and the finite without setting aside their distinction. It is the unity of God with his creature which is grounded in the fact that the divine love eternally affirms the creature in its distinctiveness and thus sets aside its separation from God but not its difference from him.[118]

We first note that the phrase "not in that of its fusion with thought" makes it clear that, for Pannenberg, the spirit which provides this "leap" is not the concept of "spirit" (*Gheist*) as found in German idealism, particularly as developed by Hegel. Instead Pannenberg is referring to the Spirit of God. This in turn suggests, at least indirectly, that Pannenberg views the paradox he sees in Hegel's understanding of the Infinite as not only a *logical* paradox but even a *metaphysical* paradox and one which Hegel's basic theme of the dialectic between thesis and antithesis, and their synthesis through *Gheist,* cannot solve. Second, we note that such infinite divine love must specifically be understood within the doctrine of the Trinity. Here, then, such love can manifest the structure of the true Infinity, subsuming the finite without expunging its difference with the finite. Finally, this understanding of divine love is the ground of God's unity with God's creation, a unity which once again reflects the structure of the true Infinite. We will return to and take up his discussion of Infinity in chapter 3.[119]

F. ESCHATOLOGY: THE CAUSAL PRIORITY OF THE FUTURE AND THE PROLEPTIC CHARACTER OF ESCHATOLOGY

We now turn to two themes in Pannenberg's overall treatment of time and eternity that are related specifically to Christian eschatology. One of these is the causal priority of the future; the other is the proleptic character of the resurrection of Jesus. The first theme is that the immediate future is a partial factor (along with the past, of course) in bringing about, manifesting, determining, or causing the present to be what it is. I will refer to this as "the causal priority of the future." The role of partial causality assigned to the future by Pannenberg clearly runs up against the commonly held assumption, not only in science but seemingly required by ordinary experience,[120] namely that the present is entirely determined by the efficient causality of the past.[121] The second theme is that the resurrection of Jesus is proleptic: it is an appearance, manifestation, and anticipation within history of the eschatological future referred to both as the Kingdom of God and the New Creation. The eschatological future of the New Creation provides the ultimate context of meaning for the still-ambiguous present world, but this meaning is already disclosed to us because it reaches back and appears in the proleptic Easter event, the resurrection of Jesus. Conversely Pannenberg tells us that proleptic eschatology starts with God's radically new act at Easter and continues forward in time until the world as a whole is transformed by God into God's New Creation.[122]

1. Anticipation as the Arrival of the Future: The Causality Priority of the Future

"Futurity is fundamental": For this pivotal theme in Pannenberg's writings I start with his groundbreaking work of forty years ago, *Theology and the Kingdom of God*.[123] The context is the debate over the relationship between the future and the presence of the Kingdom of God in the ministry of Jesus. In chapter 1 Pannenberg argues against three views: 1) God's rule is not so distant that it has no present impact (cf. Jewish eschatology); 2) it is not so imminent that its futurity can be dismissed (cf. Bultmann); and 3) it does not merely begin with Jesus and end in the distant future (cf. Cullmann). Instead, Pannenberg claims that the Kingdom Jesus proclaims is already imminent in the present and yet the Kingdom that is present is the *coming* Kingdom. This leads Pannenberg to assert the startling idea that "it is more appropriate to reverse the connection between present and future, giving priority to the future . . . . In this way we see the present as an effect of the future, in contrast to the conventional assumption that past and present are the cause of the future."[124] Taking the lead from Pannenberg I will refer to this as the "causal priority of the future."[125]

To support his claim Pannenberg proposes we develop a philosophy of the natural sciences for which our experience of the ambiguity of the future is due not just to our lack of full knowledge of the present but rather an "essential indeterminateness" or contingency in nature. "It is God whom we confront in the contingency of events which, before they actually enter our world, share in the ambiguity of the future. . . . Those contingent events are in fact acts of God from whose future they spring."[126] Then Pannenberg explicitly states his revolutionary theme of the causal priority of the future:

> Yet when events which we anticipated in anxiety and/or hope do occur, the ambiguity of the impending future congeals into finite and definite fact. In every event the infinite future separates itself from the finite events which until then had been hidden in this future but are now released into existence. The future lets go of itself to bring into being our present. And every new present is again confronted by a dark and mysterious future out of which certain relevant events will be released. *Thus does the future determine the present.*[127]

At first blush this claim seems to contradict ordinary experience, let alone the scientific picture of efficient causality in which the past alone determines the present. Nevertheless, Pannenberg argues that new events, as they arise to constitute the present, do not emerge solely from the imme-

diate past but also from the future in conformity with the present world—
and in a way that results in our ordinary view of efficient causality:

> Thus the continuity of nature is no longer understood as the irre-
> sistible dynamic of the already existing pushing forward, but as the
> building of bridges to the past that save the past from getting lost. In
> this way the familiar picture of a continuous process moving from
> the past through the present and into the future becomes possible.[128]

In almost all cases this conformity reduces to repetition, thus giving rise
to our experience of ordinary physical causality and enduring objects in
nature.[129]

Finally Pannenberg relates these ideas of the causality of the future
to Aristotle's concept of entelechy: "the yet unattained goal is present in
an anticipatory way in the moved, and this indwelling of the goal effects
the movement toward the goal." According to Pannenberg, the futurism
of Aristotle was undercut by Aristotle's use of the Platonic idea that the
entelechy is really a germ that already exists in the present and out of
which the goal unfolds. "This inner teleology, which reverses the relation of the present and future, has robbed evolutionary thought until our
day of the possibility of seeing what is new in each event as something
really new."[130]

2. The Proleptic Character of Eschatology

We now come to the second theme of this section: the proleptic character
of eschatology. This theme returns us to the issue that we found Pannenberg debating above, but it brings to it additional complications for a
new ontology of time. The debate is this: Is the resurrection of Jesus proleptic merely in that it represents a truth claim made in history about the
eschatological future, or is it also proleptic in the much more important
sense of an anticipation, appearance, and concrete manifestation within
history of the reality of the eschatological future? Pannenberg clearly defends the latter, as suggested in the following brief discussion of several
key texts.[131] His defense will lead him to make four distinct but interrelated claims about prolepsis. 1) The eschatological future is the basis for
its proleptic appearance within history. 2) Our openness to the future is

grounded in the proleptic character of the resurrection of Jesus. 3) The eschatological future as the hermeneutical context for understanding the history of the world is revealed proleptically in the resurrection of Jesus. 4) Testimony about its proleptic appearance in history takes the form of a hypothesis.

1) In chapter 4 of TKG Pannenberg once again takes up the exegetical debate over the relationship between the themes of futurity and presence of the reign of God in the ministry of Jesus. Although some New Testament interpretations stress its presence while others emphasize its futurity, Pannenberg finds the "uniqueness of the (New Testament) message precisely in the juxtaposition of [these] seemingly opposing sayings." To understand this juxtaposition Pannenberg offers the following radical claim about the relation between *appearance* and *that which appears:*

> In the ministry of Jesus the futurity of the Reign of God became a power determining the present. . . . [The] presence of the Reign of God does not conflict with its futurity but is derived from it and is itself only the anticipatory glimmer of its coming. Accordingly, in Jesus' ministry, in his call to seek the Kingdom of God, the coming Reign of God has already appeared, without ceasing to be differentiated from the presentness of such an appearance.[132]

This statement goes beyond Pannenberg's previous claims about the primacy of the imminent future over the present. Here it is the eschatological future that is given primacy over its appearance in the present. In fact Pannenberg actually makes three closely related claims: i) the eschatological future determines the present; ii) the reality of the eschatological future is the ontological basis for the reality of the present appearance;[133] and iii) the appearance of the eschatological future in the present does not undercut the difference between the eschatological future and the present.[134] To reiterate, in all of these, "future" does not mean the imminent tomorrow as discussed in the previous section but the eschatological future of the New Creation. It is the eschatological future that is self-grounding and it is the eschatological future that reaches back into the present where its manifestation is based on its eschatological reality.

2) Pannenberg also connects the assertion of the openness to the future, as constitutive to the human person, with the proleptic character of the resurrection of Jesus:

> The proleptic character of the Christ event . . . [signifies that] the resurrection of Jesus is indeed infallibly the dawning of the end of history . . . . [T]he onset of the end had occurred only in a preliminary way, happening in Jesus himself . . . . [F]or the rest of us, the resurrection of the dead, which already happened to Jesus, is still outstanding. . . . The proleptic character of the destiny of Jesus is the basis for the openness of the future for us. . . . And, conversely, without this proleptic character, the fate of Jesus could not be the ultimate revelation of the deity of God, since the openness of the future belongs constitutionally to our reality.[135]

3) Pannenberg applies contemporary hermeneutics—that context determines meaning—to interpret the eschatological future as offering the widest possible context for the meaning of the world and its future. His understanding of prolepsis then leads him to assert that the meaning as found in the future eschatological context is already given retroactively in the historical events of Easter:

> Then the essence of a man, of a situation, or even of the world in general is not yet to be perceived from what is now visible. Only the future will decide it. . . . But through the resurrection it is decided, not only so far as our knowledge is concerned, but with respect to reality, that Jesus is one with God and retroactively that he was also already one with God previously.[136]

4) Finally, because the eschatological future is still future and only proleptically revealed in history, testimony about its truth takes the form of an hypothesis: We could be wrong about the future eschatological Lordship of God. This means that the present act of faith is at most an "intelligent anticipation prior to secure comprehension."[137]

Here, I believe, we find anticipation to be a real instance of something as occurring in advance from the eschatological future. The anticipated future is already present in its anticipation—though only given the presupposition that the eschatological future of God's Lordship and the resurrection of the dead will actually occur. If this future does not occur, then its anticipation will have been only a tragic prophetic enthusiasm. Anticipation is therefore always ambiguous; its true significance depends upon the future course of what Pannenberg calls universal history.[138]

In sum, Pannenberg's writings on the theme of this volume, time *in* eternity, when viewed through the lens of his understanding of eschatology, deploy two massive conceptualities: the causal priority of the immediate future and the proleptic character of the eschaton emerging from the New Creation to have its normative efficacy in transforming present history, starting at Easter, toward the coming of the New Creation. Together, these interrelated conceptualities take the problem of "time *in* eternity" from the more frequently studied context of the relation in philosophy and philosophical theology between the time of creation and the divine eternity and ground it in a thorough-going theological treatment of the transformation of creaturely time, and its relation to eternity, into the eternal temporality of the eschatological New Creation. All of this is governed by Pannenberg's unique understanding of the Trinitarian God as an intrinsically differentiated unity.

Far more can and should be said about all this, of course. Nevertheless the task at hand requires that we move now from this brief, but somewhat detailed, outline of Pannenberg's work on "time *in* eternity" to the task of the current volume: a reformulation of Pannenberg's theology on God in light of contemporary mathematics, physics, and scientific cosmology (chapters 2–4), and the role it could play in offering new suggestions for research in both the philosophy of physics (chapters 5–6), specifically the philosophy of time, and in current programs in fundamental theoretical physics.

CHAPTER 2

Co-presence and Prolepsis in Light of Mathematics, Physics, and Cosmology

The most fundamental question in the philosophy of time is whether a static or a dynamic conception of the world is correct.

—Michael Tooley, *Time, Tense, and Causation*

There is no part of our time which is not as such also in [God's]. It is, so to speak, embedded in His eternity.

—Karl Barth, *Church Dogmatics*

At the outset of this volume, I suggested that the theme of time and eternity can be seen as including two distinct but interrelated and well-discussed issues.[1] The first is the relation between time as we currently know it (i.e., creaturely time) and the eternity of God. This issue can be found in the literature on the philosophy of religion, philosophical theology, and systematic theology. The second issue is the relation between creaturely time and the eternity of the eschatological New Creation as anticipated proleptically in the resurrection of Jesus, an eternity that is in some ways already present to Christians now. This view of eternity is discussed typically when systematic theology deals explicitly with eschatology. In this chapter, I explore Pannenberg's key insights into both issues, focusing on the first in section B and on the second in section C.

To characterize Pannenberg's unique contribution to the first issue, I have coined the term "co-presence." We begin with the everyday, sequential flow of temporal events in daily experience—each event a present moment with its own specific past and future. "Co-presence" is meant to signify the special kind of unity that these ordinary temporal events are given as they are taken up, even now, into the "supratemporal" eternity of God. In chapter 1, we found this idea arising vividly in several places in Pannenberg's *Systematics* (see section A.1) To characterize Pannenberg's unique contribution to the second issue, I have adopted his term "prolepsis" to signify the theological interpretation of the resurrection of Jesus as the appearance and culmination in history of the eschatological future consummation and reign of Jesus the Christ in the New Creation. This concept appeared in Pannenberg's *Basic Questions* roughly two decades before the *Systematics*.

To explore these fundamental themes of co-presence and prolepsis, however, we must keep in mind two facts. First, as we have already seen, Pannenberg articulates a complex metaphysical theory of being and time in critical dialogue with Continental philosophy, with its commitments to phenomenology, hermeneutics, and historical consciousness, as well as with classical philosophical sources such as Plato, Plotinus, and Boethius.[2] Pannenberg's metaphysics includes such concepts as temporal extension (i.e., "duration"); the present as the "arrival" of the immediate future; and prolepsis as the appearance in the present of that which is eschatologically future and which offers the ultimate context of meaning of the present. This metaphysical theory of time underlies Pannenberg's theology in general and his work on time and eternity in particular. Second, even with such a rich temporal ontology Pannenberg uses ordinary language about time in discussing his ontology. This language reflects a conventional understanding of time as found both in human experience and in nature. Here time includes a momentary present moment with a distinct and unique past and future. In addition, the present constantly changes as what was a set of possible and indeterminate future events becomes a unique, determinate present event, and then immediately becomes an event in the ever-receding and irretrievable past. Such a view of time is often called "flowing time" by physicists and philosophers.

The problem is that the reality of flowing time, at least as physics understands it, is widely disputed by Anglo-American analytic philosophers while it is taken for granted in the rich temporal ontology of Continental metaphysics. Many Anglo-American philosophers opt instead for a timeless view of nature and a tenseless view of language, claiming that flowing time can be reduced to *merely* subjective experience.[3] Where is Pannenberg in all this? As I stated in chapter 1, it is not entirely clear whether Pannenberg believes that the dynamics of flowing time and, more to the point, duration, are merely subjective phenomena with no basis in the physical world. Nevertheless, I will make the assumption here that he does in fact hold that the physical world is dynamic and characterized by flowing time and duration.

One of the most important arguments in defense of a timeless view of nature is based on physics: Einstein's special theory of relativity. Clearly I must address relativity's challenge to flowing time. In chapter 5 I suggest that Pannenberg's theology, reconstructed in light of special relativity, can in fact lead to a new argument for the *philosophical* defense of flowing time in light of relativity, based on the relational and inhomogeneous temporal ontology articulated here in chapter 2. With this in mind, we can take up the general philosophical defense of flowing time before drawing on our mathematical and scientific resources to explore Pannenberg's views of co-presence and prolepsis.

A. A DEFENSE OF PANNENBERG'S ASSUMPTION OF FLOWING TIME AND ITS IMPLICIT ONTOLOGY

1. The Debate over Static and Dynamic Conceptions of the World in the Philosophy of Time: A Brief Background

In *Time, Tense, and Causation,* philosopher Michael Tooley provides a sweeping panorama and detailed analysis of the historical and contemporary issues in the analytic philosophy of time.[4] Tooley notes that our ordinary view of time differs sharply from the view most philosophers take. In daily life we take it for granted that the past is settled while the future is undetermined, and, while we move in time toward the open future, the past recedes ever further from us. Philosophers, however, are

Table 2.1. Two Theories of Time in the Philosophical Literature

Types of theories	Interchangeable terminology	Who supports the theory
A-theories	Tensed/dynamic/flowing time	Minority of philosophers and ordinary usage
B-theories	Tenseless/static/eternalist/ block universe	Majority of philosophers

sharply divided between a tensed, dynamic conception of time some-what akin to that of our ordinary view, and a static, tenseless concept of time. These alternatives are frequently referred to as A-theories and B-theories (table 2.1).[5]

A minority of philosophers support an A-theory of time as tensed, flowing, and dynamic, maintaining that "time does flow, that time does have an intrinsic direction, and that there are significant ontological differences between the past, the present, and the future."[6] A-theorists disagree, however, about what these ontological[7] differences are. Some claim that only the present is real, a view known as presentism (see table 2.2, "Ontology I").[8] Others claim that the past as well as the present is real, although the future is not real, a view often called the "growing universe" or "growing block" (table 2.2, "Ontology II").[9] Still others agree with the B-theorists' static view that past, present, and future are all real, but they qualify this by claiming that the future is only potentially real, unlike the present and past that are actually real (table 2.2, "Ontology III").[10] Some philosophers support this claim by adding to the ontology of real events special tensed properties that are acquired or lost in time (table 2.2, "Ontology III-1").[11] In one version, events first lack a property called "presentness." They then acquire it momentarily only to lose it forever. In another version, these properties are called "pastness, presentness, and futurity," and they are held sequentially in time. Still other philosophers view the ontology of the universe as growing in time. The future is a super-position of possible but only potentially real states, while the present and the past are each an actual, single, and determinate real state (table 2.2, "Ontology III-2").[12] Finally, tensed theories of time offer a tensed grammar that allows us to view propositions about time as truth claims. For example, the statement "U.S. elections in 2020 will be a landslide for the Democrats" could be true or false. Moreover, the veracity of temporal

Table 2.2. Differing Ontologies Assigned to Flowing Time by A-theorists

	Ontology I: *presentism*	*Ontology II:* *growing universe with* *the future as not real*	*Ontology III:* *growing universe with the* *future as potentially real*	
			III-1: *acquired* *properties* *of one state*	*III-2:* *an ontological* *transition from many* *states to one state*
Future	not real	not real	"futurity"	many potential states
Present	real	real	"presentness"	one actual state
Past	not real	real	"pastness"	one actual state

propositions can change as time flows. So in the year 2021 this proposition will be either true or false, whereas at the present it is indeterminate.

The majority of philosophers, however, support a B-theoretic, static, tenseless conception of time. According to this view, the past, present, and future are equally real. Temporal designations such as past and future are asymmetric, but there is no asymmetry intrinsic to time itself. Talking about "now" and "then" is strictly analogous to spatial language about "here" and "there." There is no ontological difference between being here and being there or any property that distinguishes here from there. As Tooley notes, "In both cases, such statements are true or false simply because of some relation—in the one case spatial, and in the other, temporal—between two entities, one of which is picked out by an indexical or demonstrative term."[13]

Closely related to the debate between A- and B-theorists of time is the debate over dynamic and static conceptions of the world. This debate, in turn, is intimately related to differing concepts of change.[14] According to the static conception of the world, what facts or states of affairs exist does not depend upon what time it is. Change is simply a matter of objects possessing different properties at different times. But according to the dynamic conception of the world, change includes more than objects having different properties at different times. What facts exist does depend on what time it is. Consequently, change includes different facts about a given object at different times. With this distinction in place, Tooley claims that "the fundamental thing that separates tensed and tenseless accounts of the nature of time is . . . . the acceptance, or rejection, of a dynamic conception of the world."[15]

2. Pannenberg and Ontology III-2

As I suggested at the outset of this chapter, Pannenberg takes for granted a rather conventional understanding of flowing time as found both in human experience and nature: a momentary present distinguishes between a unique past and future. What was future and undetermined becomes a determinate but fleeting present and then an ever-receding, irretrievable past. Thus, flowing time has first of all an essential creaturely characteristic, the association of a unique past and future to each present moment, and this is to be preserved in eternity. But flowing time has also a second, tragic characteristic that is to be overcome in eternity: the loss of the present to the past and the unavailability of the future to the present.

Pannenberg next complexifies this understanding of time enormously. We have already discussed such complexification in part. It includes at least five key points: 1) the momentary present actually has temporal extension, or "duration," but here understood in a quite specific way; 2) duration is the means of unifying past, present, and future by relating them to, and allowing them to participate in, the differentiated unity of the eternal Trinity; 3) the structure of duration is based on what I call the "co-presence" of all events in eternity. In eternity, all present moments of flowing time, each with its distinctive past and future, are held together "simultaneously." They are mutually co-present so that their distinctions as unique moments of flowing time (time's first creaturely characteristic) are maintained, but their separation into the irretrievable past and the unavailable future of flowing time (and thus time's second, and now tragic, characteristic) is overcome; 4) the anticipation of the future in the extended present is in actuality the appearance and "arrival" of the future with its attendant ontological priority. (At the underlying level of physics we will see that this "arrival" of the future will entail what can be called "retroactive causality"; see chapter 6); 5) the resurrection of Jesus as the culmination of history is the proleptic realization in time of the global eschatological future and the consummation of history in the New Creation.

Pannenberg sets this understanding of time within the complex metaphysical and theological debates over the nature of eternity and its relation to ordinary time in dialogue with both classical sources (e.g., Plato, Aristotle, Plotinus, Boethius, Augustine) and contemporary sources (e.g., Hegel, Heidegger, Bergson, Whitehead). Nevertheless, he does *not* locate himself within the recent debates among analytic philosophers, mentioned above, over the philosophy of time. Thus, I can only speculate how he would respond to the challenges raised by analytic philosophy to defenders of flowing time.[16]

So which of the ontologies described above offers the most appropriate understanding of Pannenberg's view of ordinary, creaturely time? I would certainly rule out ontology I (presentism) since its view of the past as not real is inconsistent with Pannenberg's commitment to the reality of the past as part of temporal duration. Because temporal duration also includes the potential reality of the future, I would also rule out ontology II (growing universe with the future as not real). This leaves ontology III: a growing universe with the future as potentially real either because of its variety of realizable properties (ontology III-1) or because of an ontological actualization of one out of many potentially real future states (ontology III-2). On reflection, I believe that ontology III-2 best characterizes Pannenberg's assumptions about time in nature. Its view of an *ontological* transition from a multiple potentiality of future states to a unique actuality in the concrete present most adequately addresses Pannenberg's concept of the "becoming real" of the present out of the future.[17] Thus my task will be to defend ontology III-2 against the challenges raised by B-theorists.

3. Defending Ontology III-2 in Light of Pannenberg's Theology: Flowing Time as a Relational and Inhomogeneous Temporal Ontology

To defend what I take to be Pannenberg's implicit assumption of ontology type III-2, I argue for a relational and inhomogeneous temporal ontology. Here, the relations between events as well as the events themselves are given ontological significance. Such an approach will help us avoid some of the problems that arise for the usual A-theory of flowing time and, in turn, will enhance our conversations about Pannenberg's understanding of the divine attribute, eternity.

The Underlying Problem for Flowing Time: Contradictions Generated
by Multiple Ontologies Assigned to the Same Temporal Event

From the brief background to the debate over the philosophy of time
provided above in section A.1, it is already clear that most A-theories
assign different properties to the same event according to whether it is
past, present, or future. Since these different properties point to ontological differences for the same event regarded as past, present, and future, an ontological contradiction arises. An event considered as past is
actually real and determinate but irretrievable. Conversely, an event as
future is potentially real and indeterminate.[18] Only when the event is present is it actually real and determinate. But then arises the contradiction:
How can an event be simultaneously past, present, and future since its ontological status must then be at the same time actually real/determinate/
irretrievable (past); actually real/determinate (present); and potentially
real/indeterminate (future)? Questions such as this lie at the foundation
of the A-theorist's philosophy of time, as McTaggart argued strikingly
in 1927:

> Past, present, and future are incompatible determinations. Every
> event must be one or the other, but no event can be more than one.
> If I say that any event is past, that implies that it is neither present
> nor future, and so with the others. And this exclusiveness is essen-
> tial to change, and therefore to time. For the only change we can get
> is from future to present, and from present to past. These character-
> istics, therefore, are incompatible. But every event has them all. . . .
> How is this consistent with their being incompatible?[19]

The same questions underlie crucial issues in theology such as the relation between free will and divine foreknowledge (with such proposed
solutions ranging from Molinist middle knowledge to Aquinas's concept of divine presentiality); God's action in the world (particularly as
deployed in a non-interventionist account of objective divine action);[20]
and, centrally to this book, the relation between time and eternity.

In light of these issues, I explore a *new* form of ontology, type III-2,
that offers a helpful way out of this ontological contradiction. In posing
the contradiction above, I spoke as though time consists of events with
properties, reflecting an Aristotelian-type ontology. Instead I want to shift

in part to a Leibnizian-type ontology in which the tenseless relations between events—past and future—carry their own distinctive ontology along with the ontology carried by events in their state as present.[21] My claim is that this shift will avoid the contradiction as stated by McTaggart. In addition, if ontological weight is carried by the temporal relations of past and future, a Leibnizian-type ontology will provide in part for what A-theorists view as the "flow" of time. Moreover, the difference between these relations—past and future—accounts in part for what A-theorists view as the "direction" of time's flow. Finally, in keeping with this new type III-2 ontology, I will disagree with those A-theorists who hold presentist and growing universe views insofar as they claim that future events are simply not real.[22] I will agree, however, with growing universe views that the reality of past and present events differs markedly from the potential reality of future events—even though like all events, "future events" as present to themselves are actually real and determinate.

In sum, in this new form of ontology type III-2, I will suggest that an event A as future to event B as present is in a crucial sense ontologically different from the same event A as present to itself due to the ontologies carried by the temporal relations between A and B (a similar claim holds for event B as present and event C as past to B). I will also suggest that the future event becomes present precisely as an actualization of one of its many potential states and, in this way, represents a part of what Pannenberg calls the "arrival of the future." It is ultimately this "arrival" of the momentary present that Pannenberg attributes to the *creative action of God,* the divine action that provides the basis for the "flow" of "flowing time." It is thus a *theological* claim, and not a strictly philosophical one, that delivers the categorical difference between, and the requirement of, an A-theory versus a B-theory of time. Such a theological claim about God gives meaning to the ontological difference between an event as a cluster of indeterminate potential future states and an event as a concrete, unique, and actually real present moment.

Before proceeding further, let us visually represent the problem and my response to it in more detail. In figure 2.1, time t is treated as a continuous one-dimensional variable (e.g., the time axis, t):

Figure 2.1. Time as a One-Dimensional Continuous Variable

In figure 2.2., events, represented by points on the line, are labeled in relation to a single event chosen as the present. The present event (denoted "pr") is B, with event A in the past of B ("p"), and event C lying
in B's future ("f"):

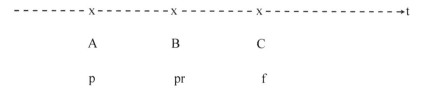

Figure 2.2. The Present Event B with a Future Event C
and a Past Event A Relative to B

where:

pr stands for the present event, B, as actually real and determinate
p stands for the past event, A, as actually real and determinate but irre-
    trievable
f stands for the future event, C, as potentially real and indeterminate

Figure 2.3 portrays the ontological contradictions that flowing time
seems to raise: *As a future event becomes the present and then the past,
the ontology assigned to this event changes. What once was an event in
the potential and indeterminate future becomes the actual and determinate present moment, and then it becomes an actual, determinate but per-
manently irretrievable past moment.* To emphasize this change in the status of an event,[23] I include some additional notation to the figure:

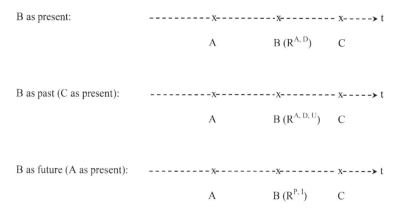

Figure 2.3. The Paradox of Flowing Time: Which Ontology Is Correct?

where:

pr, the present event, is real, actual, and determinate: $R^{A, D}$
p, the past event, is real, actual, and determinate but unavailable: $R^{A, D, U}$
f, the future event, is potentially real and indeterminate: $R^{P, I}$

But how can an event such as B have the ontological status of "present" if it is also to have the ontological status of "past" when C is the present and the ontological status of "future" when A is the present? Which ontology should we assign to event B? And how can it change "in time"? I call this predicament "the tangle of conflated ontological assignments." The conflation arises because we assert that the event B is somehow both present (actually real and determinate), future (potentially real and indeterminate), and past (actually real, determinate, and unavailable/irretrievable).[24]

A-theorists have a variety of responses for this problem, as we have already touched upon briefly. For example, they might counter by asserting that the three tenses—past, present, and future—are not assigned to the same event *at the same time.* Instead, the tenses represent properties of events that change in time. Note how C. D. Broad articulates such a response to McTaggart:

> I cannot myself see that there is any contradiction to be avoided. When it is said that pastness, presentness, and futurity are incompatible predicates, this is true only in the sense that no one term could have two of them *simultaneously* or *timelessly.* Now no term ever appears to have any of them timelessly, and no term ever appears to have any two of them simultaneously. What appears to be the case is that certain terms have them *successively.* Thus there is nothing in the temporal appearances to suggest that there is a contradiction to be avoided. . . . To sum up, . . . the fallacy of McTaggart's argument consists in treating absolute becoming as if it were a species of qualitative change.[25]

Broad's response might be taken as laying to rest McTaggart's argument. While I believe it does take us a long way toward this goal, it founders at a crucial point. Hence, I want to suggest an additional argument to help complete the response.

In my view, Broad's argument founders when he adopts a metalevel form of time as indicated by the crucial appearance of the term "*successively*" in the quotation. The problem is rooted in what I am calling the Aristotelian assumption that tenses are properties of events. Given this assumption, the only way out would be to appeal to yet another higher level of time that defines the time that defines "successively" just as Broad does, and so on ad infinitum. Instead, I will suggest an alternative view of time that understands tenses not as properties of events but as relations between events. This view avoids the infinite regress entailed implicitly in Broad's approach and thus it will be crucial in making the case against McTaggart in favor of flowing time. With such a relational view in place, we can deploy an ontological argument about $R^{A, D}$, $R^{A, D, U}$, and $R^{P, I}$, such as what I am calling a type III-2 ontology. This in turn is the final element needed to complete the case against McTaggart's contradictions because temporal ontologies are assigned to relations between events and not only to the events themselves.[26]

A Modified A-Theory of Time with Tense as Relational

The key to overcoming at least some of the problems in the ontological conflation just described will be a very specific A-theory of time. As indicated above, it involves the idea that tenses are better treated as *relations between events* than as *properties of events,* and that these relations carry an ontology in addition to the ontology of the events themselves.[27] I will refer to this proposal as a "tense as relational A-theory of time."[28] In general, the term "relational theories of time" typically refers to theories that assume that time does not exist without events and their relations. By contrast, "receptacle theories of time" assume that time exists independently of actual events. The distinction between relational and receptacle theories of time is analogous to the distinction between Leibniz's relational theory of space and Newton's receptacle theory of space that we saw above in discussing Pannenberg's analysis of omnipresence (chapter 1, section A.2). In relational theories of time, there is no meaning to time without there being temporal events. In receptacle theories of time, the idea of time is meaningful whether or not there are temporal events.[29]

My opting for a relational theory of time follows from Pannenberg's agreement with Leibniz over Newton, his reading of Einstein as continuing Leibniz's project, and his support for Augustine's view of creation not as an event *in* time but as the creation *of* time. My proposal that tense is relational is informed in even larger measure by Pannenberg's treatment of the relationality of the Persons of the Trinity (chapter 1, section D). Finally, I would add to this the idea that the world, being the creation of the Trinitarian God, in some way bears witness to its being so created. One can see this as an inversion of the traditional theological argument of the *vestigia Trinitatis* in which certain threefold features in the world are taken as analogous to, and as a separate and independent source of knowledge about, God as Trinity, knowledge that we know normatively from special revelation.[30] Thus, I will explore the idea that temporal relations between events carry some of the ontological factors normally attributed to the individual events themselves, factors typically seen as properties of the event. Instead of viewing tenses such as past, present, or future as properties of events, I propose a tense as relational A-theory of time, where the tenses "past" and "future" are relations between events and an event taken as present. This leads us to a complex temporal ontology that includes both a given event as present, where "present" is the property of that event, and other events as "past" and "future" in relation to that event as "present." Together this complex vista comprises a key element in my proposal for the ontology of time.

The first step in developing such a modified A-theory of time is to define tense as relational in three senses. We first choose an event taken to be present and that bears the ontology of real, actual, and determinate ($R^{A, D}$). In relation to this event, we then identify an event that is past in relation to this event and an event that is future in relation to this event. The following is a convenient notation for referring to these relationships:

[event related to the present event] [relation] [the present event]

Using this convention, we can write down the two relations to event B taken as the present event to events A and C (figure 2.4):[31]

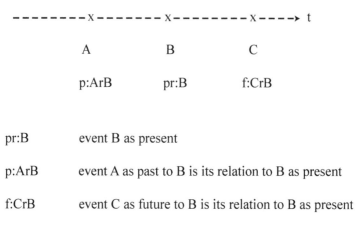

Figure 2.4. Tense as Relational for a Given Event as Present

We should immediately note two points: 1) To consider an event as present is to view its ontology as of a different type than the ontology of the relation of an event to other events. To consider the event as present is to view it as real, actual, and determinate. 2) To consider an event as past or future is to view it in relation to another event considered as present. The relations past and future are to be viewed as relations between events from the perspective of the event considered as present. The past is a relationship of an event to the present event that is actually real and determinate but inaccessible. The future is a relationship of an event to the present that is potentially real and indeterminate.

It is this difference between an event as present and events considered in past and future temporal relation to the present event that thus allows me to attribute distinctive ontologies (1) to an event as present and (2) to the past and future temporal relationships of events to the present event. It is also important to note that these relations, though tensive relations, are themselves tenseless: for example, C's relation to B as present is always future. B-theorists capitalize on this fact when they claim that *all* tensed statements can be reduced to tenseless statements, and that dynamic time is therefore merely subjective while time is in reality static. A-theorists typically reply that such a reduction is incomplete and/or inadequate. Time's flow cannot be reduced completely and without remainder to tenseless properties or relations. Much of the debate turns on this point. In addition, flowing time will require an in-homogeneous ontology including both present events and temporal past and future relations to present events such as described here. *In sum, I agree with A-theorists in support of flowing time, but with this critical modification. Within my modified A-theory of time I agree with B-theorists who claim that tenseless relations carry some ontological weight but I do so without conceding to them an eliminative temporal reductionism.*

Excursus: Representations of Tense as Relational
How then should we think of relations such as I am describing? An initial way to represent the tenseless relations ArB and CrB is by vectors[32] that extend *from* the events A and C *to* event B considered as present (fig. 2.5):

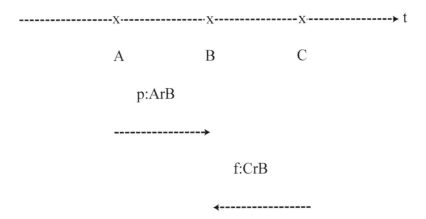

Figure 2.5. Vectors Representing the Tensed Relations

We can extend this to include the analogous relations from the perspectives of A and of C when they are considered to be present events:

f:BrA B as future in relation to A as the present
p:BrC B as past in relation to C as the present

All four of these relations are shown in figure 2.6:

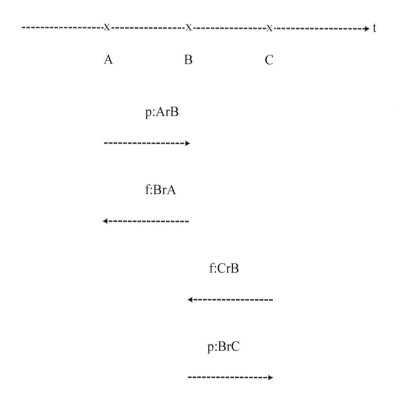

Figure 2.6. Four Tense-Relational Vectors

The relational model allows us to separate the tenses past and future that otherwise seem to conflict with each other if they are viewed as properties of one and the same event, B. It does so by distinguishing between vectors ArB and BrA (and CrB and BrC) and asserting that these vectoral relations carry the ontology of tense, not the events themselves. In this way, the tangle of conflated ontological assignments can be "untangled" into distinct ontological relations: B is not assigned "present," "past," and "future" properties or assigned properties that "vary in time," etc. Instead, the ontologies of tense are now carried by the relations between events—relations ArB, BrA, CrB, BrC—represented here by vectors. The key point is threefold: there are two different relations between each pair of events A and B, namely ArB and BrA; the relations ArB and BrA are not equivalent; and the relations ArB and BrA separately carry the tenses otherwise assigned to the same event B as conflicting properties.

Using vectors in this way can be somewhat misleading, however. The reason is that the temporal relations they represent here are *not* extensions *in* physical time t as their vector representations might suggest, especially when they are pictured as in figures 2.5 and 2.6 above.[33] To avoid this problem I turn to more abstract representations of relations that are not vectors.

One example is a typical "family tree" diagram depicting such relations as spouse, parent, child, etc. Figure 2.7 is a family tree for a fictional character, Lucas Grey, and his descendants:

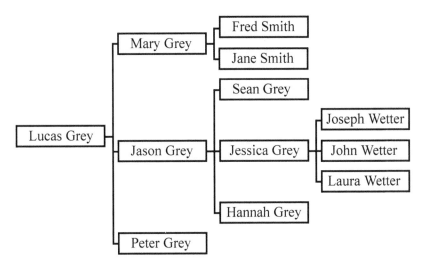

Figure 2.7. A Typical Family Tree Diagram

The space of relationships represented in figure 2.7 is obviously not the real world in which family members live out their lives. Nevertheless, it conveys crucial facts about these relations in an easily visualized form. In a similar way, the space of temporal relations between events is not the space of the timeline (t-axis) but the space of the intrinsic logic of temporal relations. In short, temporal relations are not "in time."

Another representation of relationality is a game tree diagram, such as this one for tic-tac-toe depicting the first two levels of decisions (fig. 2.8):

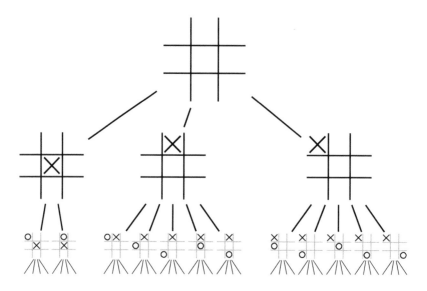

Figure 2.8. A Game Tree Diagram

Again, the space of the game tree is not the gameboard with its red and white chips on which the game is actually played. Instead it is the abstract mathematical space depicted in figure 2.8.

Yet another representation of relationality is an "entity relationship model" such as shown in figure 2.9 for the relationship between an artist and a song the artist performs, with the term "performs" standing for the relationship.

Figure 2.9. An Entity Relationship Diagram

Once again, the abstract space showing the relationship between artist, performance, and song is not the physical space and time in which the artist lives and performs her song.

A subtle example from particle physics illustrates the idea that relations, here thought of as *interactions* between particles, carry ontological weight along with the weight carried by the intrinsic properties of these particles. According to particle physics, matter is made up of quarks and leptons (such as electrons), while the interactions or forces between them are carried by bosons (including photons). In the 1970s, Sheldon Glashow, Steven Weinberg, and Abdus Salam constructed a theory, now called the Standard Model, that unified the two forces—the weak interaction and the electromagnetic interaction—into the "electroweak" interaction. One of the problems with the Standard Model is that, because of the mathematical symmetry of its equations, it did not account for the differences in the masses of quarks and leptons. The solution came via the "Higgs interaction" between the weak and electromagnetic forces that gives differing masses to quarks and leptons. The upshot is that the masses of quarks and leptons are not entirely an intrinsic property like charge or spin; instead, they arise through the interaction between quarks and leptons via the Higgs field.[34]

As a final example—and one that is particularly fitting here—I draw on Karl Rahner's pivotal discussion of the Trinity and his analysis of the meaning of the Trinitarian relations.[35] Rahner presents five "notional relations" that express the distinctions between the persons: unoriginatedness, fatherhood, sonship, active spiration, being spirated. The crucial point here, however, is that instead of thinking of the three persons, Father, Son, and Spirit, as existing independently of their relationships with each other, it is helpful to think of the persons as ontologically constituted by their relations.[36] One way to represent this is to describe these relations through their order of generation: the Father as unoriginated, the Father generates the Son, and the Father and the Son together spirate the Spirit. Clearly these relations are abstract theological concepts that, while referring to the divine Persons, do not "exist" within the Trinity as such.

From these examples of abstract relations we can think of tensed relations between events in time as carrying some of the ontology of time without thinking that the relations between events exist *in* time or imply additional temporal dimensions in the world along with ordinary, flowing time, t. This insight will be of crucial importance as we extend the argument for flowing time in the following section.

Extending the Argument to a Relational and Inhomogeneous
Temporal Ontology for Flowing Time

An additional step toward untangling the problems related to an A-theory
of time is to adopt what I call a relational and inhomogeneous temporal
ontology for flowing time. We start with the previous one-dimensional
representation of time, including the temporal relations for event B considered as the present (fig. 2.10):

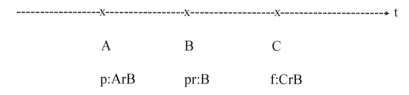

Figure 2.10. Temporal Relations for a Given Event as Present

Next we draw two more such representations, each with a different view
of the present and labeled t_1 (A is present) and t_3 (C is present), and then
sandwich B as present between them (fig. 2.11):

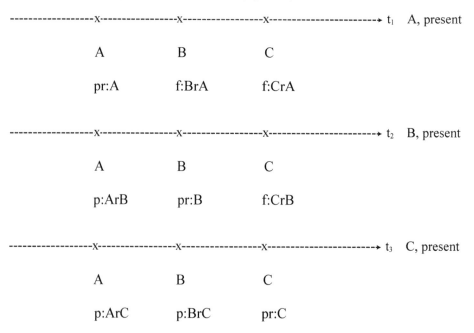

Figure 2.11. Temporal Relations for Three Events
Considered Successively as Present

Now the critical step is to view flowing time as the temporal axis, t, that connects the present events on t_1, t_2, and t_3 (fig. 2.12 below).[37] Note that each event A, B, and C considered as present has a unique set of past and future temporal relations to the other events. As I suggested above, these temporal relations carry the ontological "weight" of "past and future" associated with the present event along each of these axes. But this, in turn, means that the ontology of the two-dimensional representation in figure 12 is *inhomogeneous:* event B is considered present to itself on axis t_2, future in relation to event A considered as present on axis t_1, and past in relation to event C considered as present on axis t_3. Nevertheless this inhomogeneous ontology is not contradictory: the relations of future and past between B, A, and C are ontologies assigned to different temporal axes t_1 and t_3, associated with different events— A, B, and C—considered successively as the present moment. To underscore this point I add the indicators of ontological status to the past and future relations and to the events as present in figure 2.12.

It is crucial to keep in mind that the temporal axis, t, that cuts across the three axes t_1, t_2, and t_3, represents *physical flowing time*. The axes t_1, t_2, and t_3 are *not* meant as additional physical dimensions, but as an *abstract representation* of the temporal relations between events A, B, and C that lie in physical time, t. Drawing them separately is a way of emphasizing that the tensed relations do not lie along the axis of time, as earlier figures (figs. 2.1–2.5) might misleadingly suggest, and that in turn seems to imply that the tensed relations are relations in time and not time-less temporal relations between events in time. To emphasize this, I remove the arrowheads on the axes in figure 2.12. This point underscores the B-theorist's claim that temporal relations are timeless. I agree with this claim; still it does not commit us to a static view of time, as the B-theorist does. To make this point more forcefully I would draw on our discussion of relations as abstract, and not spatial, forms in the previous section on the family tree, etc. Likewise, temporal relations are abstract forms lying within a logical space of relations, and not temporal forms lying within physical time. If we keep the point in mind, we can continue to picture events and their temporal relations with diagrams such as here in figure 2.12 without needing to repeat the cautionary note that axes t_1, t_2, and t_3 are not additional physical dimensions, but abstract and time-independent representations of the tensed relations between events A, B, and C in physical time, t.

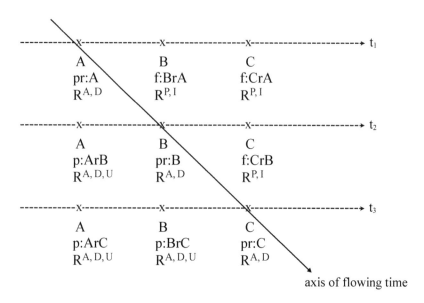

Figure 2.12. Flowing Time as the Axis Connecting
Multiple Presents in t_1, t_2, and t_3

With this move to a relational and inhomogeneous temporal ontology we have made considerable progress toward resolving the contradictions associated with flowing time.

Tense as Relational: A Way to Represent Time's Irreversibility?

One of the most intense debates both within physics and within the philosophy of physics is over the direction of time: Is the flow of time irreversible, as it clearly is in ordinary experience?[38] So, for example, we remember the past but not the future, a cup of coffee sitting by itself always cools and never spontaneously heats, we see waves receding outward after the diver hits the water and not converging from the distance to the point where the diver hits the water, and, since all spectra from distant galaxies are red-shifted and not blue-shifted, we conclude that the universe is expanding, and not contracting, in time. Are each of these phenomena separately grounded in, respectively, psychology, thermodynamics, hydrodynamics, and cosmology? Or is one of them truly fundamental, perhaps the expansion of the universe, and the rest arise from it? And how do we reconcile this with the fact that in theoretical physics time is reversible?[39]

While these questions continue energetically, the preceding discussion might offer a new and helpful terminology to the debate. In essence, if we believe that time in nature is irreversible, our discussion of tense as relational can provide a fresh basis for representing this directionality in terms of the difference between temporal relations such as f:BrA and p:ArB. So, for example, if time is irreversible, then A will be past in relation to B on all three time axes in figure 2.12 (above); we will never have f:ArB (or p:BrA).

Completing the Project by Generalizing to
a Continuum Model of Time

In the examples above I chose discrete moments in time, such as events A, B, and C, to simplify the discussion. However, we should generalize these models of time further by treating time as a continuous variable with an uncountably infinite set of events along the temporal axis. This means that the three temporal axes t_1, t_2, and t_3 (above) should be replaced by a continuous two-dimensional surface, T. Flowing time in figure 2.12, then, is a one-dimension line that "cuts across" this surface marking out a continuous sequence of temporal events as the sequence of subsequent present events (fig. 2.13):

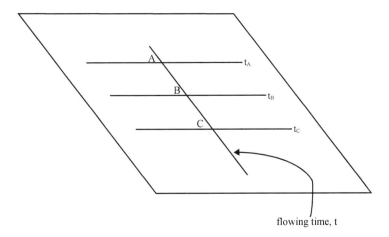

Figure 2.13. Two-Dimensional Surface T Showing the Axis of Flowing Time, t, as a Continuous Sequence of Present Events, Each with Its Own Axis of Events in Relation to the Present Event as Past and Future

Exploring the Relational Ontology of Time in Depth:
Flowing Time as Self-similar and Fractal-like

The relational ontology of time proposed above reveals the underlying complexity in the structure of flowing time: namely that this structure is self-similar, even fractal-like. To explore this claim let us return to the three time axes t_1, t_2, and t_3, and the discrete events A, B, and C, in figure 2.12.

First consider the axis t_2 and recall the temporal relations represented there: event B in relation to itself is the present moment (i.e., actual and determinate). For event B as present, event C in relation to B is future (i.e., potentially real and indeterminate), and event A in relation to B is past (i.e., actual, determinate, and inaccessible). Let us press the analysis further. Suppose we consider event A as present to itself (axis t_1). Here events B and C, in relation to event A as present, are both future events (i.e., potential and indeterminate events). This means that the *future* of event A as *past* in relation to B is different from the *future* of event A as *present* in relation to itself. Putting this starkly, the past had a different future when it was present than it does now. This may not be all that surprising, but what can easily be overlooked is that it actually leads to an *endless, iterative* process that reveals the underlying complexity of flowing time's structure.

There are two helpful mathematical terms to describe the results of this iterative process, "self-similarity," and its closely related concept, "fractal." Self-similarity is a mathematical term for an object that has, at least approximately, the same shape as its parts.[40] Many examples of self-similarity can be found in nature: a coastline, a cloud, a tree fern, the circulatory system and the villi that line the intestines in animals, DNA, snowflakes, forked lightning, a spiral galaxy, and so on.[41] Fractals are geometrical objects with two properties: self-similarity and fractional, or "fractal," dimension. Euclidean objects, such as a cube, are self-similar (they can be divided endlessly into smaller and smaller cubes), but they have integer dimensions (three for the cube). By contrast, the Sierpinski triangle, when endlessly reiterated, has a dimension calculated to be 1.585 (fig. 2.13).

Let us now explore in more detail the iterative process that reveals the self-similar structure and, possibly, fractal-like structure underlying the concept of flowing time. We can represent these ideas more formally as

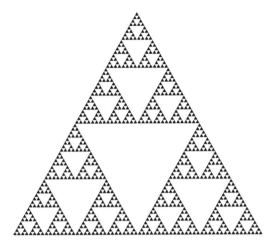

Figure 2.13.1.
Sierpinski Triangle
under Construction

Note: The white regions are cut out of the triangle successively. Four stages of extraction can be seen in this figure. To learn how to construct this fractal and calculate its dimension, see "Sierpinski Triangle," *Wikipedia,* last modified August 13, 2011, http://en.wikipedia.org/wiki/Sierpinski_triangle.

Source: "Sierpinksi Triangle," *Wikipedia,* October 21, 2007, http://en.wikipedia.org/wiki/File:Sierpinski_Triangle.svg.

follows: According to axis t_2, B is the present moment in relation to itself and is thus $R^{A, D}$. The relation of A to B is past and thus $R^{A, D, U}$ (p:ArB; actual, determinate, inaccessible). However, according to axis t_1, A is considered as the present moment in relation to itself and is thus $R^{A, D}$, while the relation of B to A is future $R^{P, I}$ (f:BrA; potential, indeterminate). To capture this fact we draw a *copy* of axis t_1 *below* t_2 in figure 2.14.[42]

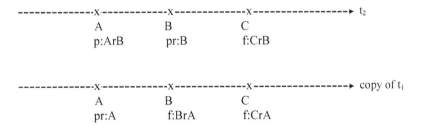

Figure 2.14. Copy of Axis t_1 below Axis t_2

Note: According to axis t_2 in figure 2.12, event B is the present moment and event A is past in relation to event B. But along axis t_1, event A is the present moment. Here we draw a copy of axis t_1 below axis t_2 such that event A on axis t_1 is directly below event A on axis t_2.

Figure 2.14 shows explicitly that from the perspective of B as present, only C is future in relation to B, but from the perspective of A as present, both B and C are future in relation to A. So the event A that is past in relation to B as present has a future that includes not only C in relation to A, as does the future of B, but also B (f:CrB). Thus the axis t_1 not only lies parallel to t_2 as in figure 2.12, where the relation between them specifies the axis of flowing time, t. In addition, in figure 2.14 we have placed a copy of axis t_1 beneath t_2 to represent the future temporal relations associated with event A considered as the present in addition to the future temporal relations associated with event A considered as the past in relation to event B.

This fact will not be significant for B-theorists. For example, if all events are $R^{A, D}$ (i.e., actual and determinate), then the relation between events B and C to A would not differ between t_2 and t_1. For ontology III A-theorists, however, events that are future in relation to A as present are both indeterminate and only potentially real in relation to A (see table 2.2 above).[43] Thus both B and C can be indeterminate in relation to A as present (i.e., along axis t_1), but only C can be indeterminate in relation to A as past (i.e., along axis t_2), because, in this case, B is present and thus determinate.

We can repeat the process still further as indicated in figure 2.15, where a copy of t_0 is included. To simplify the figure, the events are not lettered and the temporal relations between events are simply indicated as p, pr, and f (but bearing in mind that these indicate relations, not properties):

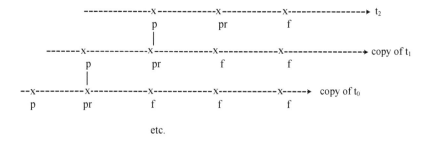

Figure 2.15. Copies of Temporal Axes t_2 and t_0, Each Displaying a Distinct Set of Temporal Relations for the Events of Axis t_2

Finally, this process can be reiterated endlessly, as indicated in figure 2.16:

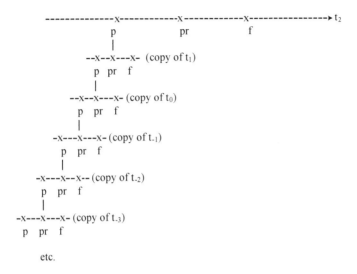

Figure 2.16. A Temporal Self-similar and Fractal-like Structure Underlying
Event A as an Endless Series of Temporal Axes, Each Capturing
a Distinct Set of Past Temporal Relations for the Events of Axis t_2

A similar argument can be made for the temporal fractal underlying
C along t_2 (fig. 2.17):

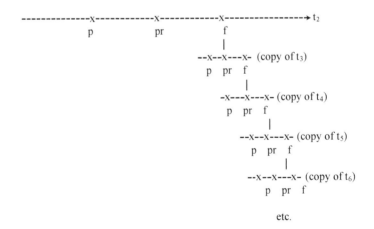

Figure 2.17. A Temporal Self-similar and Fractal-like Structure Underlying
Event C as an Endless Series of Temporal Axes, Each Capturing
a Distinct Set of Future Temporal Relations for the Events of Axis t_2

We have now arrived at a key insight: this structure, infinitely self-replicating downward from the temporal axes t_2, is self-similar and can be given a fractal-like interpretation. For this reason, I call this structure the "fractal-like character of time."

From here we can generalize to every temporal axis t_1, t_2, t_3, . . . , and to each event A, B, and C along every axis. Each event is the "tip of an iceberg," an underlying temporal self-similar and fractal-like structure. Thus, the two-dimensional surface depicted in figure 2.13, being a continuum, is the upper boundary of an uncountably infinite set of unbounded underlying self-similar and fractal-like structures, one for each moment A, B, C, . . . , along each temporal axis t_1, t_2, t_3, ...

Finally, as before, we can generalize from discrete temporal events A, B, and C, with their respective fractal-like substructures indicated above, to time as a continuous variable. Because a continuous variable like this is dense,[44] we can conclude that there are an uncountable infinite number of temporal events as in the temporal plane sketched in figure 2.13. But this, in turn, means that we can generalize the set of fractal substructures by recognizing that each level of the fractal is itself composed of a continuous temporal axis and not just the discrete events noted in the previous figures. Thus, each level of the fractal for each temporal event on the surface is itself the start of a new downward uncountably infinite temporal fractal.

In sum, we have an uncountably infinite set of temporal events along each finite temporal axis and an uncountably infinite set of such temporal axes all lying in parallel in the temporal surface across which an uncountably infinite set of "present" moments form the axis of flowing time. Lying below this surface is an uncountably infinite set of fractal-like structures each starting for every event in the surface and each, in turn, generating an uncountably infinite set of fractal levels each generating its own uncountably infinite fractal-like structure. Following Georg Cantor's terminology for transfinites,[45] I call this entire structure a "transfinite temporal self-similar and fractal-like structure" (fig. 2.18).

B. CO-PRESENCE: TIME IN ETERNITY

In the preceding discussion I began with Pannenberg's assumption that time, as we experience it in life and as it is found in history and in nature,

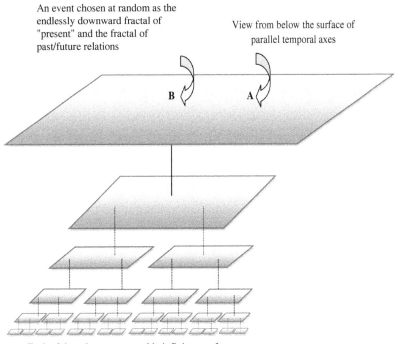

An event chosen at random as the
endlessly downward fractal of
"present" and the fractal of
past/future relations

View from below the surface of
parallel temporal axes

B   A

Each of these has an ucountable infinite set of
fractal levels not indicated here

Figure 2.18. A Transfinite Temporal Self-similar and Fractal-like
Structure Lying below an Event O Considered to be
the Present along the Axis of Flowing Time

is "flowing." I described, although briefly, the intense and complex debates in analytic philosophy surrounding this assumption. I suggested
that Pannenberg's understanding of flowing time is an A-theorist view
with what I call type III-2 ontology. Still, flowing time is subject to the
contention that it involves contradictions that arise when multiple tenses
are assigned to the same event. To avoid these contradictions, I suggested
that we make an additional assumption about time, namely that tense
should be treated as a relation between events and not as a property of
events. With this assumption I claimed that the contradictions could be
undone. I also indicated how truly complex flowing time is by suggesting
that a self-similar and fractal-like temporal structure "underlies" every
event in flowing time.

With ontology type III-2 in place we can return to the first of two issues in this chapter, namely what I am calling "eternal co-presence." Recall that, according to Pannenberg, our lived experience of flowing time, as well as time in nature, is taken up and given a kind of temporal unity in the divine eternity. For us, each present moment is related to a past that is lost and a future that is inaccessible. But in the divine eternity, all of the moments of our life, each with its unique past and future, are available to us "simultaneously." While they each retain their unique pasts and futures, all present moments are ready for us to experience "at once" and forever. I refer to this aspect of Pannenberg's concept of eternity by the term "co-presence." Pannenberg unpacks his concept of eternity by using two key ideas: distinction and separation. Here I want to explore several metaphors that illustrate Pannenberg's claim that while temporal events in ordinary experience and in nature are both *distinct and separate,* in eternity temporal events, while still *distinct,* are *no longer separate.* In this way I hope to show what is wonderfully offered us for our own theological research by the concept of eternal co-presence.

To do so, it is important to understand what Pannenberg means by "distinction" and "separation," since they frequently characterize the difference and relation between time and eternity. In my construal of Pannenberg, these terms mean the following:

Distinction. In ordinary experience and in nature, each momentary present has associated with it its own unique past and future, a past that once was and a future that is not yet. The past and the present are determinate, and the future is indeterminate. I refer to this as the "ppf structure" of time (i.e., "the past, present, and future structure"). This ppf structure is essential to flowing, creaturely time. It is part of what distinguishes each moment from all others, making it unique. It is part of the goodness of creation, and it is to be preserved eschatologically in eternity.

Separation. In ordinary experience and in nature, each present moment not only has a unique past and a unique future (i.e., its ppf structure) as part of its flowing character. In addition the present separates or divides its past from its future. Because of this separation, the "flow" of time takes on a second meaning: each moment is experienced as present only once. Each present moment immediately sinks forever into the past, while a new moment from the future takes its place as the present. Now the following is of crucial importance: this separation and division of time by the momentary present is not essential to creaturely flowing time. Instead, this temporal feature is part of the broken or fragmented character of flowing time in nature and ordinary experience, and it is to be overcome eschatologically in eternity.[46]

By using the terms "distinction" and "separation" in this way we can gain a clearer understanding of Pannenberg's view of eternity. As we saw in chapter 1, Pannenberg develops his idea of eternity in contrast to two prevailing ideas in philosophical and systematic theology: eternity as timelessness and eternity as endless flowing time. In my reading of Pannenberg, he does so through two claims about distinction and separation:

Distinction. In eternity, as in time, the distinction between events, that is, the ppf structure of each present moment, is preserved. Every present moment retains its own past and its own future. Thus eternity is not a "timeless now" in which all moments are stripped of their unique ppf structures and conflated into a single, all-encompassing present.

Without separation. In eternity, unlike our fragmented experience of creaturely flowing time, the separation between events is overcome. Instead, all events will be held together in an eternal present that comprehends them all and that preserves the unique ppf structure of each event. Thus eternity is not an unending form of what we now experience as flowing time, a sequence of separated and unrepeatable present moments each with its unique past and future. Instead, time in eternity is flowing time without a separation between events.

It is the combination of these two claims about distinction without separation that forms the basis for Pannenberg's specific concept of eternity. I refer to this concept as co-presence or, more specifically, eternal co-presence. Co-presence holds for flowing time in eternity but not for our experience[47] of flowing time in life and in nature. Co-presence is the way genuine creaturely temporality—flowing time without its fragmentation but with its rich ppf structure—is *embedded* in the divine eternity both "now" and in the eschatological New Creation. As such, it is in clear contrast with concepts of eternity as timeless and as unending creaturely time, and is a new way for us to talk about the temporality of eternity. It is, in essence, "time *in* eternity."

We are now ready to explore this very specific idea—eternal co-presence as temporal distinction without separation—through three models: 1) an "open stacks" library; 2) a non-Hausdorff manifold; and 3) "temporal entanglement." The first is meant to be intuitive and playful, offering the reader an initial glimpse into what otherwise might feel like the daunting task of understanding co-presence. The second model moves directly into some very abstract mathematics, but hopefully in a way that makes the venture more doable even for a nonscientist and one that rewards the patient reader with deeper insights into co-presence than could be found without mathematics. The third model draws on a central theme in the dependably counterintuitive world of quantum mechanics: the simultaneous correlation of properties of elementary particles at enormous distances. Here I extend the concept of entanglement drawn from quantum mechanics to what I call the entanglement in time between the different present moments in flowing time. The result of such exploration points in new directions for further study of co-presence in theology.

1. Co-presence in Eternity as an "Open Stacks" Library

The Books of Your Life

Imagine a vast library residing in the eschatological New Creation that contains all the books of your life. By "books of your life" I mean that each book in the library details your entire life from the point of view of a moment in time, perhaps a particular day or even hour or minute. So, for example, one book might describe your life from the point of view of the day you first fell in love (day B), long after your first birthday (day A), and long before your retirement (day C) (fig. 2.19a):

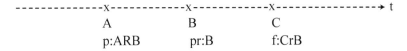

Figure 2.19a. Temporal Relations for Day B as the Present

Other books describe your life from days A and C considered respectively as the present (fig. 2.19b):[48]

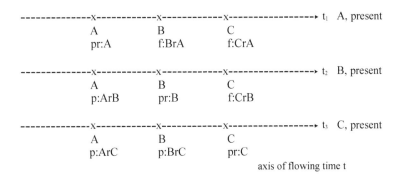

Figure 2.19b. Temporal Relations for Three Distinct Days, A, B, and C
Considered Successively as the Present Day

The key to relating time and eternity will be in the way we access these books in life now and in the eschatological future, the New Creation.

Creaturely Time: The "Closed Stacks" Library Represents
Both Distinction and Separation in Time

In life our experience of flowing time can be interpreted through the model of a "closed stacks" library. A librarian is in charge of getting books for us upon request, and, being a closed stacks library, the rules are quite strict:

- You can only withdraw one book at a time.
- You can only read books in temporal order.
- You can never read a book about a future day as the present until you have read all the books between the one you're reading now and the book about that future day (i.e., no "skipping ahead").
- Once you have read a book you can never read it again.

So you begin: you withdraw one book at a time, read it cover to cover, then return it to the librarian, requesting the next book to read. Say, for example, you read the entire book written from the perspective of day A, your first birthday. Only one year is past and definite, and the rest of your life extends throughout this book as mostly unknown. After reading "book A" you close it and return it to the librarian, who locks it away where it can never be reopened and reread. Now suppose that you would like to fast-forward to the book written from the perspective of the day when you first fell in love, book B. After reading it, you would close it, return it, and then you might want to skip ahead and read the book written on the day you retired, day C, and so on. However, this is all impossible. After reading book A you have to read all 15 x 365 = 5475 books (if you first fell in love at age sixteen and there is one book per day of your life) before getting to book B, and so on before reading book C.

Here, then, is a "closed stacks" interpretation of creaturely flowing time: each book records the past, present, and future from the point of view of a given present day; thus each book keeps the temporal distinctions associated with a particular present intact (i.e., its ppf structure). But now for the crucial point: The rule that you can read only one book at a time separates the days of your life from each other and gives them to you piecemeal, *ad seriatim,* epochal. The rule that each book must be read in temporal order without skipping ahead or turning back and the rule that a book can never be read a second time together lead to the "not yet and inaccessibility" of our experience of the future and the "never again and irretrievability" of our experience of the past. Finally the ontology of type III-2 gives to the determinate past its irretrievability and to the indeterminate future its inaccessibility. Here we have the abstract structure of flowing time overflowing with all its beauty reflecting its character as God's creation *ex nihilo* and laden down in all its sorrow in what I once called the "talons of time."[49]

Time in Eternity as Co-presence: An Open Stacks Library
Allows for Distinction without Separation

Now for the good news: in the eternity of the eschatological New Creation, the library is an "open stacks."[50] There is no librarian to gate-keep the books of our life from ourselves. Here instead the stacks are open, all the books are accessible, and the rules are wonderful:

- You can withdraw as many books at a time as you wish.
- You can read them in any order you like.
- You can read any book about the past as present or the future as present whether or not you have read all the books between that book and the book you are now reading.
- You can re-read a book about the past as present or the future as present as many times as you want.

In this model of the open stacks library in the New Creation, all the days of your life are available to you in any order, and you can take out as many at a time from any part of the library. The book for each day carries with it its own unique past and future intact. All lie before you to read at your leisure any way you wish—forever.

Here, then, is the "open stacks" interpretation of eschatological flowing time: each book records the past, present, and future from the point of view of a given present day; thus each book retains the temporal distinctions associated with a particular present intact (i.e., its ppf structure). And now for the crucial point—the rule that you can read as many books at a time as you wish overcomes the separation of the days of your life that the "closed stacks" library enforced and removes the piecemeal, *ad seriatim,* and epochal experience of flowing time. The rules that each book can be read in any temporal order, that you can skip ahead or turn back, and that a book can be re-read endlessly overcomes the "not yet" of our experience of the future and the "never again" of our experience of the past. Finally, the ontology of III-2 gives to the determinate past its irretrievability and to the indeterminate future its inaccessibility in a given present, but this no longer produces the piecemeal, *ad seriatim* character of creaturely flowing time because of the availability of all your life's books at once. Here we have the abstract structure of eschatological flowing time overflowing with all its beauty (reflecting its character as God's creation *ex nihilo*) and delivered from all its sorrow (the "talons of time" are no more).

The Endless Hermeneutical Richness of the Open Stacks Library

Recall Pannenberg's use of philosophical hermeneutics, drawing on Dilthey, Gadamer, Heidegger, and others to make the case that the meaning of a text (including life experience and historical consciousness along

with written texts) is found in its context. The key is the fusion of original context with the context of the interpreter as the wider hermeneutical
horizon of meaning. We can use the analogy of an open stacks library to
express this idea of the eternal re-experiencing and re-interpreting of our
lives by "layering" the reading process in a variety of ways.

One way is to imagine reading a specific selection of books, then
reading a different selection, and then comparing the two readings by
reading the first selection in light of the second and the second in light
of the first, and so on. This process can be continued indefinitely by reading more and more selections and comparing them with each other. With
each layer of readings in this process we generate an increasingly complex set of interpretations of our lives resulting from the effects of the
most recent selection that we have read on the meaning of the selection
read just prior to it, and this in turn from the readings of its prior selections and so on.

We can reframe the process of selecting successive readings to reflect Pannenberg's insight that the meaning of the individual moments of
life is given by their future in which the whole of life is experienced eschatologically by aligning the order of selection with the temporal order
of the books so read. Thus I re-read today's book in light of tomorrow's
book, and so on, finally approaching an understanding of today that could
never have been found just in today. We can reframe the process again
to reflect Pannenberg's insight about prolepsis by reading a spiritually
normative experience in life, say, conversion or confirmation or adult
baptism, as the last book, having read all the other books of one's life in
historical order. Other reframings are possible to reflect Pannenberg's
understanding of the present as the arrival of the future and not just its
birthing grounds.[51]

Such a hermeneutical process plays a fundamental role in all fields
of study. A basic definition of hermeneutics is the art, theory, and practice of interpretation. According to hermeneutics, the interpretation of
texts and data is an irreducible component of the quest for meaning in all
fields. Moreover, the process of hermeneutics leads to the unlimited discovery of unpredictable and radically new insights that inevitably arise
through continuous, multi-lensed readings of "fixed," given texts such as
the Bible—or, as in the above analogy, the books of our life.[52] Once life
is over and all the books recorded, it might not seem possible for some-

thing profoundly new to arise merely by re-reading them. To alleviate this undervaluation of the fecundity and necessity of hermeneutics, let us start with an example drawn specifically from biblical hermeneutics, since the closure and fixity of the biblical canon is often the reason for its dismissal as a source of genuine novelty by those interested in what they consider to be wide open, new knowledge and experience. Historically the canon of the New Testament arose in the Western Latin tradition through conciliar decisions of the early church as to which texts would be included and which excluded in the canonical documents.[53] These discussions culminated for the most part by the end of the fourth century with the Councils of Rome (382) and Carthage (397). Nevertheless, the closure of the canon—itself still the subject of controversy long after these councils[54]—in no way brought an end to the hermeneutical process of discovering rich new worlds of meaning in the New Testament texts. Particularly over the past two centuries, biblical-critical approaches to source, form, redaction, etc., have produced often radically different perspectives on such perennial theological issues as the historical Jesus in relation to the Christ of faith, the relation between faith, justification, and sanctification, the demythologization and remythologization of the New Testament, the meaning and significance of miracles in the Gospel of John, the historicity of the resurrection of Jesus as reported in the Synoptic Gospels, and so on.

Another example of hermeneutics can be taken from the dawn of Western liturgical music in Gregorian chant, dating back to the tenth century and codified in the twelfth and thirteenth centuries. In its earliest form Gregorian chant was a type of plainsong, a monophonic chant consisting of a single, unaccompanied melody sung in a fluid, rather than fixed, rhythm. By the late Middle Ages, polyphonic chants were frequent, with two or more independent voices sung together either at the octave or at a more interesting musical interval, principally the fifth (an interval of seven semi-tones, such as C to G). Gregorian chant can also include a "recitation," or reciting tone, a single musical note that is sung continuously during part or all of the chant, one that colors and grounds the entire piece. Here the independent voices can be understood as interpreting each other as they form an emergent whole: first the simultaneous harmonic sound heard at each moment and then the entire piece that, though heard sequentially, is remembered as a synthetic whole. A closely related musical example of the hermeneutics within music is the fugue, a contrapuntal form developed in the Renaissance and Baroque periods (ca. 1600–1760) perhaps most famously by J. S. Bach. Here, two or more voices are written that sound quite distinct but when played together are harmonious, especially when three or more voices are included and musical chords arise in each moment as notes are sounded simultaneously among the voices. Bach's contrapuntal music was famous for massively deploying a harmonic progression in which each piece cycles through part of the "circle of fifths"—the twelve tones of the chromatic scale arranged by fifths—before returning to its key.[55] Once again one thinks of the previous discussion of the "open stacks" library and the multiple readings of the books of one's life, particularly as the wholeness of eternal duration provides an ontological unity within which to experience the endless hermeneutical structure generated by the individual readings as a self-transcending whole.

Co-presence within Duration: Multidimensional
and Environmental Ecological

Finally, there are three questions—and probably others—about co-presence that should be addressed even if very briefly here. The first question is this: What accounts for our ability to integrate the co-present reading of so many life books together simultaneously in an experience of the wholeness of life in eternity? Clearly the taste of the co-presence of the moments of our life experience in the present world is tragically fragmented, serialized, and "once-off." To answer this, and to complete the concept of co-presence, we must turn with Pannenberg and his source, Augustine, to include our internal experience of temporal duration. Then we must extend this to project the possibility of its objective grounding in nature.

Thus, it is duration that provides the "superstructure" within which co-presence operates. In life our subjective experience of duration allows us to remember our past and anticipate our future as a whole in a given moment, even though this happens in a limited, fragmentary, and seemingly illusive way. So in the eternity of the New Creation we will be able to hold together—in the temporal extension called duration—the multitude of life books we read, but then we will hold all of them at once. The temporal extension of our New Creation experience will be true duration, bridging limitlessly across the entire plethora of the books of our lives and all their combinations in a feast of hermeneutic abundance. In it we will be able to experience our life history "all at once" as a whole, without separation yet with the essential temporal distinctions which give each moment its unique temporal and historical character. In eternity we will at last experience the true co-presence that structures duration and that is experienced in part in this world through memory and anticipation.

The second question is in regard to the multidimensionality of co-presence. How do we incorporate the insight that each moment of the past and future has its own past and future moments and that this reiterates endlessly? From the preceding discussion we saw that each moment of time is not only connected to its past and future by temporal relations but that such connections diverge toward infinity because time is a continuum and therefore dense. More important, we saw that each moment of time has a fractal-like substructure in which each past moment has its own future and past, as does each future moment, and that this extends endlessly into the infinite. One way to image what I am calling the fractal-structure of this account of time is through an expansion of the previous analogy of an open library. Now each book in the library of our lives is the "tip of the iceberg," a library of books representing a fractal-like temporal substructure. And each book in this substructure is its own starting point for a fractal-like substructure of books in life libraries. This process continues endlessly downward into endlessly increasing temporal complexity.

A third question arises as we place our reading process within a larger context: How does each book in the library of our lives refer and connect to the books of all those with whom we have interacted, especially those in our inner community of family and friends? Our open stacks library links[56] us to an ever-growing circle of people whose libraries contain their own downward fractals and relate us to yet a wider and wider context of life experience and hermeneutical detail. From here our own library reaches out to the wider ecumenical and global communities, and outward to the entire ecological and environmental community of life, which in turn reaches backward in time to all of life as it evolved in planet Earth (and perhaps elsewhere, if we have exchanged living materials with Mars, etc.). Thus in a very definite but incredibly complex sense each moment of our life links with an ever-burgeoning wealth of libraries of the life experience of others—both human lives and the wider ecological world of life—each of which has an endless repository of life experiences linked with others in a flowering diversity of graceful and beautiful connections. And all this, in turn, could and should be extended further into the endlessly opening horizon of meaning provided by the proleptic context of the eschatological future in which all life will be made whole, all experience healed, forgiven, and made available for endless enjoyment, and in which all life can lift its voices in worthy praise of the God of life, the Christ of redemption, and the Spirit of consummation.

2. Co-presence in Eternity as a Non-Hausdorff Manifold

Examples of Non-Hausdorff Manifolds in Physics and Cosmology

I next explore the idea of co-presence in eternity, which I have drawn from Pannenberg's work, by using what is called in mathematics a non-Hausdorff manifold. In particular I draw on the concept of non-separability offered by non-Hausdorff manifolds to show what I take to be an analogous meaning of non-separability in my interpretation of Pannenberg's view of co-presence in eternity.

Recall what I consider to be Pannenberg's fundamental claim about time and eternity: 1) In the creaturely time of this fragmented world, distinct events in their state as present events are *distinct events;* each present event has its own distinct past and future, neither of which are available to that present moment. The past is real and determinate, but distinct from and irretrievable to, the present, while the future is a realm of indeterminate possible realities only capable of being anticipated, but not experienced, in the present moment. 2) In the creaturely time of this fragmented world, distinct events are also *separable.* The past is irretrievable to the present, and the future is unavailable to be experienced as present. Thus, even though this separability is neither necessary nor essential to the temporal character of the present moment, it is inevitable to the temporal character of the present world.

Our hope is that eternity, both "now" and "eschatologically," will overcome this separability while preserving temporal distinctions. Thus in eternity, the separability of past, present, and future will be overcome but without eliminating the distinct relations that we refer to as the past and future of each present. Thus in eternity all present events together with their respective pasts and futures are *non-separable:* all present events as present events will be co-present to each other, held together in an extended, simultaneous whole that nevertheless retains their unique, determinate pasts and unique, indeterminate futures (co-presence). It is the separability of the past and future from the present in creaturely time that, I believe, produces our lived, existential situation in which we cannot experience the past or the future as present in and along with our present moment—and not just the fact that the present has a unique, determinate past and a realm of indeterminate possible futures. I represented this claim above by the analogy of the open and closed stacks library. Can mathematics offer us another, more formal and rigorous way to represent this claim, even if it, too, has inherent limitations and problems? I believe it can, and one way to do so is to represent time and eternity by Hausdorff and non-Hausdorff manifolds, respectively.

The simplest example of a manifold is Euclidean space: a line is a one-dimensional Euclidean manifold, a surface is two-dimensional, a volume is three-dimensional.[57] A manifold of n-dimensions is a mathematical abstraction from Euclidean space in which the space near every point in the manifold is either exactly or approximately Euclidean: intuitively, it is locally flat and continuous. Nevertheless, the manifold as a whole may have a more complicated shape and curvature, such as a sphere, a torus, a Möbius strip, and so on. These spaces are still locally flat, or Euclidean, if one looks at "small regions" but they are globally curved or non-Euclidean.[58] What is important for us, however, is that even these non-Euclidean spaces are still Hausdorff. We must now explore what this means, and how non-Hausdorff manifolds differ from them.

To do so we need to define terms such as "open" and "closed" sets on a manifold. Let us start with the x-axis, an example of a one-dimensional manifold. We can define an open set intuitively as follows: a set of points U along the x-axis containing the point u is said to be "open" if for every point u we can move a short distance from it and still remain in U. An example of such an open set is the interval $0 < x < 1$. Clearly we can start from any point P within the interval $0 < x < 1$ and move a non-zero distance in either direction repeatedly without reaching the points $x = 0$ and $x = 1$.[59] Now compare U with the set V composed of the open set U plus the points $x = 0$ and $x = 1$. We define V as a "closed" set, with $x = 0$ and $x = 1$ as its boundary. Again, intuitively, if we start from $x = 1$, *any* finite movement in the positive x-direction, no matter how small, takes us out of the set V. Additional terms can be defined. Two sets are said to be "disjoint" if they have no points in common (intuitively, they do not overlap), and a "neighborhood" of P is an open set containing P. Note that these definitions can be generalized to open and closed sets in a two-dimensional Euclidean manifold (such as the x-y plane), to higher-dimensional Euclidean manifolds, and to non-Euclidean manifolds (such as the sphere or the torus).

With this terminology in mind we can probe one of the most impor-tant distinctions between a Hausdorff and a non-Hausdorff manifold: namely, the former is "separable" while the latter is "non-separable." We start by imagining two distinct points P and Q in a space H (fig. 2.20). The neighborhood of P is the open set N_P, and the neighborhood of Q is the open set N_Q. If H is Hausdorff, it is always possible to find neigh-borhoods N_P and N_Q, which are disjoint. A Hausdorff manifold is called "separable" to signify this fact.[60] What, then, would a non-Hausdorff manifold look like?

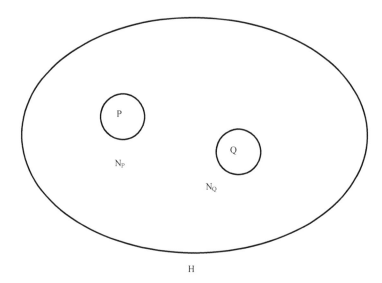

Figure 2.20. A Hausdorff Manifold H Containing Points P and Q
and Their Disjoint Neighborhoods N_P and N_Q

A non-Hausdorff manifold includes at least one pair of distinct points that cannot be separated, that is, they do not have disjunctive neighborhoods. A well-known example is the "line with two origins."[61] We start with two copies of the real line, R_1 and R_2, and their respective origins, O_1 and O_2 (fig. 2.21):

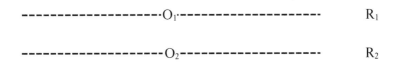

$$R_1$$

$$R_2$$

Figure 2.21. Two Copies of the Real Line, R_1 and R_2, with Their Respective Origins, O_1 and O_2

Next, we identify all the points of R_1 and R_2 except the origins O_1 and O_2. The origins O_1 and O_2 are now non-separable. To see this, one need only consider an open set U_1 that includes the origin O_1 and compare it with an open set U_2 that includes the origin O_2. Clearly U_1 does not contain O_2, but it *will* contain points also contained by U_2 because all points in R_1 and R_2, except the two origins, have been identified. In short, the two origins O_1 and O_2 cannot be separated even though they have not been identified. The resulting manifold is non-Hausdorff (fig. 2.22):

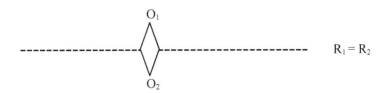

$$R_1 = R_2$$

Figure 2.22. Non-Hausdorff Manifold: Line with Two Origins

Note: All points along R_1 and R_2 are identified except for the origins O_1 and O_2.

A second example of a non-Hausdorff space is the "branching line."[62] Once again consider two copies of the real line R designated R_1 and R_2 whose coordinates are x_1 and x_2, respectively (fig. 2.23a). Next identify all points $x_1 < 0$ along R_1 and $x_2 < 0$ along R_2 (fig. 2.23b). This produces the branching line manifold in which the two origins $x_1 = 0$ and $x_2 = 0$ are distinct but non-separable: each open neighborhood of the origin $x_1 = 0$ in R_1 contains points $x_1 < 0$ that are also contained in an open neighborhood of the origin $x_2 = 0$ in R_2, even though the origins O_1 and O_2 and the branches $x > 0$ in R_1 and $x > 0$ in R_2 are not identical (fig. 2.23a–c):

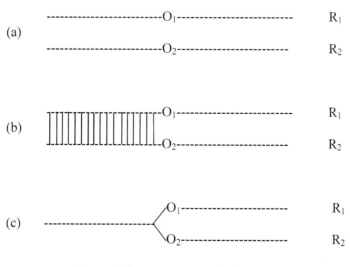

Figure 2.23a–c. Non-Hausdorff Manifold:
The "Branching Line" Space Constructed in Three Steps

Finally, a third, closely related example of non-Hausdorff space is the "complete feather" obtained by creating a branch at every point along the real line R, and then at every point along each branch, and so on indefinitely.[63] In figure 2.24, three branches have been grafted onto the real line R, the first with three branches grafted onto it, and the first of these with three branches grafted onto it. The process is repeated without limit, giving the real line an infinity of branches, each of which has an infinity of branches and so on endlessly.

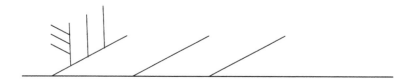

Figure 2.24. Non-Hausdorff Manifold:
The "Complete Feather" under Construction

With this in mind, let us now return to temporal events lying along ordinary, Euclidean time t. Consider two events, P and Q, each surrounded by a neighborhood of events N_P and N_Q. Because Euclidean time is Hausdorff, we can always find neighborhoods N_P and N_Q that are disjunctive. So, for example, if at noon I have lunch, at 3 p.m. I have a snack, and at 6 p.m. I have dinner; then there is an open neighborhood surrounding lunch (say, 11:45 a.m. < lunch time < 12:15 p.m.), another neighborhood surrounding snack time (say, 2:45 p.m. < snack time < 3:15 p.m.), and yet another neighborhood surrounding dinner (say, 5:45 p.m. < dinner time < 6:30 p.m.), all of which are disjunctive. This reflects the fact that a Euclidean manifold, here the t-axis, is separable and thus Hausdorff. Thus, I propose that the Hausdorff character of time that underlies the separability of time is a factor in what we experience as creaturely flowing time.[64] More precisely, I propose that the Hausdorff character of flowing time gives rise to what is at the heart of the tragic element of time's flow, namely that being in the present isolates us radically from the unavailable past with which we can no longer be connected and the indeterminate future that we can never experience in the present and that thus remains mere anticipation. This also means that the Hausdorff character of time contributes to the transcendental preconditions for the possibility of true becoming by rendering distinct temporal events separable in the mathematical sense. If so, then the recognition that ordinary time appears Hausdorff in character, and that there may be non-Hausdorff ways of conceiving of time, offers a new and promising insight into Pannenberg's understanding of time in relation to eternity.

Implications for Co-presence in Eternity:
Temporal Distinctions without Separation

Thus we come to a crucial question regarding time and eternity: Is the Hausdorff character of time essential to time in eternity, or can it be overcome without undercutting what *is* essential to time, that is, what is essential to the nature of temporal existence and to the possibility for us to experience ourselves as historical beings? According to my reading of Pannenberg, the eternity of the New Creation, with its Boethian mode of universal simultaneity, will be such that the distinct temporal relations, which are associated with every event (i.e., its own particular ppf structure with an indeterminate future and determinate past) and which contribute to an event's constitutive, treasured character as temporal, will remain in place in eternity. However, the *irretrievable loss* of the present into the determinate past that can no longer be experienced in its presentness and the *unavailability* of the indeterminate, potentially real future that, as future, can never be experienced in its presentness will be eliminated in eternity. In essence, these features—temporal isolation consisting in past irretrievability and future unavailability—are *not,* in fact, essential to the fully temporal character of our lives and of the natural world. Instead, that which is good, right, and beautiful about the flow of time as the creation of God will be ours forever in eternity because eternity retains the distinctive ppf temporal relations constitutive to each event. But God's eternity will not maintain the separation between events as present and with it the inevitable, but ultimately unnecessary, irretrievable loss of the past and unavailability of the future.[65] Because events in time are no longer cleaved apart as separable, although they remain distinct, the temporal distinctions that contribute to the created goodness of the flow of time will remain forever in eternity, to be enjoyed endlessly as seen by the "open stacks" analogy. But because events in eternity will not be separable, the isolation of our temporal experience, that is, the isolation resulting from the irretrievability of the presentness of the now-past event and unavailability of the presentness of the non-yet future event, will be overcome. In this way a key factor that contributes to the tragic element in the flow of time—its separability and resulting isolation, or what I once called the "talons of time"—will be healed.

The next question is whether there is a way to employ the idea of a non-Hausdorff manifold to capture this relation between time and eternity. I want to suggest that there is: namely that the temporality of eternity, unlike that of ordinary, this-worldly time, is at least non-Hausdorff. In essence, in the non-Hausdorff temporal manifold of eternity, distinct times (such as lunch at noon and dinner at 6 p.m.) retain their unique character through having their own temporal neighborhoods (i.e., their own futures and pasts) and distinct identities (i.e., noon is never 6 p.m.), but they are nevertheless inseparable: they cannot be cleaved apart in eternity as they can be in ordinary Hausdorff time. It is this *distinction-preserving temporal non-separability,* offered by a non-Hausdorff concept of the temporality of eternal time, in contrast to *distinction-preserving temporal separability,* offered by a Hausdorff conception of ordinary time, that captures part of what I believe Pannenberg means by (what I am calling) the co-presence of all events in eternity.

Let us start with a concrete example using a "branching line" representation for non-separability of co-presence in eternity. In figure 2.25a, we see the ordinary time axis, t, with events A, B, and C that represent the events of lunch, snack, and dinner as discussed above, and with event D as a late night snack.

$$\text{------------------x------------------x------------------x--------------x----} \rightarrow \text{ real time t}$$
$$\qquad\quad A \qquad\qquad B \qquad\quad C \qquad\qquad D$$

Figure 2.25a. Ordinary Time Axis, t, with Events A, B, C, and D
Representing Lunch, Snack, Dinner, and Late Night Snack

In figure 2.25b we represent times of events with separate temporal axes of relations. So for axis t_B, event B is present while event A (lunch) is past in relation to B as present (p:ArB), and events C (dinner) and D (late night snack) are future in relation to B as present (f:CrB; f:DrB), and for axis t_C, while event C (dinner) is present, both A and B are past in relation to C as present (p:ArC, p:BrC), and D is still future in relation to C (f:DrC). For reference, I have labeled the events A_p, B_{pr}, C_f, D_f, and so on in figure 2.25b.

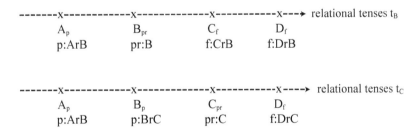

$$\text{-------x-------------x--------------x-----------x---}\rightarrow \text{ relational tenses } t_B$$
$$\quad A_p \qquad\qquad B_{pr} \qquad\qquad C_f \qquad\qquad D_f$$
$$\quad p{:}ArB \qquad\quad pr{:}B \qquad\quad f{:}CrB \qquad\quad f{:}DrB$$

$$\text{-------x-------------x--------------x-----------x----}\rightarrow \text{ relational tenses } t_C$$
$$\quad A_p \qquad\qquad B_p \qquad\qquad C_{pr} \qquad\qquad D_f$$
$$\quad p{:}ArB \qquad\quad p{:}BrC \qquad\quad pr{:}C \qquad\quad f{:}DrC$$

Figure 2.25b. Two Axes of Temporal Relations for Event B
as Present and Event C as Present

Now in figure 2.25c (compare with the construction of the branching line in figure 2.23b), we make the crucial move: i) we identify all events $t_B < B_{pr}$ and $t_C < C_{pr}$:

i) $t_B < B_{pr} = t_C < C_{pr}$

but ii) we keep B_{pr} along axis t_B and C_{pr} along axis t_C distinct:

ii) $B_{pr} \neq C_{pr}$

To clarify,

> i) by "identify" I mean that the time of each event $t_B < B_{pr}$ is put into a one-to-one relation with the time of each event $t_C < C_{pr}$. I do not, however, mean that the contents of these events (e.g., what happened in people's lives at each event) are conflated. Thus each event $t_B < B_{pr}$ remains past in relation to B_{pr} as present; B_{pr} retains its unique set of past relations. Similarly, each event $t_C < C_{pr}$ remains past in relation to C_{pr} as present.

> ii) By "distinct" I mean that the events B_{pr} and C_{pr} are *not* identified. They are, instead, the first events along t_B and t_C to retain their full, individual temporal identities.

In this way we can say, with a certain degree of plausibility, that event B as B_{pr} is non-separable from event C as C_{pr}: each event retains its unique past and future even while the two events cannot be disjoined into non-overlapping open sets of past and future events. In essence, the past, according to event C as present, is present "in" and co-present "to" event C as present.[66]

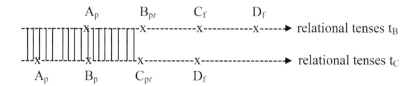

Figure 2.25c. Co-presence of Event B_{pr} with C_{pr} Representing
Conditions i and ii

This argument captures at least part of what the idea of eternal co-presence means for a specific set of moments, events B and C. I am suggesting that by using the mathematics of a non-Hausdorff manifold, we can indicate how the event B_{pr} as present can be brought into a temporal relation with event C as present, C_{pr}, without being equated ontologically with C as C_{pr}. Put less technically, in eternal co-presence I will experience the reality of my past (in its presentness) in my present moment without conflating the past as past with the present as present. This is because I have "added" the set of temporal relations associated with the event B as present to the set of temporal relations associated with the event C as present without equating B as present with C as present. This interpretation is represented by figure 2.25d using the "branching line" discussed (compare fig. 2.23c).

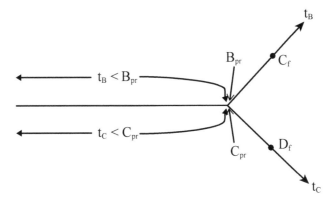

Figure 2.25d. Non-Hausdorff Co-presence: Branching Line Representation Resulting from Imposing Conditions i and ii on the Axes of Temporal Relations for Events B and C

Extension

This branching centered on event C_{pr} can be thought of as one thread among an uncountable infinity of such threads each related to each moment in time (which, recall, is dense). This larger perspective offers what I will call "the infinitely complex, non-Hausdorff tapestry of life." In essence, we can repeat the preceding procedure for all events in the past of C as well as for all events in the future of C, in each case, generating a different non-Hausdorff branching line in which each temporal moment, as truly present to itself, is brought once again into a set of temporal events in which it is past and future in relation to other events. Now the uncountably infinite set of all such branching lines, seen as a whole, casts a partial but highly illuminating light on creaturely co-presence in eternity as suggested by Pannenberg using the mathematical language of a non-Hausdorff topology.

But there is more: we can repeat the process for each event by moving "below" it into its underlying, fractal-like structure as described above. We then come to gaze upon an uncountably infinite-dimensional fractal-like manifold of events associated with each event as present and, beyond this, upon a set of such uncountably infinite-dimensional fractal-like manifolds, one for each event in the endlessly dense set of temporal moments along the ordinary temporal axis.

Hints for Part Two: Implications of Co-presence in Eternity
for Temporality in This World

The non-Hausdorff model of eternity in comparison with the Hausdorff model of ordinary time offers a mathematical language to illuminate Pannenberg's idea that in creaturely, eschatological eternity, all the "present moments" of our lives can be united in an overriding eternal present, which, while respecting their distinctions (i.e., their individual pasts and futures), nevertheless unites them in a "co-presence" without separation. In this way, eternity is seen to overcome time's ruthless passage.

The non-Hausdorff model also illuminates two other theological concerns. First, it suggests that what is "fallen" about creation is the apparent (i.e., phenomenological) loss of the non-Hausdorff property of authentic, created temporality. Second, it suggests that the eschatological transformation of creation into the eternal New Creation will include the retrieval and perpetual guarantee of time's non-Hausdorff character and lead to a temporal "fractal of fractals" endlessly generating complexity and beauty. Finally, the non-Hausdorff model suggests possible topics or directions for scientific research to which I return in part two of this volume where I consider programs represented as TRP → SRP (Theological Research Program → Scientific Research Program).

3. Co-presence in Eternity as "Temporal Entanglement"

Entanglement in Quantum Mechanics

Entanglement is one of the central features of quantum mechanics.[67] It refers to the fact that identical particles once united in a single physical state retain something of that original unity as suggested by the surprising correlations between their properties even when they are separated at enormous distances and when their properties are measured simultaneously.[68] Particles showing this kind of subtle unity are "entangled."

Entanglement is actually a complex form of a general feature of quantum mechanics known as superposition. This feature lies at the heart of quantum mechanics and is, arguably, its most nonclassical characteristic. Superposition shows up most clearly and famously in the "double slit" experiment. Here individual electrons fired at a sheet of metal with two narrow, parallel slits seem to pass through *both* slits before landing on a detector (fig. 2.26).[69] The result is quantum interference, in which bands of light and dark build up on the detector as huge numbers of electrons are fired at it. Classically we would see two piles of electrons, one below each slit, evidencing their particle-like nature. Quantum mechanics predicts interference bands due to their wave-like nature, suggesting water waves passing through a breakwater.

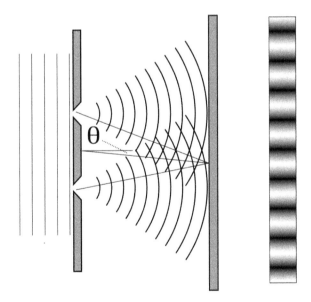

Figure. 2.26. Double-Slit Experiment

Note: The electrons are represented as waves emerging from the source at the left of the figure and passing through two slits in the metal plate in the middle. Interference bands occur on the detector to the right of this figure.

Source: "Doubleslit," *Wikipedia,* May 17, 2011, http://en.wikipedia.org/wiki/File:Doubleslit.svg.

We can represent electron superposition mathematically. If ψ_1 represents the possibility that the electron passed through the first slit, and ψ_2 the second slit, then the state of the electron just before its detection is ψ_T, the linear superposition of these two possibilities:

$$\psi_T = \psi_1 + \psi_2 \tag{1}$$

According to quantum mechanics, the probability of finding an electron anywhere along the detector is given by the square of ψ_T:

$$|\psi_T|^2 = \psi_1{}^2 + \psi_2{}^2 + 2(\psi_1)(\psi_2) \tag{2}$$

It is the third term in (2), the product of the two states, that describes the interference effects along the detector, effects that are entirely contrary to the classical world and ordinary life. Compare this interference pattern with the results that we would expect if electrons behaved like classical particles. If they did, we would see two piles of electrons on the detector, one below each slit.

We can now explore quantum entanglement with the preceding discussion of superposition in mind. Let us start with an electron and focus on one of its properties, spin.[70] Measurements show us that electron spin is always either up or down along any axis of measurement x, y, or z. Suppose we put two electrons into a combined state whose total spin is zero, that is, the spins of the two electrons are opposite. Such spins are said to be anti-correlated. The mathematical description of their combined state is the entangled wavefunction ψ_T:

$$\psi_T = \psi(+)_1\psi(-)_2 - \psi(-)_1\psi(+)_2 \tag{3}[71]$$

Here ψ_T on the left-hand side of the equation is the description of the state of the two electrons. The two ψ followed by brackets ( ) on the right-hand side of the equation indicate the state of one or the other particle. They are labeled $\psi(\ )_1$ and $\psi(\ )_2$ to indicate which particle they refer to (1 or 2), and their spin is indicated by (+), meaning "spin up," and (−) for "spin down." The mathematical reason why the state of the two particles is called "entangled" is that there is no way to think of this equation as only describing electron 1 in spin state "up" and electron 2 in spin state "down," as $\psi(+)_1\psi(-)_2$ implies, without also thinking of the opposite combination of states, $\psi(-)_1\psi(+)_2$, as possible. Instead, in a highly counterintuitive way, the wavefunction ψ_T depicts each particle in both states at once.

To understand this more clearly, let us compare the quantum description of an entangled state (3), above, with the way we would describe the two electrons if we used classical physics. According to classical physics we can always distinguish between two particles by stating where each of them is at a given time, by following their individual trajectories in space, and so on. This means that their combined state can be written as a simple product of the two separate states with the first particle either with spin up (4a) or down (4b):

$$\psi_T = \psi(+)_1\psi(-)_2 \tag{4a}$$

or

$$\psi_T = \psi(-)_1\psi(+)_2 \tag{4b}$$

Here is the crucial point. The classical choice of either (4a) or (4b) is not an option here because it fails to describe such basic quantum phenomena as the Pauli principle, which states that two electrons cannot be in the same state at the same time. The Pauli principle in turn gives rise to such wide-ranging classical phenomena as the chemical properties represented by the periodic table of elements. Equation 3, however, does lead to the Pauli principle and is therefore the correct way to describe the two-electron state.[72] But it does so at a price: with (3) we get an entirely nonintuitive worldview in which each electron is in both spin states, a worldview that overthrows our everyday understanding of nature.[73]

It is no wonder that George Greenstein and Arthur G. Zajonc underscore the importance of entanglement in these words: "Few developments in the history of physics have been of comparable significance for our philosophical understanding of the world."[74] According to Greenstein and Zajonc the fact that entangled states cannot be thought of as composed of separate states in a classical sense led Schrödinger to proclaim that entanglement is "not *one* but rather *the* characteristic trait of quantum mechanics."[75]

One of the striking implications of quantum entanglement is that when two identical particles are emitted in opposite spin states, they seem to remain in a single combined state regardless of their spatial separation. Yet when either particle is measured, they both take on a definite state: one, spin up, and the other, spin down (although which particle takes which spin state is entirely unpredictable). This conclusion follows both from the now famous theorems of John S. Bell[76] and the experiments performed to test them by Alain Aspect, Philippe Grangier, Gerard Roger, and many others.[77]

Let us look at this result in more detail. Imagine that two electrons are emitted in opposite directions along the y-axis with one traveling off to laboratory A and the other to laboratory B (fig. 2.27):

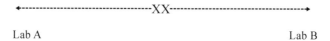

Lab A   Lab B

Figure 2.27. Two Electrons Produced at XX with Opposite Spins
Are Emitted and Travel to Labs A and B

If we measure the spin of the electron along the y-axis in lab A it might be up or down. Nothing we do seems to affect the result because if we repeat the experiment thousands of times the individual results are completely random. Taken together they are, on average, 50% up and 50% down:

Results for lab A, spin along the y-axis: + + − + − − + − − − + − + +
+ − etc.

Similarly, for measuring the spin of the electron along the y-axis in lab B:

Results for lab B, spin along the y-axis: − − + − + + − + + + − + − −
− + etc.

This holds even when the measurements are taken simultaneously in labs A and B. It is only when we compare these measurements that we find, in every case, the spins of each pair of electrons are "anti-correlated": when the electron in lab A has spin up, the electron in lab B has spin down, and vice versa. Such exact anti-correlations can be seen in a comparison of the preceeding results for labs A and B.

The natural way to represent these anti-correlations would be with one of the two above wavefunctions, 4a or 4b. Yet to correctly represent the initial combined state before the electrons were emitted, an entangled state such as equation 3 is required, as we saw above in discussing the Pauli principle and its relation to the chemical properties of the elements. But if equation 3 represents the initial, entangled state of the two particles when they are first emitted, when and why did this state "collapse" and the resulting state become either 4a or 4b? And keep in mind the fact that the two labs can be miles apart and the measurements synchronized such that even light cannot travel from one measurement to the other.

Before responding to this question we need to dig a bit deeper into quantum entanglement. It is to John Bell's lasting credit that he discovered even further complexities in spin entanglement. Bell turned from a comparison of the spins of the two electrons along the same axis to a comparison between spins measured along different axes. Let us say that lab A measures spin along the y-axis as before, but that lab B now measures spin along the z-axis. It turns out that when the results of thousands of such measurements are compared, the spins of the two electrons along different axes are anti-correlated to a much smaller amount than one would predict if spin is a classical property. It is also a victory for quantum mechanics because, if we calculate the anti-correlation for different axes using quantum mechanics, the experimental results are entirely consistent with the predictions. Thus quantum mechanics accounts for spin correlations but a classical model does not.[78]

So what are our options for understanding these results? Broadly speaking there are two options: 1) Relativity is violated by an instantaneous, "superluminal" interaction between the distant labs. This interaction communicates the results of the measurement of the electron's properties in one lab to the electron in the other, forcing it to be in the anti-correlated state.[79] 2) Relativity is not violated. Instead, the two electrons remain to some degree ontologically united in a single state. Though they are measured individually in separate labs they still form a "non-separable" unity. Most scientists find the former option unattractive given the foundational role played by relativity in contemporary physics. Yet the latter suggests a drastic revision of the ontology of physics such that the two particles, while in some ways distinct, are also profoundly inseparable, even if this inseparability is veiled beneath the experience of reality that characterizes the ordinary, classical world. In short, quantum non-separability seems to be the clear, but highly counterintuitive, inference of quantum entanglement.

Implications for Co-presence in Eternity:
Temporal Distinctions without Separation

Can quantum non-separability, as an inference from quantum entanglement, serve to illuminate Pannenberg's insights about co-presence in eternity? Perhaps so. Of course quantum non-separability is a *spatial* phenomenon; it is not about the non-separability of temporal events, per se.
Nevertheless, it provides another analogy, along with those we have
already explored, for what Pannenberg means by the way that eternity
as a whole includes all temporal events in nature and history in a non-
separable unity, a unity that still preserves the distinctive characteristics
of those events.

The analogy is the following: 1) For quantum non-separability, different events in space retain their own spatial contexts. Similarly, for eternal co-presence, different moments in time retain their temporal contexts
of pasts and futures. 2) For quantum non-separability, different events
in space are entangled, reflecting an underlying non-separable ontology.
Similarly, for eternal co-presence, different moments in time are "temporally entangled," brought into a unity by the endless duration of eternity in
which they are upheld "simultaneously" but without temporal conflation.
In this way, the quantum analogy offers an insight into the notion of co-
presence in eternity as an ontology of temporal non-separability. Moreover, this analogy is simply not available to us if we are confined to the
ordinary world of human experience, the world described by human language and taken up into classical physics and modern philosophy.

It is thus here in quantum mechanics that we now find a unique resource for interpreting Pannenberg's theological conception of the divine
eternity and its power to unify the "spread-out-ness" of ordinary flowing
time. Quantum entanglement and its inference, quantum non-separability,
offers an analogy for understanding eternal co-presence in which the distinction between events (i.e., their unique ppf structures) is maintained
and yet their apparent temporal separability—the tragic quality of flowing time—is overcome. Temporal co-presence in eternity is thus seen as
arising from the temporal non-separability of eternity. It is an ontological
unity bequeathed to the temporal processes of nature by their Creator, a
unity that gives to our experience of flowing time the sense of an underlying unity, one that extends to all temporal events in nature. More specifi-

cally, it is a "differentiated unity," to use Pannenberg's term, that can only be postulated of the eternity of the Trinitarian God.

C. PROLEPSIS: ETERNITY *IN* TIME

The goal of this section is to explore new ways to think about Pannenberg's concept of prolepsis using resources from math, physics, and cosmology. First, however, something should be said about the relation between the present world and the New Creation that will emerge by God's radical new act beginning with the Easter event and extending forward in time until "all will be well," as the fourteenth-century mystic Julian of Norwich so elegantly wrote. It will be out of this New Creation that Christ will "reach back" proleptically—not "in time from the future" but "outside of time from the future"—to the normative redemptive event, his Resurrection and Ascension. I begin with a brief review of the New Testament understanding of the transformation of the Creation into the New Creation as based by analogy on the bodily resurrection of Jesus. I then offer a rudimentary "diagram" that seeks to capture the "now yet coming" character of the eschatological transformation (section 2). Next I use this diagram as a basis for the idea of prolepsis. The key here is that the relation between the eschatological New Creation and the Resurrection events does not extend back through historical time but transcends it (section 3). To capture the idea that prolepsis involves a transcending of historical time I offer two analogies based on the topology of a "multiply connected manifold." The first is drawn from the idea of black holes as described by Einstein's general relativity. The second is taken from Andrei Linde's "eternal inflation" multiverse (section 4). I close with a look at the theological idea of multiple prolepsis based on the normative Easter event such that our own death can be thought of as related to our resurrection in the New Creation as a form of prolepsis (section 5).

1. The Transformation of the Creation into the New Creation Based by Analogy on the Bodily Resurrection of Jesus

How are we to talk about the relation between the present created world and the New Creation? In the introduction to this volume I report in some detail on current research by New Testament scholars on the bodily resurrection of Jesus (see the appendix to the introduction, section B). I note that the term "transformation" is frequently used to conceptualize the relation between Jesus of Nazareth and the Risen Jesus. This term embodies the claim that there are elements of continuity *and* discontinuity between Jesus of Nazareth and the Risen Jesus. I then use "transformation" analogously for the relation between the present universe as God's creation and the coming New Creation that God will bring about eschatologically. Here the phrase "bring about" combines two concepts dialectically: the "realized" Easter event and the "apocalyptic" future. I embed this claim in guideline 6 of CMI: Eschatology in light of the "resurrection of the body" (see the appendix to the introduction, section D), which includes two additional insights about transformation.

The first insight is about the "transformability" of the universe and the formal conditions for its possibility (the "such that" or "transcendental" argument): If God will transform God's creation, the universe, into the New Creation, it follows that God must have created the universe such that it is transformable by God. In particular, if the universe is to be transformed and not replaced, God must have created it to have precisely those conditions and characteristics that will be part of the New Creation. I refer to these conditions and characteristics as "elements of continuity." Still, if a transformation is to occur, and not just a continuous evolution of the present world into the New Creation, there must be conditions and characteristics of the present creation that we will *not* expect to be continued into the New Creation. There must also be conditions and characteristics of the New Creation that are not present in the present world creation as we know it but that *will* characterize the New Creation. Both of these can be called "elements of discontinuity" between creation and New Creation.

The second insight is about continuity within discontinuity: inverting their relative importance in the eschatological transformation. When we come to the resurrection of Jesus and thus to the eschatological hope for the future, I propose that the elements of "continuity" will be present but within a more radical and underlying "discontinuity" as is denoted by the concept of the transformation of the universe by God. This insight distinguishes eschatological transformation from the emergence in creation of new levels of complexity, including its emergence through non-interventionist objective divine action (NIODA) such as is found in my version of theistic evolution.[80]

Together, these two insights represent what I call "first instantiation contingency." As previously noted, I refer to it by the acronym FINLON, or "the first instance of a new law of nature." FINLON refers in specific to the resurrection of Jesus as the first instance of a radically new phenomenon in the world. But within the context of the coming New Creation the resurrection of Jesus will be the first instance of a regular new phenomenon, the general resurrection from the dead and life everlasting. It might better be termed "the first instance of a new law of the New Creation" (FINLONC).[81]

2. The Transformation of the Creation into the New Creation: A Diagrammatic Approach

Before we discuss "prolepsis" we first need to think about the relation of transformation between the present creation and the New Creation in more detail. Only then can we layer onto this understanding Pannenberg's concept of the prolepsis *from* the New Creation *back to* the Easter event in the present creation. So, with the discussion of the previous section in mind, I would pose this question: How might we think about the elements of continuity and discontinuity as configuring what we mean by "transformation"? Can we offer in a rough diagrammatic style a starting point for approaching the incredibly complex relation between creation and the New Creation, one that balances the simple futurity of ordinary clock time and the appearance of something absolutely new, namely the coming reign of God through God's paschal transformation of the world?[82] And how do we do this in a way that includes the contrasting biblical theologies of a realized eschatology and an apocalyptic eschatology?

Figure 2.28a is a modest first step that seeks to address the complex relation between the creation and New Creation in a highly schematic and abstract way. The leftmost line in figure 2.28a represents the history of the universe, including Easter (indicated by the symbol of the cross) up through the present and into the future. It is then transformed into the New Creation as represented by the second line. Note that the second line takes up where the first line ends to indicate two points: 1) I am rejecting those eschatologies that portray the New Creation as entirely continuous with and a mere evolutionary development of the present creation. 2) I am rejecting those eschatologies that portray the New Creation as a second "full-blown" creation completely disjunctive from the present creation (a second creation *ex nihilo*).

Figure 2.28a. Resurrection-Based Transformation of the Universe

However, by saying that the second line takes up where the first line ends is somewhat misleading. The idea of *transformation* is not reducible to that of a *transition* from a "first this" to a "then immediately that." Instead, it is a juxtaposition of two concepts held together in tension: transformation as a pervasive characteristic of the entire process of the creation of the New Creation out of the present creation and transformation as an abrupt eventlike transition from the present creation into the New Creation. This juxtaposition underlies the common assertion that the reign of God is realized by Jesus' proclamation and yet it is a radically future, apocalyptic event, a coming parousia that is not yet. How might we improve figure 2.28a to more adequately represent these myriad concepts?

Figure 2.28b is a "first order" attempt to do so. Here the theme of continuity is consistent with the pervasive characteristic of realized eschatology while the theme of discontinuity relies on the sudden character of apocalyptic eschatology. Thus the single line of figure 2.28a is actually composed of three lines: one line representing continuity and two lines representing discontinuity.

To represent the element of continuity in figure 2.28b, I use a single continuous line to stand for those elements in the present world that will continue to exist in the New Creation. These might include aesthetics, the ethics of agape love, embodiment, temporality, relationality, community, music, art, poetry, science, mathematics, dance, sexuality, food, humor, and so on.[83] When it comes to the elements of discontinuity, however, there are at least two types, as discussed above: i) I use a dashed line that terminates at the transition for those aspects of the present world that will not continue into the New Creation: moral evil, including sin as such and all its explicit forms ranging from hubris and concupiscence to racism, homophobia, economic injustice, political injustice, terrorism, violence to individuals, to communities, to other species, and to the ecosystem as a whole, etc., and natural evil including biological natural evil such as suffering, disease, death, and extinction, and physical natural evil including tsunamis, hurricanes, earthquakes, and so on.[84] ii) I use a wiggly line that begins at the transition to represent the birth of radically new elements characterizing the New Creation: authentic and unmitigated Christian life, genuine agape love of neighbor, the full flowering of the gifts of the Spirit, Augustine's *non posse peccare* (roughly, "it is not possible to sin"), the beginning of eternal life as the fully temporal, endless experience of God and each other, etc.

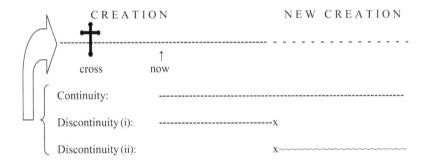

Figure 2.28b. Continuity and Discontinuity in the Characteristics of the Universe

But again, to avoid oversimplifications in speaking of a "transition," I want to suggest an additional, "second order" complication to the preceding. This would "spread out" the "transition time," or, better yet, generalize its instantiations from a moment in time to *all* of time starting at the cross or even at the original creation of the universe. The former reflects the sense in which the Easter event is the normative saving event in all of human history and the "final" revelation of God's unconditional grace. The latter offers a sense of the universality of the impact of the Easter event not only forward in time but even reaching back long before human history and including the healing of a creation "groaning" for transformation (Rom. 8). There are other possibilities, too. The choice clearly depends on one's particular interpretation of Christian eschatology. My point here is simply to suggest that there is a great deal of flexibility in an attempt like this to explore the incredibly complex conceptual landscape that is unavoidably encountered in this theological locus.

3. Eschatology as Prolepsis: A Diagrammatic Approach

The preceding discussion is meant to provide a brief background to Pannenberg's radical concept of eschatology as prolepsis. (See further chapter 1, section F, "Eschatology," part 2, "The Proleptic Character of Eschatology.") The context of his discussion is the debate over the proleptic character of the resurrection of Jesus. Pannenberg defends the radical claim that the bodily resurrection of Jesus is proleptic in being an anticipation, appearance, and concrete manifestation within history of the reality of the transhistorical eschatological future. Recall that Pannenberg explicates the latter through four distinct but interrelated claims about prolepsis: 1) The eschatological future is the basis for its proleptic appearance within history. 2) The openness of the present to the future is grounded in the proleptic character of the resurrection of Jesus. 3) The eschatological future as the hermeneutical context for understanding the history of the world is revealed proleptically in the resurrection of Jesus. 4) Testimony about its proleptic appearance in history takes the form of a hypothesis. For purposes of expediency I focus entirely on the first claim and leave the other three for another time.

We find the first claim in Pannenberg's discussion of the relationship between the futurity and presence of the reign of God in the text I cited in chapter 1:

> In the ministry of Jesus the futurity of the Reign of God became a power determining the present . . . [The] presence of the Reign of God does not conflict with its futurity but is derived from it and is itself only the anticipatory glimmer of its coming. Accordingly, in Jesus' ministry, in his call to seek the Kingdom of God, the coming Reign of God has already appeared, without ceasing to be differentiated from the presentness of such an appearance.[85]

From this text, three closely related claims emerge: 1) the eschatological future determines the present; 2) the reality of the eschatological future is the ontological basis for the reality of its present appearance; and 3) the appearance of the eschatological future in the present does not undercut the difference between the eschatological future and the present.

One way to visualize Pannenberg's ideas here is to start with the diagrammatic approach used above to suggest the transformation of the creation into the New Creation. In figure 2.29 we symbolize this transformation via a dashed line/dotted line representation as found in figure 2.28a–b. Now the crucial move is to add an additional curved line that starts in the eschatological future and arcs "back" to the Easter event two millennia ago. These two different lines connecting the cross and the eschaton should not be viewed as lying on this physical sheet of paper. Instead, they are separate, independent connections between these two events that have been represented as though they were lines on the paper. In essence, while I have drawn figure 2.29 *on* this page, the page plays no ontological role in conveying the relations between the cross and the eschaton. All there is in this conception of history and the eschaton is the combination of (i) the movement of ordinary time from the creation through its transformation into the eschatological future of the New Creation (the dashed line/dotted line symbol) and (ii) the movement of the eschatological future of the New Creation back to the cross (the curved line starting in the New Creation and pointing back to the cross).

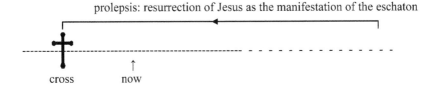

CREATION   NEW CREATION

prolepsis: resurrection of Jesus as the manifestation of the eschaton

cross   now

Figure 2.29. Resurrection as Prolepsis

4. Eschatology as Prolepsis: A Topological Approach

Are there models in mathematics, physics, and cosmology that might help illuminate Pannenberg's claims about prolepsis? In section B.2 above, we looked at non-Hausdorff manifolds as offering a way to think about co-presence in eternity. Now we turn to another idea drawn from mathematics, a "multiply connected manifold," to aid our thinking about prolepsis. Just as non-Hausdorff manifolds differ from Hausdorff manifolds in possessing characteristics such as branchings, multiply connected manifolds differ from Hausdorff manifolds in the way they connect regions within manifolds.

Think of a circle drawn on a Euclidean plane such as this page. You could imagine reducing its radius to zero as the circle vanishes to a point. Now imagine drawing a circle on a sphere. Again, you could reduce its radius to zero, even though the surface of a sphere is non-Euclidean.[86] Now think about the surface of a doughnut (fig. 2.30). You can draw two kinds of circles on it; neither of which can be reduced to a point, nor can they be made congruent with each other.[87] The doughnut's surface is not only non-Euclidean like the sphere, it is multiply connected.

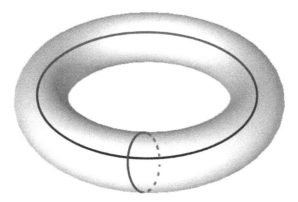

Figure 2.30. Surface of a Doughnut: Non-Euclidean and Multiply Connected

Note: The surface of a doughnut (i.e., torus) with two circles, neither of which can be reduced to a radius of zero or made to "fit onto" (i.e., be congruent with) the other.

I want to explore briefly two examples of the way in which multiply connected manifolds can illuminate Pannenberg's idea of eschatology as prolepsis. The first example comes from singularities in Einstein's general theory of relativity. The second comes from the multiply connected spacetimes found in Andrei Linde's "eternal inflation" cosmology. The first represents an anomaly *within* the universe, the second an anomaly *between* distinct universes (or portions of the multiverse).

Singularities in Spacetime

The field equations of Einstein's general theory of relativity (GR) allow for multiply connected spacetimes in the form of singularities known colloquially as "black holes." The simplest example is the Schwarzschild black hole produced by the mass of a nonrotating, noncharged object. If we let a Schwarzschild hole represent the curvature of the spacetime of the solar system caused by the sun, we can calculate the classical orbits of the planets as well as such nonclassical effects as the deflection of starlight by the sun. The latter provided one of the key tests of GR.

The Schwarzschild hole is often depicted graphically by two flat spacetime surfaces connected by a tubular-looking structure, as in figure 2.31:

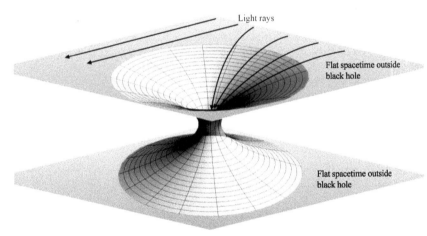

Figure 2.31. Graphic Representation of a Schwarzschild Black Hole, Which Connects Two Otherwise Flat Spacetimes

Roger Penrose has developed a diagrammatic approach that greatly helps to visualize the complexities of singularities. In figure 2.32 we see the worldlines of Ted and Sara as they travel forward in time from the infinite past denoted by alpha, α. Ted's path diverges from Sara's and crosses the "horizon" of the black hole, after which it inevitably approaches the singularity at the heart of the black hole, indicated by a double horizontal line. Sara continues into the infinite future, approaching the ultimate event, omega, Ω. Note that the double line of the singularity within the black hole is connected topologically, though not causally, to the event Ω in Sara's infinite future. In nontechnical language the endpoint of Ted's trajectory, the singularity within the black hole, and the endpoint of Sara's trajectory, the infinite future event called omega, are both the same and yet distinct events. The point of this bizarre but mathematically valid description of general relativity's predictions for the journeys of Ted and Sara is that a model universe consisting of even one singularity has to be described in what amounts to at least two different temporal perspectives. The remote future (i.e., omega) remains remote for some observers such as Sara but it becomes the immediate future (i.e., the singularity) for others such as Ted.

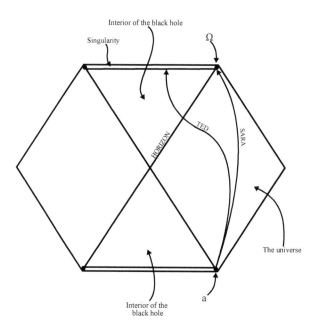

Figure 2.32. Penrose Diagram for the Schwarzschild Singularity:
Worldlines of Ted and Sara

Can the Schwarzschild black hole serve in any way as a model of Pannenberg's view of prolepsis? It might suggest that the relation Pannenberg describes between the eschatological future and the present is not a "path" in the ordinary spacetime of this world but a separate spatio-temporal connection between two different realms of this world. If so, we can improve on this model by turning to more complex spatio-temporal connections. These include charged (Reissner-Nordstrom) and charged and rotating (Kerr) black holes.[88] In the Penrose diagrams for these two types of black holes there are many portions of our universe, or perhaps even different universes, connected by black hole singularities. Moreover, travel through these black holes is physically possible compared to the impossibility of travel through the Schwarzschild black hole. Perhaps the Penrose diagrams for the Reissner-Nordstrom and Kerr black holes can further clarify Pannenberg's idea of the eschatological future as appearing and being manifested in the concrete reality of the first Easter event.

Finally, if we really believe that the end times have already appeared at least in part in the events of Easter, we ought to insist that the universe must have a more complex temporal topology than that of the linear time of ordinary experience and classical physics. If this is so, we might find this prefigured in some way by the musings of theoretical physics and cosmology. We will return to this topic in chapter 6.

Andrei Linde's "Eternal Inflation"

Another example drawn from physics and cosmology that might convey a sense of the proleptic relation between the cross and the eschaton in Pannenberg's work is that of the "multiverse." Here we will focus on the multiverse as explored through research into quantum cosmology by Andrei Linde. In figure 2.33, we see a computer visualization of an endless expanse of individual universes connected by branching tubes.[89] The dimensions, laws of physics, and natural constants can vary from universe to universe. As before, the space on which the diagram is drawn

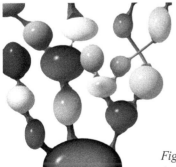

Figure 2.33. Computer Visualization of Andrei Linde's "Eternal Inflation" Multiverse

(i.e., this piece of paper) is a mere convenience for illustrating Linde's ideas and has no ontological significance of its own.

5. Multiple Prolepses as Topological Connections between Individual Death and the Eschaton

Finally, I want to suggest a generalization of the proleptic relation between the eschaton and the cross to include the relation between our own death and the New Creation. I then want to offer a mathematical analogy for Pannenberg's view that at the moment of our death we enter *immediately* into the presence of Christ and the eschatological New Creation. To address the issue of theodicy raised by the problem of natural evil, I will conclude briefly with a generalization on the relation between the eschaton and the cross that includes the deaths of all creatures.

First, then, I propose that the normative prolepsis that God accomplished in the Easter event creates the basis for an ongoing series of prolepses throughout the entire future history of the universe until its full transfiguration into the New Creation. In figure 2.34a–b, I schematically represent some of these prolepses by the addition of new lines from the eschatological realm back to individual moments in historical, post-Easter time. Figure 2.34a repeats the previous diagrammatic representation in figure 2.29 of the relation between the proleptic event of Easter and the eschatological New Creation. In figure 2.34b this relation is repeated twice and suggests how it can be thought of as repeated endlessly. Thus Easter is not only the normative prolepsis; it is also the prototype of proleptic events for all times in human history.[90]

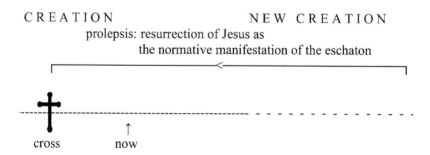

Figure 2.34a. Resurrection as Prolepsis

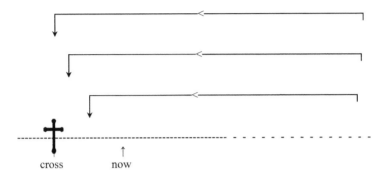

CREATION   NEW CREATION
Multiple prolepses: the manifestation of the eschaton at
all events in the future of creation

cross   now

Figure 2.34b. Multiple Prolepses and the Continuity
of Personal Identity through Death

Such an approach to the proleptic relation between the eschaton and the cross also allows us to address a crucial question about the continuity and temporality of personal identity between our individual deaths in time and the general resurrection of the dead at the end of time. This continuity is represented in figure 2.34b by the lines reaching "back" from the New Creation to the moment of each individual's death, and representing Christ's particular proleptic presence to us at that final moment. We can now address this crucial question: Is the "length" of the paths between our death and the general resurrection finite or zero? A finite length could represent the possibility of something like purgatory through which, while saved and destined for paradise, individuals after death must pass in order to be purged of their sins. A path of zero length could represent the possibility that at the moment of our death we are immediately present in the New Creation. It is not hard to find a way, analogically, to represent either possibility mathematically. Here I do so by drawing on Einstein's special relativity (see overview, chapter 4, section A). We start with the concept of a metric as a measure of the distance between points in space. Let point A be at the coordinate origin and point B have coordinates x, y, z. The distance r between them is given by the familiar metric or Pythagorean theorem:

$$r^2 = x^2 + y^2 + z^2$$

Now consider that these two points A, B are in spacetime. We can measure the spacetime "distance" (called the "interval") between them analogously to the way we measured distance in space—but with an important difference. Let event A lie at the coordinate origin O and let event B have coordinates x, t (where we suppress coordinates y, z for simplicity). The interval τ between them is now given by the Lorentz metric

$$\tau^2 = t^2 - x^2 / c^2$$

where c is the speed of light. Note the crucial insertion of a minus sign between the time and space coordinates here. In essence, this difference between intervals in spacetime and distances in space leads to the paradoxes of relativity. In specific, for most events B in spacetime the interval between A and B is finite, but there is a striking exception: for any event that lies along the path of a photon emitted at the origin, the interval τ is zero.[91] So we have the counterintuitive result that, for distinct events A and B lying along the path of a photon in spacetime, the spacetime interval between them is null even though the events are separate events.

Imagine walking outside on a crisp, clear winter's evening and gazing at the constellation of Andromeda. There you might spot the spiral galaxy M31 some 2.5 million light-years from Earth. This means that the light you are observing from Andromeda was omitted long before humankind evolved on planet Earth. Yet no time has passed for the photons in their long journey across intergalactic space. Clearly there are two very different measures and meanings of the passage of time in this simple example. The event A, when the photon you see was omitted by a star in Andromeda, and the event B, when it is captured by your eye, are "instantaneous" for the photon even though they are definitely *not* the same event for the photon or for you.

Finally, we can imagine the proleptic manifestations of the New Creation in the present world as being generalized to include all living creatures and not just humankind. This allows us to begin to address the problem of theodicy raised in the introduction to this volume: How do we deal with the enormity of "natural evil," in this case, death and extinction, sweeping throughout the nearly four billion years of life on Earth?

Allowing such a proleptic redemption for all creatures, or at least to all who are capable of sentience, avoids a horrific and anthropocentric "means-end" theodicy, namely that the evolution of species, with its enormity of suffering and death, was all "worth it" or at least "required for" (i.e., justified by) the eventual evolution of humankind. Instead, proleptic redemption offers a vision of the mercy of God the Creator and of Christ's suffering on the cross as extended to all of life through the power of the Holy Spirit, bringing all living creatures, all the "sparrows who fall" (Matt. 10:31), immediately into the endless joy of the eschatological New Creation.

CHAPTER 3

From Hegel to Cantor

*New Insights for Pannenberg's Discussion
of the Divine Attributes*

*No one shall expel us from the paradise Cantor had
created.*

—David Hilbert, "Über das Unendliche"

*The concept of the Infinite is perhaps the most significant
example of a philosophical concept which, without already
being an idea of God in the actual sense, still functions as
a criterion for the appropriate formulation of theological
assertions about God: whatever stands in contradiction to
the notion of the infinity of God cannot be a component of
a rationally demonstrable notion of the one God.*

—Wolfhart Pannenberg,
Metaphysics and the Idea of God

In this chapter I explore the ways in which the concept of infinity in
modern mathematics offers insights for reformulating Pannenberg's doctrine of God. My specific focus will be on Georg Cantor's work on infinity in formal set theory in relation to Pannenberg's use of the concept

of infinity in his discussion of the divine attributes. The choice to focus on Cantor's mathematics arises because of Cantor's breakthrough in our traditional way of thinking about infinity in mathematics, in which the infinite is in sharp contrast to the finite dating back to the ancient Greeks: the infinite is the *apeiron*—the unbounded, the unlimited, the formless. This understanding of infinity led theology to follow the *via negativa:* a way of coming to know God through an utter separation between the finite world and God its Creator. Recent developments in mathematics starting in the nineteenth century, though foreshadowed by Galileo Galilei's discoveries in the seventeenth century, have shed provocative light on infinity. And Georg Cantor in particular has given us a new mathematical conception of infinity in which there are an endless variety of infinities, which he called the "transfinites," and yet, lying beyond them, there is an unreachable Absolute Infinity. In effect, because of Cantor's breakthrough in our traditional way of thinking, we can now say a lot more about infinity than merely that it contrasts with the finite—indeed, a potentially endless amount more. These revolutionary discoveries in mathematics lead to exciting new possibilities for the concept of infinity in Pannenberg's explication of the divine attributes.

A. CANTOR'S SET THEORY, TRANSFINITE NUMBERS, ABSOLUTE INFINITY, AND THE THEOLOGICAL REACTION DURING CANTOR'S LIFE

1. A Historical Note on Infinity in Greek Philosophy and Mathematics

The concept of infinity in mathematics, philosophy, and theology from early Greek thought through the European Enlightenment is both extraordinarily complex and endlessly fascinating.[1] In a masterful survey of this period, Orthodox theologian David Bentley Hart states that "there is not—nor has there ever been—any single correct or univocal concept of the infinite."[2] Granted Hart's point, what I want to lift up here is a very basic distinction between that *aspect* of the concept of the infinite that by and large dominated this history, namely the infinite as merely the negation of the finite, and the new concept of the infinite found in modern mathematics, specifically starting with the work of Cantor in which the infinite includes, even while negating, the finite. To portray this distinction in its simplest form, while recognizing that this might be an oversimplification, I will touch briefly on a few key examples in this rich history.[3]

By defining the concept of infinity as *apeiron,* the unbounded, the ancient Greeks strictly contrasted the infinite with the finite: something is infinite if it has no boundary or end, if it is chaotic and lacking structure or order. The infinite was also seen as opposed to those things that make up the finite experience of everyday life from rocks to people. Thus underlying the world of finite entities is a formless and indeterminate infinity. In addition, finite entities are good precisely because they are formed, bounded, and determinate in contrast with the infinite. Paradoxes such as the famous race between Achilles and the tortoise, which date back at least to Zeno of Elea (490–430 BCE), were understood to arise precisely because of the sharp contrast between the finite and the infinite. In fact, over a century before Zeno, Anaximander (610–547 BCE) wrote of the infinite as a spherical, limitless substance: eternal, inexhaustible, lacking boundaries or distinctions. By contrast, the world of finite entities arises out of that which is infinite. More severely, Pythagoras (570–500 BCE) rejected the infinite as having anything to do with the real world. He contended that all actual things are finite and thus they can be represented by the natural, or whole, numbers. Pythagoras taught that the geometrical forms in one, two, and three dimensions are the result of imposing a mathematical limit on the underlying infinite structure of nature. Plato (428–348 BCE) believed that the good must be definite rather than indefinite, and therefore it must be finite rather than infinite. According to Plato, God as the demiurge imposes limitations (i.e., intelligible forms) on preexistent, inchoate matter, giving rise to the structured world around us as an ordered whole instead of a formless, unintelligible, infinite chaos. In all these cases the infinite took on a negative quality compared with the finite.

With Aristotle (384–322 BCE), however, we find a rather different conception of infinity, and one that continued in Western thought without serious challenge until the revolution in mathematics in the nineteenth century. Following his predecessors Aristotle began by considering infinity as a negative quality, that is, as something lacking: "Its essence is privation."[4] But he altered the concept of infinity to mean an unending and incomplete process: "The infinite has this mode of existence: one thing is always being after another, and each thing that is taken is always finite, but always different."[5] Here the only infinity considered possible is potential infinity: something capable of being endlessly divided or added to, but never fully actualized as infinite. For example, consider the series of unending succession of natural numbers, the endless succession of the seasons, or the continuous division of a line interval into smaller and smaller portions.[6] Aristotle thought of these as potentially infinite, since the series can be continued endlessly while remaining finite; these series are never actually infinite. Going through successive elements in a series, we never get beyond the finite steps in the series to reach their limit, the actually infinite. We never view the infinite series as a whole.[7]

A number of early Christian writers were informed by this more "positive" conception of infinity as they developed the doctrine of God.[8] Augustine (354–430), through his debt to Plotinus, believed that since God is infinite God must know all numbers and must have limitless knowledge of the world.[9] Writing a millennium after Augustine and drawing on the philosophy of Aristotle, Thomas Aquinas (1225–74) believed that only God is infinite and all creatures are finite. Yet even God, whose power is infinite, "cannot make an absolutely unlimited thing."[10]

2. The Modern Understanding of Infinity in the Mathematics of Cantor

The modern understanding of the mathematical concept of infinity began with Galileo Galilei (1564–1642), although its roots lie in a variety of earlier scholars, perhaps most notably Nicholas of Cusa.[11] In a delightful section of *Two New Sciences,* Galileo argues that, as far as the natural numbers are concerned, infinite sets do not obey the same rules that finite sets obey. Let us join the conversation between Galileo's three interlocutors—Salviati, Sagredo, and Simplicio—about the incomprehensibility and indivisibility of infinity. The problem they face is how to think about the construction of a continuous, divisible line segment out of an infinity of indivisible points. As Simplicio points out, a longer line segment would seem to contain a greater infinity of indivisible

points than a shorter one. Still, he admits that "this assigning to an infinite quantity a value greater than infinity is quite beyond my comprehension."[12]

Salviati responds that the problem arises "when we attempt, with
our finite minds, to discuss the infinite, assigning to it those properties
which we give to the finite and limited; but . . . this is wrong, for we cannot speak of infinite quantities as being the one greater or less than or
equal to another." To prove his claim, Salviati turns to the unending sequence of the natural numbers: 1, 2, 3, . . . and the unending sequence of
their squares: 1, 4, 9, ... Intuitively we think that there must be more
natural numbers than there are squares, since the set of natural numbers
contains the squares along with other numbers, such as 2, 3, 5, 6, 7, and
8, which are not the squares of natural numbers. Yet the squares can be
put into a one-to-one correspondence with the natural numbers:

  natural numbers: 1 2 3 . . . n . . .
  squares of natural numbers: 1 4 9 . . . n^2 . .

by the simple rule: $n \rightarrow n^2$. And so, against our intuition, we now see that
the series of squares 1, 4, 9, . . . is *equivalent* to the series of natural numbers 1, 2, 3, ... In essence, there are as many squares as there are natural numbers! Given this paradoxical result, Galileo, through the words of
Salviati, states that "the totality of all numbers is infinite, . . . the number of squares infinite, and . . . the number of squares is [not] less than the
totality of all numbers, or the latter greater than the former; and finally
the attributes 'equal,' 'greater,' and 'less,' are not applicable to infinite, but
only to finite, quantities." Returning to different line segments, Salviati
concludes that all such segments contain an infinity of points, none more
nor less than the others.

Galileo's insights into infinity remained relatively dormant until the
nineteenth century. While the simultaneous discovery of the calculus in
the seventeenth century by Sir Isaac Newton (1642–1727) and Gottfried
Wilhelm Leibniz (1646–1716) involved complex questions about the status of infinitesimals as minute extensions in space and durations in time,
even Newton, Leibniz, and Carl Friedrich Gauss (1777–1855) adhered to
the view that infinitesimals are only abstractions or potential infinities. It
was only with the discoveries of Bernard Bolzano (1781–1848), Richard

Dedekind (1831–1916), and Georg Cantor (1845–1918) that a radically new mathematical conception of infinity was born.

A biographical caveat will give us some sense of the person behind the discoveries and this, in turn, will be helpful as we explore some of the theological issues underlying Cantor's mathematical research (section A.4). Georg Ferdinand Ludwig Philipp Cantor was born in St. Petersburg, Russia on March 3, 1845. His mother was a devout Roman Catholic, his father a Jew who had converted to Lutheranism. Raised in a musical family who relocated to Germany, Cantor was an accomplished violinist. Cantor received his doctorate at age twenty-two on number theory and accepted a position at the University of Halle two years later in 1867. In 1873 he proved that the rational numbers are countable and, a year later, that the real numbers[13] are uncountable. Starting in 1883 he began a series of attempts at proving the "continuum hypothesis." In 1895 and 1897 he made his last major contributions to set theory, including finding the first of a series of paradoxes, or antinomies, in set theory. During much of his lifetime Cantor's views were supported by some mathematicians, principally J. W. Richard Dedekind (1831–1916), but they were opposed by many others, notably Leopold Kronecker (1823–91). This opposition may have contributed to Cantor's declining health in later years, including a series of nervous breakdowns that began in 1884. Cantor died on January 6, 1918, in a mental institution.[14]

Cantor began with the deployment of the crucial concept of a set as a "collection into a whole of definite distinct objects of our intuition or of our thought."[15] Part of its attractiveness is that in Cantor's view set theory could study a vast range of objects—"a pack of wolves, a bunch of grapes, or a flock of pigeons."[16] In an elegant comment, Cantor describes the move from the richness that can be found in the diverse kinds of elements that can be thought of as a set to the sparse idea of what he called its cardinal number. The move takes the form of a double abstraction from "the nature of its various elements . . . [and] the order in which they are given. We denote the result of this double act of abstraction the cardinal number" of the set.[17]

Cantor then developed the basic mathematical concepts and principles of finite sets, including addition, multiplication, proper subset,[18] sets as countable and as uncountable, and so on.[19] The idea of a set being countable is of particular importance for Cantor's work, and it harkens

back to Galileo's insight about natural numbers and their squares. When
we count a finite set we place each element in the set in a one-to-one
correspondence[20] with a subset of the natural numbers 1, 2, 3, . . . Cantor coined the term "countable set" for a set with the same number of elements as a subset of the natural numbers. The number of elements in the
set is called its *cardinal number*. A "well-ordered set" is a set for which
every nonempty subset has a smallest element and on which an overall
order is imposed. So, for example, if we consider the non-negative natural numbers 0, 1, 2, and 3, with the assumption that they can be ordered
by magnitude (i.e., that $0 < 1 < 2 < 3$), then the set $\{0, 1, 2, 3\}$ is well-
ordered. Its *ordinal number* is 4 because there is a fourth largest element, namely the element 3.[21]

In his breakthrough insight Cantor then showed how to apply these
concepts and principles to infinite sets. This in turn meant he could give
an explicit procedure for constructing different kinds of infinity. Rather
than assuming the infinite is in direct *contrast* with the finite, as had been
held to be the case for over two millennia, Cantor treated infinite sets
partly in *analogy* with finite sets.

To see this, let us return to Cantor's work, starting with the set of
natural numbers $\{1, 2, 3, . . .\}$. Cantor thought of this set as an infinite,
completed whole and not just as an unending but always finite sequence
of numbers, as Aristotle had taught. In other words, Cantor distinguished
between the sequence 1, 2, 3, . . . as potentially infinite but always actually finite and incomplete, and the set $\{1, 2, 3, . . . \}$ as actually infinite and complete, the endless sequence of natural numbers now thought
of as a whole. He called the potentially infinite sequence 1, 2, 3, . . . a
"variable finite" and symbolized it as ∞. To distinguish it from the actually infinite set $\{1, 2, 3, . . .\}$, he called the latter $\aleph_0$ ("aleph-null"). Thus
∞ never reaches completion, never becomes $\aleph_0$. To think about ∞ is to
think of an ever-increasing finite series of numbers continuing forever
without reaching an end. To think about $\aleph_0$ is to stand outside this series
sub specie aeternitatis and to consider it as a single, unified, and determinate totality. To be precise, $\aleph_0$ is the cardinal number of the actually
infinite set $\{1, 2, 3, . . .\}$. Cantor called $\aleph_0$ a "transfinite" number to stress
its status as an actual infinity.

Cantor then extended what we know about counting finite sets to
infinite sets: all infinite sets whose elements can be put in a one-to-one

correspondence with the natural numbers will have the *same* cardinal number, $\aleph_0$. He called these sets "denumerably infinite" or "countably infinite." Using these insights we can restate Galileo's paradox about square numbers. Since there is a one-to-one correspondence between the natural numbers and the square numbers, it means that the set of square numbers is countably infinite; it has the same cardinal number, $\aleph_0$, as the set of natural numbers. Similarly, the set of even numbers is equivalent to the set of natural numbers, as is the set of odd numbers. Another way to put this is that a countably infinite set is one that can be put into a one-to-one correspondence with a proper subset of itself.

In a stunning move Cantor next showed that we can generate infinities that are "bigger" than the set of natural numbers in that they have different ordinal numbers, although they all have the same cardinal number, $\aleph_0$. To see this we start again with the infinite set of natural numbers, $\{1, 2, 3, \ldots\}$ and designate its ordinal number as ω. Now let us think of this set as a whole, as complete in itself. If we do so, then it is possible to conceive of adding 1 to the end of this set, thereby forming a new set $\{1, 2, 3, \ldots, 1\}$. The cardinal number of this new set $\aleph_0 + 1$ is equivalent to the cardinal number of the original set $\aleph_0$ since they can clearly be put in a one-to-one correspondence with each other. However, the ordinal number of this new set $\omega + 1$ is *not* equal to ω. Still, if we add 1 to the beginning of the set of natural numbers, $\{1, 1, 2, 3, \ldots\}$, both the cardinal number $1 + \aleph_0$ and the ordinal number $1 + \omega$ are equivalent to the respective cardinal and ordinal numbers of the original set, namely $\aleph_0$ and ω. Note the meaning of the symbol $1 + \omega$: We start with the number 1 and then add on the endless sequence of numbers $1, 2, 3, \ldots$. We then take this as a whole, and we have the set of natural numbers $\{1, 2, 3, \ldots\}$ whose ordinal number is ω. Compared with this, the symbol $\omega + 1$ means that we add to a *given, already infinite whole,* namely $\{1, 2, 3, \ldots\}$, a new element 1, thus forming the "larger" set $\{1, 2, 3, \ldots, 1\}$. In essence, the set $\{1, 2, 3, \ldots, 1\}$, taken as a whole, is not equivalent to the set $\{1, 1, 2, 3, \ldots\}$, taken as a whole. Thus $\omega + 1$ is not equivalent to ω.[22]

Next we repeat the procedure by adding 1 to the set $\{1, 2, 3, \ldots, 1\}$. This yields another new set, $\{1, 2, 3, \ldots, 1, 2\}$, whose ordinal number is $\omega + 2$. As we continue the process, we generate a ladder of increasingly complex infinities. For example, if we add the set of natural numbers to the set of natural numbers, we obtain the set $\{1, 2, 3, \ldots, 1, 2,$

3, . . .}, whose ordinal number is $\omega + \omega$ or, equivalently, $\omega \cdot 2$. From here we continue the process, generating sets with ordinal numbers $(\omega \cdot 2) + 1$, $(\omega \cdot 2) + 2$, $(\omega \cdot 2) + 3$, and so on until we reach $\omega \times \omega$ or, equivalently, ω^2. Next comes ω^3, ω^4, and so on. This in turn points toward a summit, ω^ω— but there is still infinitely more. We can think of an infinite series of exponential powers, raising ω to the power ω infinitely many times. What is even more astonishing is the fact that the elements in any of these transfinite sets can be put in a one-to-one correspondence with the elements in the set of natural numbers, $\{1, 2, 3, . . .\}$. This means that all of these sets, even though differing radically in their ordinal number, are *denumerably* infinite: *they have the same cardinal number,* $\aleph_0$. Mathematicians express this fact by noting that

$$\aleph_0 + \aleph_0 = \aleph_0$$

even though

$$\omega + \omega \neq \omega$$

Apparently the analogy between finite and infinite sets is better seen as a simile and a dissimile: the rules that infinite sets obey are *both like and unlike* the rules that finite sets obey.

But there are still more surprises ahead, since we can imagine sets whose "infinity" is so great, as it were, that they *cannot* be put into a one-to-one correspondence with these sets: they are "uncountably infinite." In 1874, Cantor proved this for the real numbers.[23] Since the real numbers can be put in one-to-one correspondence with the points of a straight line (i.e., linear continuum), Cantor called the cardinal number of the set of real numbers "the power of the continuum," designated by the letter "c." In 1877 Cantor proposed the continuum hypothesis: the real numbers form the *first* uncountable infinity, that is, the first infinity whose cardinality is greater than the cardinality of the set of natural numbers. He denoted its cardinality as $\aleph_1$. The continuum hypothesis is the first of David Hilbert's famous list of twenty-three unanswered questions posed in 1900. Even today, the continuum hypothesis remains controversial, as evidenced by the ongoing research of mathematician Hugh Woodin.[24]

With the concept of infinity as $\aleph_1$ we discover vastly startling results. As Cantor found, there is a one-to-one correspondence between the set of points in a plane and the set of points in a line; one might well have thought the former to be infinitely greater than the latter. Cantor then extended this result to the points in a three-dimensional space, and then to a space of any number of dimensions. All these mathematical objects—the line, the plane, the volume, the hypervolume in four or more dimensions—have the *same* cardinal number, c. We can go further still. Cantor showed that one can construct an entire series of transfinite cardinal numbers starting with $\aleph_0$ and leading to $\aleph_\omega$ and beyond this to $\aleph_{\omega+1}$, $\aleph_{\omega+2}$, . . . , $\aleph_\omega \omega$, $\aleph_{\aleph 0}$, and so on. In fact, there is no end to the kinds of infinities we can construct. Yet at the same time, these infinite sets share an important feature with finite sets since, no matter how complex, they are conceivable by construction—hence the reason for Cantor to call them "transfinites."

Cantor took a final step and considered what lies beyond even the transfinite numbers: he called it Absolute Infinity and symbolized it as Ω. He expounded Absolute Infinity's relation to the transfinites in terms of what is called a reflection principle. In Cantor's context of set theory the reflection principle states that properties of the class of all sets are shared with (or "reflected down to") the properties of the sets in that class.[25] Using the reflection principle, Cantor could assert what might appear to be a contradiction: in one sense Absolute Infinity is truly inconceivable in that, by definition, it transcends all the transfinites. Yet in another sense it is not strictly inconceivable because its properties are shared with the transfinites. So, does the claim that Ω is both conceivable and inconceivable represent a contradiction?

To see that it does not, consider the converse. Suppose that Ω is as conceivable as the transfinites are. To be conceivable in this context means to possess a distinctive property, such as cardinality, through which a given set can be uniquely identified. So if Ω is conceivable it means that there must be some property P that is exclusively a property of Ω. We can then refer uniquely to Ω and to nothing else by referring to that property. Now, in order to claim that Ω is inconceivable, we merely have to show that every such property P is actually shared by both Ω and some transfinite number. If so, then there is no property P that is unique to Ω. Thus we can consistently assert the conceivability and the inconceivability of Ω. On the one hand, we can conceive of Ω because each of its properties P is shared by some transfinite. Yet because of this, we can never differentiate Ω completely from the transfinite, since we can never describe Ω as possessing a property P that it does not share with a transfinite set. In essence, we can never tell Ω from the transfinites because of the fact that all its properties are shared with the transfinites. We can never know if we are conceiving of Ω and not some transfinite set. In short, we can never conceive of Ω as unambiguously distinct from the transfinites. Ω is thus inconceivable because Ω can never be uniquely characterized or completely distinguished from some transfinite number, though infinite too, that is a lower order of infinity than Absolute Infinity. In essence, the transfinite numbers, Cantor's endless types of infinities, lead toward Absolute Infinity Ω but never reach it: Ω lies beyond all comprehension. In this way Cantor set up a threefold distinction regarding the infinite:

> The actual infinite arises in three contexts: *first* when it is realized in the most complete form, in a fully independent other-worldly being, *in Deo,* where I call it the Absolute Infinite or simply Absolute; *second* when it occurs in the contingent, created world; *third* when the mind grasps it *in abstract* as a mathematical magnitude, number, or order type. I wish to mark a sharp contrast between the Absolute and what I call the Transfinite, that is, the actual infinities of the last two sorts, which are clearly limited, subject to further increase, and thus related to the finite.[26]

3. Antinomies in the Theory of Transfinites: A Stumbling Block for a Theological Appropriation of Cantor's Work?

Before considering the relation between Cantor's mathematics and the theology of his day, we need to unpack the problem of paradoxes that arise in Cantor's work. Are there intrinsic problems with Cantor's set theory that would undercut its applicability to theology? One might well answer in the affirmative[27] because of the paradoxes, or, more specifically, "antinomies,"[28] that it generated. An antinomy involves the assertion that two mutually contradictory statements can both be shown to be true within a given mathematical theory. For the most part, these paradoxes pertain to the concept of "the set of all sets"—what Douglas Hofstadter calls "self-swallowing sets."[29] In 1895, Cantor, and independently the Italian mathematician Cesare Burali-Forti, discovered an antinomy that states that the set of all ordinal numbers cannot exist.[30] Although it was hoped that a simple cure could be found for the antinomy, in 1902, Bertrand Russell discovered an antinomy that led to a fundamental crisis in set theory, and one that spilled over into logic theory. "Never before had an antinomy arisen at such an elementary level, involving so strongly the most fundamental notions of two most 'exact' sciences, logic and mathematics."[31] Russell argued that if we move to the set of all sets that do not contain themselves we arrive at a contradiction: we have a set that, because it is the set of all sets, must contain itself, and yet because it is a set of all sets that do not contain themselves, cannot be such a set,[32] thus generating the antinomy.

If set theory were to contribute to the foundations of modern mathematics, it would need to be reformulated in light of these antinomies.[33] In 1904 Ernst Zermelo offered an apodictic way to avoid the antinomies such as Russell had found through a restricted axiomatic formalization of set theory.[34] Zermelo's work, modified by Abraham Fraenkel and others in the 1920s, issued into what is now called ZFC, the standard axiomatic set theory that combines the Zermelo-Fraenkel (ZF) axioms with the axiom of choice.[35] It should be noted that, in keeping with Galileo's discovery about the integers and following Cantor's definition of an infinite set, ZFC defines an infinite set as a set that is equivalent to its own proper subset.[36] Fraenkel and Lévy note "the reluctance of many a beginner to accept the possibility" of such equivalence and that

> some trained philosophers have taken the same reluctant attitudes may be explained by their adherence to the classical principle *totum parte maius* (the whole is greater than a part). This principle in its proper meaning is, however, limited to the domain of finite sets. . . . Unfortunately, dogmatic adherence to that principle has seriously hampered the growth of set theory.[37]

The preceding discussion gives just a hint of the complexity of the issues raised by antinomies in formal set theory. According to Ernst Snapper these issues are part of a larger class of three distinct but related crises in the foundations of mathematics generated by the apparent failures of logicism, intuitionism, and formalism and their associated philosophies, realism, constructivism, and nominalism.[38] Snapper first discusses logicism, which began with the work of Gottlob Frege (1848–1925) and continued with the work of Bertrand Russell and Alfred North Whitehead. Logicism's purpose was to demonstrate that classical mathematics is a part of logic and thus to show that classical mathematics is free of contradictions. The problem, according to Snapper, is that of the nine axioms of ZF set theory, two of them are not logical propositions: the axiom of infinity and the axiom of choice. Clearly an appeal to everyday experience with such apparent infinities as the set of natural numbers does not constitute a logical proof that the infinite set exists "outside our mind" in the way that a philosophy of Platonic realism would entail.

A second crisis in the foundations of mathematics, according to Snapper, came with intuitionism. The origins of this crisis can be traced to the work of L. E. J. Brouwer (1881–1966) in 1908, when the antinomies in Cantor's set theory were rapidly arising. Instead of starting with the axioms of logic, intuitionists sided with Immanuel Kant's view that humans have an immediate awareness and certainty for the number 1 from which we construct all the finite numbers by induction. Mathematics, then, is a constructive activity of the mind, not a discovery of extramental mathematical realities. Clearly intuitionist mathematics is free of contradictions, but, as Snapper notes, it has failed to attract many classical mathematicians for several reasons: it rejects many of the beautiful theorems of classical mathematics, its proof of many theorems seems arduous and inelegant, and some of its theorems are considered false by classical mathematicians. A third crisis came with formalism, which can be traced back to 1910 and the work of David Hilbert (1862–1943). It is an attempt to give a formal language L for an axiomatized theory T such as Euclidean geometry. In 1931, Kurt Gödel showed that no such formal language can prove that T is free of contradictions.[39] Attempts to avoid Gödel's argument by using what Hilbert called "finitary reasoning" ultimately seem to fail. The nominalist roots of formalism, in which abstract entities have neither extramental nor constructivist existence, are challenged by this failure. Snapper's conclusion is that while technical mathematical research continues in full swing even without a firm foundation, the search for such a foundation is something that many philosophers of mathematics "yearn for." The key "lies hidden somewhere among the philosophical roots of logicism, intuitionism, and formalism."[40]

I close this section by returning briefly to the question stated at its outset: Are there intrinsic problems with Cantor's set theory that would undercut its applicability to theology? I would suggest that, despite the profound problems caused by the antinomies associated with set theory and the eventual series of crises that developed in the foundations of mathematics outlined by Snapper, Cantor's work *is* applicable theologically if we are seeking to use it as a conceptual tool to enhance the way Pannenberg explicates the role of the concept of infinity in the doctrine of God. This will be the general thrust of section B of this chapter.[41] Let us first return to Cantor's theological motivation for the concept of Absolute Infinity in order to explore its potential fruitfulness for the way Pannenberg employs the concept of the infinite in his discussion of the doctrine of God.

4. Theological Reactions to Cantor's Transfinite Set Theory during His Lifetime

According to historian of science Joseph W. Dauben, Cantor argued for the acceptability of transfinite set theory both in explicit ways that appealed to a formalist philosophy of mathematics and in implicit ways that drew on his theological views. From a philosophical perspective, Cantor assumed that mathematical ideas are valid if they can be shown to have internal conceptual consistency, particularly when it comes to the definition of key concepts. His strategy was to start with the system of irrational numbers, which during the late nineteenth and early twentieth centuries was generally accepted as consistent. From there he argued that it was only a short step to the consistency of the transfinites. Yet Dauben also claims that Cantor believed in the validity of transfinite set theory for theological reasons. According to Dauben, "either one assumed the existence and reality of the actual infinite, or one was obliged to give up the infinite intellect and eternity of the absolute mind of God."[42] But Cantor kept these theological reasons relatively confidential, only mentioning them in private correspondence. It is to this correspondence, following Dauben, that we now turn.

In a letter written in 1895 to the French mathematician Charles Hermite, Cantor offered a theological reason for the existence of the natural numbers: they exist "at the highest level of reality as eternal ideas in the

Divine Intellect."[43] Cantor later was to draw on Augustine to support this claim.[44] He then extended this argument to defend the existence of the transfinites, arguing that these numbers also exist "at the disposal of the intentions of the Creator and His absolute boundless will." Cantor referred to the form of their existence in God's mind as the "Transfinitum."[45] He also wrote to various Roman Catholic scholars, urging that if his theory was understood correctly it would not be seen as conflicting with Catholic doctrine. Cantor apparently meant this to protect the Church from being seen as challenging the logic of mathematics. Still, given the opposition to the concept of transfinites from some of the outstanding mathematicians of his time,[46] Cantor was also most likely motivated by a fear of being misinterpreted himself. As we will see below, this fear led in part to his articulation of the mathematical concept of Absolute Infinity.

According to Dauben, the key to understanding the Church's response to Cantor's work can be traced back to Pope Leo XIII's earlier support of neo-Thomistic philosophy and the interest it generated about science among Catholic scholars.[47] The pope believed that the Church must engage contemporary culture if it was not to be sidestepped, and a key factor in culture was science. For the pope the correct means for engaging culture was neo-Thomistic philosophy. Leo XIII saw the social evils of his age as a result of false philosophies whose interpretations of science led to atheism and materialism. Instead, neo-Thomism offered the correct philosophical interpretation of science that succeeded in rebuking atheism and materialism. Central to the pope's revival of neo-Thomism was his 1879 encyclical, *Aeterni Patris*. It is worth quoting the pope in some detail as here we clearly see the value of Catholic philosophy in both supporting and interpreting science:

> Nor will the physical sciences themselves, which are now in such great repute, and by the renown of so many inventions draw such universal admiration to themselves, suffer detriment, but find very great assistance in the restoration of the ancient philosophy. For, the investigation of facts and the contemplation of nature is not alone sufficient for their profitable exercise and advance . . . . To such investigations it is wonderful what force and light and aid the Scholastic philosophy, if judiciously taught, would bring. . . . And here it is well to note that our philosophy can only by the grossest injustice

be accused of being opposed to the advance and development of natural science.[48]

With this as background, Dauben turns to the reaction of Cantor's contemporary, the German Catholic philosopher and apologist Constantin Gutberlet.[49] Beginning in 1886, Gutberlet used Cantor's work to support his own theological and philosophical views on infinity. Still, after his views were attacked by other German scholars, Gutberlet focused on whether mathematical infinity challenges the infinity of God. In response, Cantor wrote to Gutberlet that the transfinites actually increase theology's understanding of God's nature and dominion. In light of Cantor's argument, Gutberlet then claimed that the mind of God was the ground for the existence of infinite decimals, the irrational numbers, the exact value of π, and the transfinite numbers. Dauben summarizes Gutberlet's position in a remarkable way: "Either one assumed the existence and reality of the actual infinite, or one was obliged to give up the infinite intellect and eternity of the absolute mind of God."[50] In the process of their interaction, Gutberlet's work increased Cantor's own interest in the theological and philosophical significance of infinity.

According to Dauben, many other Catholic scholars, influenced by *Aeterni Patris,* became interested in Cantor's work. Nevertheless, a crucial question remained in dispute: Are the transfinites real only as ideas within the divine mind, or are they objectively real in nature? Dauben tells us that while Gutberlet restricted the reality of the Transfinitum to the ideational realm of the divine intellect, Cantor believed they could also be seen as concrete realities in nature. This then led Cardinal Johannes Franzelin to warn that Cantor's view pointed heretically toward pantheism. As Dauben interprets Franzelin, "any actual infinity *in concreto,* in *natura naturata,* was presumably identifiable with God's infinity, in *natura naturans.* Cantor, by arguing his actually infinite transfinite numbers *in concreto,* seemed to be aiding the cause of Pantheism." In response, Cantor added a further distinction between the eternal, uncreated, Absolute Infinite ("Infinitum aeternum increatum sive Absolutum") related to God and the divine attributes on the one hand and the created, Transfinite Infinite ("Infinitum creatum sive Transfinitum"), which can be found in nature and the actually infinite number of objects in nature on the other hand. This distinction apparently satisfied Franzelin's concerns about pantheism. As Franzelin wrote, "the two concepts

of the Absolute-Infinite and the Actual-Infinite in the created world . . .
are essentially different. . . . When conceived in this way, so far as I can
see at present, there is no danger to religious truths in [Cantor's] concept of the *Transfinitum*."[51]

With this distinction, we return full circle to Cantor's reflection principle, in which Cantor distinguishes between actual infinity in the mind
of God (the Absolute Infinite) and the actual infinity of the Transfinite,
composed of actual infinity in the world (*in concreto*) and actual infinity
as grasped by the human mind (*in abstract*) (see the end of section A.2
above). We can now ask whether Cantor's mathematics can play a fruitful role in Pannenberg's understanding of the concept of true Infinity in
the doctrine of God, particularly in light of the way Pannenberg's interpretation of the concept draws on Hegel.

B. EMPLOYING CANTOR'S THEORY OF THE TRANSFINITES AND ABSOLUTE INFINITY IN PANNENBERG'S ARTICULATION OF THE DIVINE ATTRIBUTES AND ESCHATOLOGY

In chapter 1, we saw that the philosophical concept of infinity plays a crucial role in Pannenberg's interpretation of the doctrine of God, particularly in his treatment of the divine attributes, their role in the doctrine of
the Trinity, and ultimately in Christian eschatology.[52] As I quoted in chapter 1, Pannenberg writes, "The Infinite that is merely a negation of the
finite is not yet truly seen as the Infinite (as Hegel showed), for it is defined by delimitation from something else, i.e., the finite. . . . The Infinite is truly infinite only when it transcends its own antithesis to the fi-
nite."[53] "The thought of the true Infinite . . . demands that we do not think
of the infinite and the finite as a mere antithesis but also think of the
unity that transcends the antithesis."[54] Pannenberg attributes his understanding of infinity to Hegel, and he contrasts it with its traditional meaning in Greek thought where infinity as *apeiron* is merely the antithesis of
the finite. What I will explore here is the way insights from Cantor's work
on the transfinites and his concept of Absolute Infinity can offer an at-
tractive alternative to Pannenberg's use of Hegel's concept of infinity in
the doctrine of God.[55]

1. Pannenberg's Use of Hegel's Understanding of Infinity as a Presupposition of All the Divine Attributes

If we are to compare what Hegel offers to Pannenberg regarding infinity with what might be found in Cantor, we first need to determine the texts by Hegel that Pannenberg uses in his explication of the doctrine of God in the *Systematics*. Our best clue comes from Pannenberg's discussion of the divine attributes in chapter 6 of ST1. Here Pannenberg repeatedly makes creative use of Hegel's understanding of infinity. Curiously, though, as I noted above in chapter 1, Pannenberg offers only one citation to Hegel. It is found in a note on p. 397 where he refers to Hegel's *Science of Logic*.[56] Aside from the discussion of the divine attributes, the topic of infinity in Hegel's thought rarely occurs explicitly in Pannenberg's *Systematics*.[57]

As is well known, the portion of *Science* that Pannenberg cites includes in astonishing detail Hegel's profound interpretation of the relations between the finite and the infinite. In the section entitled "Alternating Determination of the Finite and the Infinite," Hegel first writes about what can be called "false infinity":

> The infinite is; in this immediacy it is at the same time the negation of an other, of the finite. As thus . . . it has fallen back into the category of . . . something with a limit . . . The finite remains as a determinate being opposed to the infinite.[58]

Shortly thereafter Hegel writes:

> We have before us the alternating determination of the finite and the infinite; the finite is finite only in its relation to the ought or to the infinite, and the latter is only infinite in its relation to the finite. . . . It is this alternating determination negating both its own self and its negation, which appears as the progress to infinity, a progress which in so many forms and applications is accepted as something ultimate beyond which thought does not go but, having got as far as this 'and so on to infinity,' has usually reached its goal. . . . The infinity of the infinite progress remains burdened with the finite as such, is thereby limited and is itself finite.[59]

Here again Hegel is referring to a false sense of infinity understood merely in contrast with the finite: infinity as an unending series of integers or an endless line, both of which remain finite even as they are unlimited.

But Pannenberg is concerned with Hegel's *dialectical* concept of infinity. According to Hegel, "the self-sublation of this infinite and of the finite, as a single process—this is the true or genuine infinite."[60] For example, in the section entitled "Affirmative Infinity," Hegel claims that

> in saying what the infinite is, namely the negation of the finite, the latter is itself included in what is said; it cannot be dispensed with for the definition or determination of the infinite. . . . This yields the decried unity of the finite and the infinite—the unity which is itself the infinite which embraces both itself and finitude—and is therefore the infinite in a different sense from that in which the finite is regarded as separated and set apart from the infinite.[61]

This text and others represent Hegel's new metaphysical understanding of infinity. It starts with the traditional notion of infinity as the negation of the finite, but it adds to this the notion of infinity as containing the finite within itself, thus sublating its own negation (Hegel's new contribution). It is clearly this understanding of infinity that Pannenberg uses in his doctrine of the divine attributes.

I suggest that Cantor's mathematical understanding of the infinite reflects, at least in part, the logical structure underlying Hegel's metaphysical concept. I also propose that it offers additional richness to Pannenberg's discussion of the divine attributes. This richness is due to Cantor's discovery of the *three*fold distinction between the finite, the transfinite, and Absolute Infinity and because Cantor's threefold distinction, unlike that of Hegel, is nondialectical.

2. Cantor's Concept of the Transfinites and Absolute Infinity as New Resource for Pannenberg's Conception of the Infinite in the Doctrine of God

First we must ask whether Cantor's mathematical understanding of the relation between the finite and the transfinites reflects the underlying logical structure of Hegel's philosophical concept of infinity and, in addition, whether it provides advantages over Hegel's concept. We must then discuss Cantor's notion of Absolute Infinity in relation to the transfinites. Having done this we will be in a position to explore Cantor's work as a new resource for Pannenberg's discussion of the divine attributes.

The Relation between the Finite and Cantor's Transfinites Compared with the Logical Structure of Hegel's Concept of Infinity

It is noteworthy that Cantor's use of the formal language of mathematics to discuss the finite in relation to the infinite can, in itself, be considered an advantage over the inherently ambiguous medium of human language and philosophical terminology through which Hegel's concept of infinity is expressed. But far more important, the power and clarity of mathematics allowed Cantor to discover *similarities* between the finite and the infinite. In specific, Cantor showed that what he called *transfinite* infinities lie between the finite and Absolute Infinity, and that these transfinites share very important similarities with purely finite numbers. Can this mathematical discovery offer a way to represent, at least in part, the *logical* structure underlying Hegel's philosophical claim?

To answer this question, let us first recall Pannenberg's interpretation of Hegel's concept of the infinite:

> The Infinite that is merely a negation of the finite is not yet truly seen as the Infinite (as Hegel showed), for it is defined by delimitation from something else, i.e., the finite. . . . The Infinite is truly infinite only when it transcends its own antithesis to the finite. . . . We [must] combine the unity of the infinite and the finite in a single thought without expunging the difference between them.[62]

It is helpful to restructure Pannenberg's interpretation of Hegel's concept of the Infinite in the form of two claims. The first is an affirmation of the traditional definition of the Infinite. The second is Hegel's dialectical complexification of this concept:

Claim 1: The Infinite is the negation of the finite.

Claim 2: The Infinite transcends the negation of the finite by including the finite without destroying the difference between the finite and the Infinite.

Let us next consider how Cantor's mathematics can provide a formal language to express the underlying logic of these philosophical claims. We start by recalling the rules that the properties of cardinality and ordinality for *transfinite sets* obey:

Transfinite cardinality: $\aleph_0 + \aleph_0 = \aleph_0$

Transfinite ordinality: $\omega + \omega \neq \omega$

We then compare these to the rules that the cardinality and ordinality of *finite sets* obey:

Finite cardinality: $\aleph_0 + \aleph_0 \neq \aleph_0$

Finite ordinality: $\omega + \omega \neq \omega$

These two sets of rules can be simplified as follows:

The cardinality of the transfinites is not like the cardinality of the finite.

The ordinality of the transfinites is like the ordinality of the finite.

This in turn means that:

A. Cardinality represents a crucial difference between the finite and the transfinites, a difference by which they are distinguishable.

B. Ordinality represents a crucial similarity between the finite and the transfinites, a similarity by which they are indistinguishable.

Now we are ready to make the crucial move. I claim that A, taken alone, is analogous to the traditional relation of strict negation between the finite and the infinite (see the first of Pannenberg's two claims above). What about B? Actually B, taken alone, does not seem analogous to either of Pannenberg's claims. But let us consider A and B taken as a whole and represent them by the notation {A, B}. My proposal is this: in order that the similarity in ordinality, B, not undermine the difference in cardinality, A, we give epistemic priority to A over B. This means that we hold the difference and distinguishability between the finite and the transfinites as the primary relation between A and B, as in the classical tradition of infinity as *apeiron*. Yet we also hold the similarity and indistinguishability between the finite and the transfinites as a secondary role in this relation. This secondary role of similarity and indistinguishability is quite surprising to the classical tradition because it qualifies their relation: it is no longer one of sheer difference.[63] We can summarize this claim as follows:

> The finite and the transfinite are primarily *distinguishable* and yet in an important but secondary sense they are *indistinguishable.*

We can represent this claim by the symbol {A, b} where the lowercase term implies secondary, but not vanishing, significance compared with the uppercase term. In sum, the paradigm shift from the classical understanding of the relation between the finite and the infinite to Cantor's understanding of the relation between the finite and the transfinite infinite can be symbolized as a shift from {A} to {A, b}:

$$\{A\} \rightarrow \{A, b\}$$

I will use this symbol to represent this paradigm shift about the concept of infinity in what follows as we explore its fruitfulness for Pannenberg's doctrine of God, specifically his treatment of the divine attributes.

Cantor's Absolute Infinity and the Reflection Principle
as Further Resources for Pannenberg

Along with Cantor's discovery of the transfinites we have also discussed his concept of Absolute Infinity, Ω. According to Cantor, Absolute Infinity lies endlessly beyond the realm of the transfinites. It is related to, even "realized in," God and the divine attributes. But it also shares its properties with the transfinites and is thus intimately related to them, according to Cantor's reflection principle. I believe this highly fluid concept, Ω, can offer several insights into Pannenberg's work.

Before exploring this, let us recall how the reflection principle helps avoid what might appear as a contradiction about the (in)comprehensibility of Absolute Infinity in Cantor's work. We begin with Cantor's definition of Absolute Infinity as beyond comprehension, that is, as lying beyond the unending ladder of the transfinite but still comprehensible numbers. Now the transfinites are "comprehensible" in the sense that, although they are infinite, their properties can be formally described and they can be used to distinguish one transfinite from another. Hence, to ensure that Absolute Infinity is incomprehensible, we claim that if it were not so, then it, too, could be described uniquely. This means we could state explicitly at least one property that it alone possesses and that singles it out from all the transfinites. We therefore reverse this claim and insist, via the reflection principle, that all of the properties of Absolute Infinity must be shared by the transfinites. Because Absolute Infinity has no unique property, we can never point unequivocally to it by stating that unique property. In this sense, at least, Absolute Infinity is indescribable and thus incomprehensible. Yet conversely this means that we do, in fact, know something about Absolute Infinity: all of its properties are disclosed to us through the transfinites. The Absolute Infinite is in this sense knowable; each of its properties must be found in at least one transfinite number. In essence, infinity as Absolute is disclosed through infinity as relative, that is, through the transfinites.

In this way the reflection principle turns what appears to be a contradiction about the (in)comprehensibility of Absolute Infinity into a set of mutually entailing polar assertions ripe for theological appropriation. To speak somewhat metaphorically, it is as though the transfinites form an endless veil surrounding Absolute Infinity. The veil is all we can ever know about Absolute Infinity. What lies behind it, Absolute Infinity in itself, is hidden by that veil and is incomprehensible. Yet, genuine knowledge about Absolute Infinity is forever revealed in the veil that hides it, for we can endlessly learn more and more about Absolute Infinity as we continue to discover more and more about the transfinites who share its properties. Thus we can move endlessly to ever more complete knowledge of what can never be known exhaustively and which is, in this important sense, genuine mystery.

Let us now introduce this mathematical metaphor into the context of the doctrine of God and God's self-revelation. With Pannenberg we can assert that the God who makes Godself known to us as Creator and Redeemer is incomprehensible. "Any intelligent attempt to talk about God . . . must begin and end with the confession of the inconceivable majesty of God which transcends all our concepts. . . . It must also end with God's inconceivable majesty because every statement about it . . . points beyond itself."[64] Mathematically, Absolute Infinity is known through the transfinites, and yet being so, it remains unknown in itself. Theologically, the God who makes Godself known is the God who in Godself is unknowable. What God has chosen to disclose to us in history and nature is a veil, but a veil unlike any other. For it is both a kataphatic, revealing veil *and* an apophatic, hiding veil. Precisely through what God makes known to us as the veil of general and special revelation is the veil behind which the mystery of God is endlessly hidden precisely as it is endlessly revealed. To capture this theological understanding of God's hiddenness in God's self-disclosure, I suggest the metaphorical statement that God's self-revelation is "the veil that discloses God as hidden and inconceivable."[65] This in turn reflects Pannenberg's claim that the incomprehensibility of God is based on God's infinity, but now using Cantor's interpretation of infinity instead of that of Hegel.

Cantor's understanding of Absolute Infinity also gives us a way to express in mathematical language the philosophical distinction between God and creation enshrined in the theological doctrine of creation *ex nihilo*. Recall that to avoid the charge of pantheism leveled by Cardinal Franzelin, Cantor proposed the existence of Absolute Infinity and related it to the mind of God. He then sharply distinguished it from the transfinites, which he considered to be created by God since they are "limited, subject to further increase, and thus related to the finite." In doing so, Cantor offered us a profound theological insight well worth pursuing. As long as we consider the transfinites to be creaturely, despite the fact that they possess characteristics that embody infinity, we can maintain the radical distinction between God and creation built into the doctrine of creation *ex nihilo*—a doctrine that, of course, Pannenberg firmly maintains. We can now build on this theologically by turning to Pannenberg's notion of God as radically transcending the world in such a way as to be fully immanent in the world. Pannenberg saw this as possible precisely because of the structural role of infinity controlling his concept of the divine attributes, and he adopted this structural role from Hegel's understanding of infinity. My proposal is to shift the discussion from Hegel's concept of infinity to that of Cantor. I believe this shift will offer an enriched understanding of Pannenberg's overall theological agenda: that the precise form of God's transcendence of the world makes possible both God's immanence in the world and God's eschatological consummation of the world. We will explore this shift in section B.5 below.

Let me pause to summarize the results so far. We have traced some of the contours of the paradigm shift in our understanding of infinity in mathematics from the classical understanding of the infinite as *apeiron* to Cantor's discovery of the transfinites. In specific, we saw that insofar as ordinality is concerned, the finite is distinguishable from the transfinites as one would expect from traditional notions of the infinite as *apeiron*. But insofar as cardinality is concerned, the finite and the transfinites are indistinguishable, and this is truly surprising. My interpretation of these two facts is that we should view the relation between the finite and the transfinite as primarily distinguishable and yet in some important senses as indistinguishable. I represented this situation by the symbol {A, b}.

We have also found initial reasons to believe that Cantor's concept of Absolute Infinity and its relation to the transfinites can be fruitful theologically as we explore Pannenberg's exposition of the divine attributes. First, the inconceivability of Absolute Infinity reflects Pannenberg's claim that the incomprehensibility of God is based on God's infinity. Next, Cantor's work supports, rather than challenges, the ontological distinction between God and creation enshrined in the doctrine of creation *ex nihilo* as long as we consider the transfinites to be creaturely. Finally the reflection principle provides a venue for discussing Pannenberg's concept of the divine attributes such that God radically transcends the world while being immanent in the world without this immanence compromising God's transcendence.

In the following section I will explore Pannenberg's interpretation of the divine attributes in light of the relations between the finite and the transfinite in creation, symbolized by {A, b}, and between the transfinite and Absolute Infinity, as structured by the reflection principle. As we will see, {A, b} allows us to place both finite and transfinite infinity within the category of creation. Yet it allows us to give an enriched view of creation an "open" through the infinity of the transfinites to God their transcendent Creator. As we will also see, the reflection principle allows us to relate the divine attributes immanently to the characteristics of the transfinites even while they remain entirely God's creation. Together I believe that {A, b} and the reflection principle will provide the structural basis, now built on mathematics instead of on Hegel's metaphysics, for deploying Pannenberg's theological discussion of the divine attributes.[66]

Exploring Cantor's Work in Pannenberg's Interpretation of the Divine Attributes

We can now return to Pannenberg's theological response to the concept of the true Infinite in sections 6 and 7 of ST1, chapter 6. In ST1, section 6, Pannenberg argues that the concepts of God's holiness and of God's Spirit are both based on the structure of Hegel's true infinity. God's holiness is truly Infinite because, even while opposing the profane world, it nevertheless enters into it and, eschatologically, makes it holy.[67] Similarly, God's Spirit is truly Infinite because, while the Spirit opposes the world, it works in it, giving life to all creatures and sanctifying them by bringing them into fellowship with God.[68]

In my view Pannenberg's argument is not only about the true Infinity of God. I believe it must also entail a crucial point that lies implicitly within Pannenberg's writings. This is the claim that God created the world in such a way that God's entering into the world to make it holy and sanctified without divinizing it is possible. But this claim, in turn, entails two additional points and takes us to the importance of Cantor's work on infinity.

I believe that Pannenberg's view of eschatology requires that, first, the world must be created as *both* finite and transfinite. Specifically it must be characterized by what I am calling the {A, b} relation between the finite and the transfinite. The second point is that the world's transfinite infinities must stand in relation to the Absolute Infinity of God such as we find in Cantor's reflection principle. In essence, Pannenberg's understanding of the divine entering into the world to transform it while remaining transcendent to it requires a combination of the two points. If this is correct, then Pannenberg must either be making the following assumption implicitly, or if he is not, we need to add it to his argument:

The world as created by God is both finite and transfinite as denoted by {A, b}. Moreover the transfinites stand in relation to Absolute Infinity as described by the reflection principle.

Additional points should also be noted. First, when I construe Pannenberg's argument in this way, it can be taken to lie within the tradition
of the *vestigium Dei,* namely that the world bears a "mark" of its Creator. (We will discuss this "mark" below in relation to God's eternity and
creaturely time.) In fact, the entire "time and eternity" problematic of this
volume can be seen as a *vestigium*-style argument. Note, however, that
this is *not* the traditional *vestigium* argument of natural theology that seeks
to move from the so-called vestiges of the Trinity in nature to the Trinitarian nature of God. Instead, my argument moves from Creator to creation and not from creation to Creator. Thus, it is radically different from
the traditional argument.[69]

Second, even if we consider the transfinite infinities as aspects or
descriptions of the created world, the doctrine of creation *ex nihilo*
should be sufficient to undermine any charge of pantheism such as was
leveled at Cantor. The charge stemmed from the traditional distinction
between the world as finite and God as infinite. But with Cantor even a
transfinite world does not lead to pantheism because the distinctions between finite and the transfinite on the one hand as created and Absolute
Infinity on the other as divine can be used within the doctrine of creation
ex nihilo to erect an unbreachable ontological gulf between God and
God's creation.[70]

Finally, viewing the creation as both finite and transfinite overcomes
a closely related potential challenge to Pannenberg's claim about God's
entering the world in holiness and through the Spirit. The challenge arises
from the traditional view of finitude as *apeiron.* Arguably a "fully finite"
world could not, in principle, be open to God's holiness and divine Spirit
as proposed by Pannenberg without leading to pantheism and the divinizing of the world or to atheism and the secularizing of God. But if the
world that God did, in fact, create is both finite and transfinite, it is a
world which can be infused with God's holiness and life-giving Spirit
while remaining creaturely. In my mind this is one of the most wondrous
gifts God gave the world in creating it {A, b} and in relating to it, as
suggested by the reflection principle.

Let us turn now to a more detailed discussion of Pannenberg's argument. In ST1, section 6, Pannenberg illustrates his argument by claiming that the proper meanings of eternity and omnipresence are only achieved when they are grounded on "the structure of the true Infinite." Eternity, when correctly understood, offers a "paradigmatic illustration and actualization of the structure of the true Infinite which is not just opposed to the finite but also embraces the antithesis."[71] In comparison, eternity, as timeless, presupposes an "improper" concept of the infinite for, if defined simply by its opposition to the finite (i.e., as the *apeiron*), eternity would in fact become finite. Omnipresence, in turn, when based on the structure of the true Infinite, includes both divine transcendence and divine immanence. Thus the transcendence of God makes it possible for God to be immanently present to the world in such a way that God's immanence does not undercut God's transcendence of the world. Pannenberg then grounds this claim explicitly on the doctrine of the Trinity: "The trinitarian life of God in his economy of salvation proves to be the true Infinity of his omnipresence."[72] Each of these claims leads me to believe that Pannenberg is implicitly assuming that the created world is what I am calling {A, b} and that the transfinites are related to Absolute Infinity via the reflection principle. Pannenberg's view of creation must have what Cantor would call transfinite as well as finite characteristics, because this would support Pannenberg's theological conception of the proper relation between the true infinity of God's eternity and omnipresence and their relation to the created realities of time and space. And again Pannenberg's acceptance of the doctrine of creation *ex nihilo* guards him from pantheistic misinterpretations that divinize the transfinites.

In ST1, section 7, Pannenberg describes the relation between the infinity of God and the finitude of creation in terms of Trinitarian divine love:

> Divine love in its trinitarian concreteness . . . embraces the tension of the infinite and the finite without setting aside their distinction. It is the unity of God with his creature which is grounded in the fact that the divine love eternally affirms the creature in its distinctiveness and thus sets aside its separation from God but not its difference from him.[73]

Here again, as above, I believe that Pannenberg's view presupposes that creation can be embraced by God precisely because God created it to be embraceable by God. This, in turn, entails that God created the universe as both finite and transfinite while the belief that God created it *ex nihilo* ensures us that its transfinitude does not lead to pantheism.

Cantor as Resource for Pannenberg's Discussion of Time and Eternity in the Context of Eschatology

We close this chapter with one final insight from Cantor for Pannenberg's discussion of time and eternity in the context of eschatology. Pannenberg writes that "in the renewed world that is the target of eschatological hope the difference between God and creatures will remain, but that between the holy and the profane will be totally abolished (Zech. 14:20–21)."[74] How might we use the concept of {A, b}, and the reflection principle, contextualized so far in the doctrine of creation, to explore Pannenberg's understanding of the eschatological New Creation? *I propose that the {A, b} relation of the present creation will shift to an {a, B} relation characterizing the New Creation.* Nevertheless, in the transformation from creation to New Creation the reflection principle will remain in place. The former shift represents what I call, following CMI, guideline 6, an "element of discontinuity," while the latter represents what I call an "element of continuity." Together the shift to {a, B} and the continuation of the reflection principle offers, I believe, mathematical language for expressing at least a portion of Pannenberg's conception of eschatology.

By a shift from {A, b} to {a, B}, I do not mean that the finitude of the world will be diminished in relation to its transfinite character, or that the latter will be privileged over the former. Pannenberg is very clear on this: finitude remains, and it is endlessly celebrated in the New Creation. For example, as we saw in chapter 1, creatures in the New Creation remain finite in space and time. Even though they do not die, their endless living is a finite form of temporality, since endlessness is still finite. Rather, I believe that the shift from {A, b} to {a, B} means that the transfinite character of the world will take on increasing significance in the eschatological New Creation as it embraces the world's finitude.

The key points are: first, that through the reflection principle the transfinites share in the properties of Absolute Infinity even while remaining entirely creaturely, as consistent with the doctrine of creation *ex nihilo.* Second, it is through the {A, b} relation in the present creation that the transfinites and finitude co-mingle some of their properties. Third, in the New Creation we can hope that our language about the transfinite character of God's creation, as it is now transformed eschatologically by God's new act starting at Easter into {a, B}, can take on a more intense role in describing the world, depicting the world as ever more deeply open to God's immanence as shown metaphorically in the reflection principle. In essence, Cantor's mathematical language for the modern concept of infinity thus gives us a mode of expressing what Pannenberg so eloquently tells us theologically: "in the renewed world that is the target of eschatological hope the difference between . . . the holy and the profane will be totally abolished."

In sum, Pannenberg's vision captures much of what is essential to the Christian doctrine of God and to Christian eschatology. My hope is that Cantor's mathematics, despite its formal problems, offers Pannenberg an alternative language to that of Hegel's philosophy in expressing the underlying structure of the divine attributes in the doctrine of God and in the intricacies of Christian eschatology.

Covariant Correlation of Eternity and Omnipresence in Light of Special Relativity

Relativity physics is a puzzling case for my thesis, the most puzzling indeed of all. . . . What is God's "frame of refer-ence," if there is no objectively right frame of reference for the cut between past and future?

—Charles Hartshorne,
A Natural Theology for Our Time

Michele has preceded me a little in leaving this strange world. This is not important. For us who are convinced physicists, the distinction between past, present, and future is only an illusion, however persistent.

—Albert Einstein

Human kind cannot bear very much reality.

—T. S. Eliot, "Burnt Norton"

In this chapter I turn to Albert Einstein's theory of special relativity (SR) and the task of reformulating Pannenberg's treatment of eternity and omnipresence in light of the spacetime interpretation of SR. This interpretation of SR, given by the mathematician Hermann Minkowski just two years after Einstein published SR, is that space and time cease to be the entirely separate physical realms as they are in ordinary experience and as they are enshrined in classical physics. Instead, they form a single, four-dimensional geometry called "spacetime." Here I offer a reformulation of the theological relation between eternity and omnipresence based on the incorporation of the physics of SR in which space and time are combined into spacetime. Once this is in place, we will explore several new theological insights into the interweaving of the divine eternity and omnipresence in relation to the world as understood in terms of SR. We will also explore the way in which the divine omnipresence offers a response to the philosophical problem, based on the debates between Newton, Leibniz, and Clarke, of a concept of the unity of the world.

First, however, I want to stress that SR poses a striking challenge to any theology that incorporates what is called its "spacetime interpretation." This is because the spacetime interpretation of SR is *itself* subject to an interpretation known as the "block universe." According to the block universe interpretation of spacetime, all events in the world—past, present, and future—are equally present and real. There is no objective distinction between what we subjectively call past and future. Instead all events in life, history, and cosmology are just "there" in the frozen geometry of spacetime, and the flow of time that is so deeply given to our personal experience is "an illusion, however persistent."[1] Such an interpretation of spacetime obviously undercuts the kind of "flowing time" interpretation of time that I defended in chapter 2 as applying not only to our personal experience but to nature itself. And flowing time is, arguably, essential to most forms of religious experience and their theological systematization. In chapter 2, we saw how this would be true especially for Pannenberg, even as he adds subtle layers of nuance in his interpretation of distinction versus separation in the relations between events in time—and their being taken up into the divine eternity. Hence, the challenge of the block universe interpretation of SR must be met if we are to incorporate the spacetime framework of SR into our theological discussion of the divine attributes.

As previously indicated, my approach in this volume is to challenge the block universe interpretation of spacetime by deploying a novel flowing time interpretation of spacetime (chapter 5). If this is successful it will allow us to appropriate the spacetime interpretation of SR for theological purposes without having to "buy into" the block universe interpretation of spacetime. Integral to the construction of this new flowing time interpretation will be the results of this chapter's theological reformulation of eternity and omnipresence within a spacetime framework. This construction, in turn, can help demonstrate the fertility of theology, when reformulated in light of science, in pointing to new directions in the philosophy of science and in scientific research (chapters 5 and 6).

In this chapter I develop the reformulation of theology in light of physics following the agenda of part one of this volume, namely what I denote by SRP → TRP (Scientific Research Program → Theological Research Program). I ask the reader to suspend judgment temporarily about whether I can successfully address the challenge raised by the block universe interpretation of SR until chapter 5. There, as appropriate for part two of this volume, I reverse the direction of discussion, following the notation TRP → SRP (Theological Research Program → Scientific Research Program), and suggest how this reformulated theological interpretation of the relation between eternity and omnipresence leads to new philosophical insights, specifically into a "flowing time" interpretation of SR.

A. OVERVIEW OF SPECIAL RELATIVITY

A note to the reader: The technical material in this section, including the mathematics and the use of spacetime diagrams, may be unfamiliar even to those readers already well conversant with the "theology and science" literature, let alone religious studies and theology readers new to the field. In fact, except for readers with a background in physics, even those with degrees in the other natural sciences may find the mathematics and its interpretation here to be challenging. I would like to suggest that such readers consider one of the following strategies. The first is to read just the summaries found at the beginning of each part of this overview. The trusting reader may simply take them for granted and move on quickly to section B. The second one is to read the summaries and then browse quickly through the discussions that follow each summary. Frankly, while either of these strategies may work for parts 1–6, I urge every reader to delve more fully into parts 7–9 as these will play such a pivotal role in the theological discussions that follow in this chapter (sections B and C) and the discussion of a flowing time interpretation of the spacetime interpretation of SR in chapter 5. The third strategy is to dive carefully into all the material in this overview in detail. While this is the most challenging tack, it is clearly the one that holds the most promise, for only by working through the mathematics can one gain one's own critical insights into what SR tells us about nature and assess SR's competing interpretations. Obviously I hope most readers will chose this approach. Finally, there are a number of books that offer introductions to relativity at the college level both to humanities majors and to physics undergraduates that are helpful sources.[2] In addition I will provide links to several excellent online resources.[3]

1. Two Postulates of Special Relativity

Summary

In his original 1905 paper on what we now call the special theory of relativity (SR),[4] Albert Einstein offered two postulates that hold for all inertial observers, that is, for all observers in uniform relative motion: 1) the fundamental laws of physics take the same form for all inertial observers[5] (the principle of relativity);[6] and 2) the speed of light in free space has the same value, c, for all inertial observers.[7] Equivalently we can say that the fundamental laws of nature and the speed of light in free space are invariant for all observers in uniform relative motion due to the covariance of the Lorentz transformations.

Discussion

There are a variety of ways to derive the theoretical consequences of these postulates and compare them with the wealth of experimental evidence that supports them. Einstein himself recognized that the so-called Lorentz transformations (see section 5 below) for shifting "covariantly" between the coordinates of observers in uniform relative motion (and as

opposed to the Galilean transformation of classical physics) would leave
the key equations of electromagnetism (i.e., Maxwell's equations) "invariant."[8] Here I will take a somewhat unusual approach and start with the
way identical "clocks" behave when they are in uniform relative motion
compared with one taken to be at rest. By "clocks" I mean something as
simple as elementary particles, such as muons, which decay after a set
amount of time called their "half-life." Their behavior is called time dilation since clocks in motion relative to a clock at rest will seem to run
more slowly. In the case of elementary particles, those moving will take
longer to decay than the ones at rest; their half-life is "dilated." This is
surely one of the most striking and counterintuitive aspects of SR, yet it
lies at the heart of contemporary theoretical physics and it has been verified in countless experiments over the past century.[9] To illustrate time dilation I will next introduce what are called "spacetime diagrams."[10]

2. Spacetime Diagrams

Summary

A helpful teaching device for describing SR is the spacetime diagram,
which represents the movements of objects in space and time. In principle, such a diagram has four dimensions: three for the spatial dimensions x, y, and z (think width, depth, and height of a room or building),
and one for time t (think clock, watch, metronome). Typically we restrict
the diagram to two dimensions to make it easier to picture on a sheet of
paper: one dimension for the x-axis, the other dimension for the t-axis.[11]
The diagram is constructed by someone at rest at the origin, and the lines
and curves in the diagram represent the motion of people, cars, protons,
etc., from the observer's point of view. A spacetime diagram can also be
used to show the path of photons emitted at the origin, creating what is
called a "light cone" (see diagrams 4.2, 4.3).

Discussion

Let us imagine that an observer, Ted, is standing at the side of a street
that extends to his left and right. His friend, Ann, is parked in her car
where Ted is standing. Let us lay out the x-axis along the street and

mark the place where Ted is standing as the origin. When Ted gives the signal, Ann begins to drive to the right then slows down and stops at a traffic light at the location x = a. Meanwhile Ted uses his stopwatch to measure the time t for these events, including when Ann starts to drive (t = 0) and when she has stopped again. Spacetime diagram 4.1 shows the "path" for these events: the car at rest at the origin, then slowly moving to the right, then coming to rest at a stoplight. We routinely call this path Ann's "worldline." Note that the term "worldline" can also be used for the paths of photons such as in diagrams 4.2 and 4.3, below.

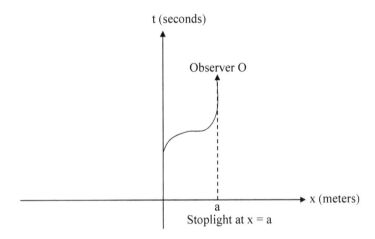

Diagram 4.1. Ann's Worldline: Ann's Car Moves to the Right
along x, Then Stops at the Stoplight at x = a

A note about units of measurement: Because the speed of light plays such a central role in SR, and because the speed of light is so enormous— approximately 3 x 10⁸ m/s (300 million meters per second)—if we stayed with units of meters and seconds the worldlines for light in diagrams like 4.1 would be almost horizontal. This in turn would make it very hard to visualize the physics associated with SR. It is therefore routine to adjust the spatial scale so that the worldlines of photons lay at a 45° angle to the t-axis. We do so by measuring distance in the unit known as a "light-second": the distance light travels in one second (i.e., 3 x 10⁸ m).[12]

Diagram 4.2 shows the worldlines for photons emitted by a light bulb at the origin and moving to the left and right along the x-axis.

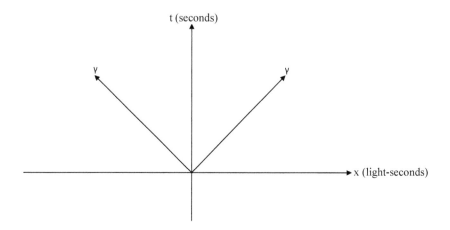

Diagram 4.2. Photon Worldlines at 45° to the t-Axis and x-Axis

If we include in the spacetime diagram the y-axis along with the x-axis we obtain the two-dimensional light cone structure of photon world-lines (diag. 4.3):

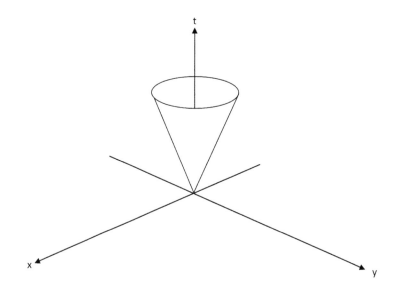

Diagram 4.3. Light Cone Structure Produced by Photons Emitted
from the Origin in the x-y Plane

3. Time Dilation

Summary

Moving clocks (i.e., clocks in uniform relative motion) run more slowly than clocks identical to them at rest. Using a spacetime diagram, we can easily depict this counterintuitive phenomenon called time dilation.[13] In this presentation of SR I take time dilation to be an experimental fact and develop its theoretical implications. These include the "downfall of the global present" (section 4), the choice of the Lorentz transformations over their classical counterpart the Galilean transformation (section 5), and such related phenomena as length contraction and causal invariance (section 6). I will then turn in detail to a particularly illuminating "spacetime paradox," the pole-in-the-barn paradox (section 7). Following this, we will return to the arguments for four-dimensional "spacetime" over the classical worldview, "space plus time" (section 8), and the real significance of the ambiguity of the global present (section 9).

Discussion

Let us start with a simple example of time dilation. Many elementary particles, such as muons, decay in the lab with a precise half-life, and because of this they can serve as a "clock" in our exploration of SR.[14] The half-life for muons at rest in the lab is 2.2×10^{-6} seconds. However, when produced in the earth's upper atmosphere by cosmic rays, muons typically travel at near light speed toward the earth. At these speeds they take over one hundred times as long to decay, often reaching sea level some 60 kilometers (6×10^4 m) below—an impossible feat even traveling at the speed of light (3×10^8 m/s) if their half-life at such speeds was still 2.2×10^{-6} seconds:

> 6×10^4 m/3×10^8 m/s $= 2 \times 10^{-4}$ seconds, which is about one hundred times longer than their half-life of 2.2×10^{-6} seconds.

This is an example of time dilation.[15] It contradicts what we would intuitively expect to happen: Why should moving clocks run slow? Nevertheless, time dilation is a fundamental fact about the physical world.

Now let's probe deeper into time dilation. Imagine we have six identical clocks that tick once per second. Keeping one at rest with us, we throw the other five clocks to the right or the left along the x-axis at different velocities v_1, v_2, etc. Next we plot the coordinates (x, t) when each clock ticks once on a spacetime diagram, including the clock at rest at the origin, O. If classical physics and our common sense were correct, the plot would be a straight line, t = 1 second. This would be the ordinary global present, one second later than when our experiment began. In essence, the x-axis or global present simply shifts up one unit along the t-axis (diag. 4.4).

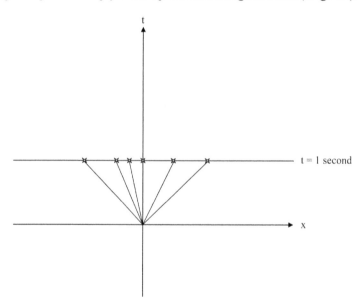

Diagram 4.4. Expected Results for One-Second Ticks of Six Identical Clocks Thrown to the Left and Right from the Origin

But experiments with clocks such as muons moving at relativistic velocities (velocities near c) give a radically different result. The "one-second ticks" in spacetime do not lie along the straight line t = 1 second. Instead, they form a hyperbola:[16]

$$\tau^2 = t^2 - x^2/c^2 \tag{1}$$

where, by convention, we call the time measured by a clock at rest its "proper time" τ. Equation 1 is often called "the hyperbola of one-second events," since its shape is that of a hyperbola (diag. 4.5).

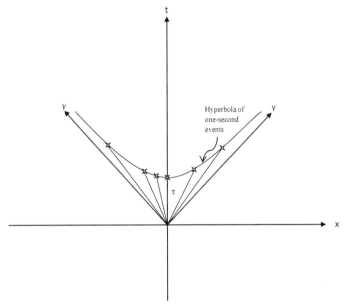

Diagram 4.5. Actual Result for Six Clocks Showing the Hyperbola
of One-Second Events

Diagram 4.5 makes it abundantly clear why this phenomenon is called
time dilation: against our best intuition, it takes the clocks longer to tick
when they are moving relative to the clock at rest. This is evident in the
diagram where τ is at the "bottom" of the hyperbola and all the other
events lie further along the time axis. With a little algebra we can turn
equation 1 into equation 2, which explicitly shows that t, the time of the
moving clocks, is larger than τ, the proper time of the clock at rest:

$$\tau^2 = t^2 - x^2/c^2 \tag{1}$$
$$= t^2 - v^2t^2/c^2 \text{ since } x = vt$$
$$= t^2 (1 - v^2/c^2)$$

or

$$\tau = t (1 - v^2/c^2)^{1/2}$$

Finally, we rewrite this result as:

$$t = \gamma\tau \tag{2}$$

where $\gamma = (1 - v^2/c^2)^{-1/2}$. Since $0 \leq v < c$, $\gamma \geq 1$. This means that $t \geq \tau$.[17]

4. The Downfall of the Present

Summary

The immediate implication of the fact of time dilation is "the downfall of the present," that is, the loss of the unique, global present that we take for granted in ordinary experience and in classical physics. We can see this in at least two ways: as the direct result of time dilation and, a bit more indirectly, as the inability to uniquely synchronize clocks in relative motion.

Discussion

a. Time dilation and the downfall of the present. Recall that in classical physics and everyday experience the present is a universal, global "now" (diag. 4.4). The global present is a straight line, t = 0, that divides all of time for all observers *regardless of their motion* into the same global past and the same global future. More important, the global present moves uniformly into the future such that one second from now as measured by any one clock will be one second from now for all clocks.

Instead, we just saw the experimental evidence for time dilation (diag. 4.5). It proves that events one second from now as measured by clocks in relative motion to a clock at rest is a curve, namely the hyperbola of one-second events, and not a straight line. Identical observers in relative motion do not proceed uniformly together into the future. Because of this, the events we expect to be mutually present are not those that identical, moving clocks mark out. Time dilation does indeed imply the loss of the traditional global present.

b. Clock synchronization and the downfall of the present. Another way to get at the downfall of the present is by asking the following operational question: How do we synchronize two clocks A and B in relative motion? Intuitively the answer would be to synchronize a third clock C at rest with respect to A, then move C to B and set B's time to match it— and thus to match A. Now A, B, and C are synchronized. Although in relative motion, they define a common present between them. The problem with this method is that, because of time dilation, when we move

clock C from A to B its rate slows down and it is no longer synchronized with A! So moving a clock between two others in relative motion
fails as a way to synchronize them.

Let us try another way to synchronize clocks in relative motion
based on a simple method for synchronizing clocks *at rest* with respect
to each other. Set two identical clocks, A and B, a distance x_0 apart and at
rest with respect to each other. Being identical and at rest we know that
they tick at the same rate, but how do we know which event along B's
worldline corresponds to a given event on A's worldline, that is, how do
we determine which events are simultaneous? The simplest way is to at-
tach a mirror to clock B, project a photon from A to B, and note the event
C where it bounces off the mirror and the event D with time t where it
returns to A (diag. 4.6). Divide the lapsed time t in half. That marks
event E along A, which we *define* as simultaneous with C. The line from
E to F is the "axis of simultaneity," representing all the events that observers at rest with clocks A and B consider to be mutually "present."
Equivalently, it is the x-axis at time t/2 for clocks A and B.

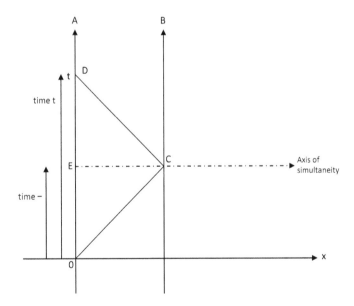

Diagram 4.6. Axis of Simultaneity for Clocks A and B at Rest

Note: Event E of clock A at the axis of simultaneity (dashed/dotted line) is defined as simultaneous with Event C of clock B.

We are now ready to tackle the problem of synchronizing two sets of clocks, A–B and F–G, in relative motion (diag. 4.7). We repeat the process above to determine the axis of simultaneity, or x-axis, for the clocks A–B at rest. But here we also let some photons go past the mirror at C and on to the moving clock G, bounce off the mirror at event H, and return to the moving clock F. If we use the same formula for determining which event P along F is simultaneous with H, we will immediately see that the line from P to H is not parallel to the line from E to C: the axes of simultaneity for clocks A–B (their x-axis) and F–G (their x'-axis) are different. Hence for clocks in relative motion there is no mutually agreed upon universal present.

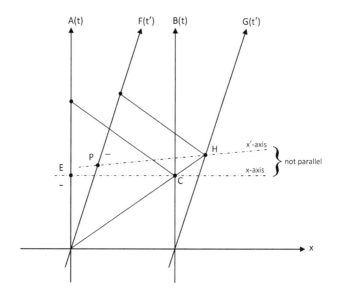

Diagram 4.7. Axes of Simultaneity (x-Axis, x'-Axis) Differ for
Clocks A–B and F–G in Relative Motion. Event P for Clock F
Is Simultaneous with Event H for Clock G

To repeat, for clocks in relative motion there is no mutually agreed upon universal present.

c. Conclusion: The "downfall of the present" is complete. The results are clear: according to SR there is no physically meaningful way of determining a universally agreed upon present. The implication for most scientists is that there is no universal, unique physical meaning to the "present," only a "present" as defined operationally by each moving observer.[18]

5. Einstein's Principle of Relativity and the Lorentz Transformations

Summary

We now return to Einstein's postulates that lie at the heart of SR and ask what it would require mathematically to make the experimental phenomenon of time dilation consistent with them: Are the equations that describe time dilation invariant for observers in uniform relative motion? Do they take the same form for all such observers? As we shall see, the answer is that they do not take the same form if we use the classical Galilean transformations, which take us from the coordinates of one observer covariantly to those of another in relative motion. Instead, we must adopt what are called the Lorentz transformations.[19] It is these transformations, and not the classical ones, that support the invariance of time dilation. We shall see in part 6 below that they also lead to such physical phenomena as "length contraction" (the "inverse" of time dilation), the invariance of the speed of light (Einstein's second postulate), and causal invariance (all observers in relative motion will agree on the temporal and thus causal relations of past, present, and future for events along the worldline of any observer).

Discussion

We began our discussion of time dilation with muon decay and then moved to the more general problem of a set of identical clocks thrown from the origin along the x-axis at various velocities. The observer O plots the x, t coordinates of the events where each clock ticks "one second" in a spacetime diagram. The observer then discovers that these events lie along a hyperbola as described by equation 1:

$$\tau^2 = t^2 - x^2/c^2 \tag{1}$$

where τ is the proper time of the clock at rest at the origin with the observer O. But what about another inertial observer, O', moving along the x-axis at a velocity v with respect to O? For simplicity we can assume that when O and O' coincide at the origin they each set their clocks to zero (i.e., $t' = 0$ when $t = 0$). Will observer O' write down the same equation (1) for the coordinates x', t' that she measures for the "one second" ticks?

Let us begin with the transformations used in classical physics and reflecting ordinary experience, the Galilean transformations:

a) Galilean transformations
$$x' = x - vt$$
$$t' = t$$

Notice that the traditional idea of a universal, global present is represented clearly by the second equation in which t' is set equal to t. In physical terms this means that all clocks tick at the same rate regardless of their relative velocities. But if we consider time dilation this should make us suspicious of the Galilean transformation.

Our suspicions are confirmed because the Galilean transformations do not, in fact, preserve the hyperbola $\tau_2 = t_2 - x_2/c_2$, as is easily seen using the inverse transformations from x', t' to x, t:

$$x = x' + vt'$$
$$t = t'$$

in equation 1:

$$\tau^2 = t^2 - x^2/c^2$$
$$= t'^2 - (x' + vt')^2/c^2$$
$$= t'^2 - (x'^2 + 2x'vt' + v^2t'^2)/c^2$$
$$\neq t'^2 - x'^2/c^2$$

Suppose, however, that in place of the Galilean transformations we use the Lorentzian transformations between coordinates (x, t) and (x', t'):

b) Lorentz Transformations[20]
$$x' = \gamma (x - vt)$$
$$t' = \gamma (t - vx/c^2)$$

whereas before, $\gamma = (1 - v^2/c^2)^{-1/2}$. A bit of algebra shows what happens when we start with the hyperbola $\tau^2 = t^2 - x^2/c^2$ in the O frame and transform to the coordinates of O'. The result is that the hyperbola is *invariant* as required by Einstein's postulates.

Again we use the inverse equations from x', t' to x, t:

$$x = \gamma\,(x' + vt')$$
$$t = \gamma\,(t' + vx'/c^2)$$

Substituting them into the hyperbola yields:

$$
\begin{aligned}
\tau^2 &= t^2 - x^2/c^2 \\
&= \gamma^2\,(t' + vx'/c^2)^2 - \gamma^2\,(x' + vt')^2/c^2 \\
&= \gamma^2\,(t'^2 + 2vx't'/c^2 + v^2x'^2/c^4) - \gamma^2(x'^2 + 2vx't' + v^2t'^2)/c^2 \\
&= \gamma^2\,(t'^2 + 2vx't'/c^2 + v^2x'^2/c^4 - x'^2/c^2 - 2vx't'/c^2 - v^2t'^2/c^2) \\
&= \gamma^2\,(t'^2 + v^2x'^2/c^4 - x'^2/c^2 - v^2t'^2/c^2) \\
&= \gamma^2\,(t'^2 - v^2t'^2/c^2 + v^2x'^2/c^4 - x'^2/c^2) \\
&= \gamma^2\,(1 - v^2/c^2)\,(t'^2 - x'^2/c^2) \\
&= t'^2 - x'^2/c^2
\end{aligned}
$$

Note that the events where the clocks tick "shift" along the hyperbola as we transform between observers in relative motion even while the shape of the hyperbola remains invariant. To see this explicitly let us plot the events A, B, C, ... F, first from the perspective of observer O, for whom event C is a one-second tick by the clock at rest, and then from the perspective of observer O', for whom event B is a one-second tick by the clock at rest (diag. 4.8):

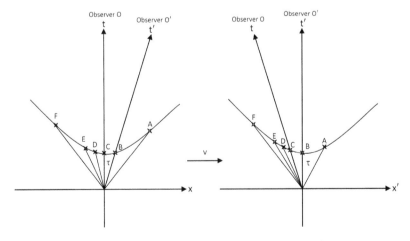

Diagram 4.8. Events A ... F from the Perspective of Observer O
and Observer O' in Relative Motion with Velocity v

Clearly the individual coordinates of A . . . F in the (x′, t′) frame differ
from the coordinates of these same events in the (x, t) frame of O. Nevertheless, these events all lie along the same invariant hyperbola of one-
second events that both O and O′ draw:

$$\tau^2 = t^2 - x^2/c^2 = t'^2 - x'^2/c^2$$

We will soon see that the invariance of the hyperbola of one-second
events provides an important step in allowing us to move from dependence on two specific frames of reference, one for O and the other for O′,
to a single generalized spacetime diagram that includes both observers'
points of view.

6. Consequences of the Lorentz Transformations:
Length Contraction, the Invariance of the Speed of Light,
and Causal Invariance

Summary

We have seen that the Lorentz transformations provide a mathematical
underpinning for the physical fact of time dilation. Three immediate consequences of the Lorentz transformations are length contraction, the invariance of the speed of light, and causal invariance.

Discussion

According to *length contraction* a moving object appears shortened
along the axis of motion. To see this, consider an observer O′ carrying a
ruler of length L_0 (called the "proper length" because it is at rest for O′)
and moving to the right at a velocity v along the x-axis of observer O. For
simplification, in diagram 4.9 below we draw only the worldlines for the
two ends of the moving ruler. Note that the moving observer's x′-axis is
rotated up from the stationary observer's x-axis by an angle θ with respect to the x-axis because of the way we previously defined observer-
dependent simultaneity. Note too that the moving observer's worldline,
or t′-axis, is rotated down from the resting observer's worldline by the
same angle θ. Finally, note that for observers moving to the left, their
x′-axis is rotated down from the x-axis by the same angle. This handy
fact makes drawing the x′-axis easy.

The ruler lies along the x'-axis. Its proper length L_0 is the distance from the origin to event A.[21] The observer at rest, however, measures the length of the ruler to be L, the distance from the origin to event B along the x-axis (diag. 4.9):

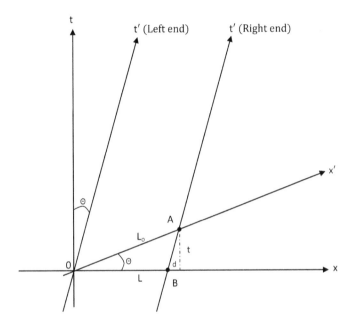

Diagram 4.9. Length Contraction of a Moving Ruler
According to a Ruler at Rest

The first step in relating L and L_0 is to determine the coordinates of event A in the moving frame and in the rest frame:

moving (ruler) frame: $(x', t') = (L_0, 0)$
rest frame: $(x, t) = (L + d, t) = (L + vt, t)$

where d is the distance that x claims the ruler has moved to the right in the time t it takes for its right end to arrive at event A, or $d = vt$.
Next we use the Lorentz transformations to change the moving coordinates (x', t') into the rest frame coordinates (x, t):

$x = \gamma (x' + vt') = \gamma L_0$
$t = \gamma (t' + vx'/c^2) = \gamma v L_0/c^2$

Finally, we substitute these results into the equation $x = L + vt$:

$$x = L + vt$$
$$\gamma L_0 = L + v\,(\gamma v L_0/c^2)$$
$$= L + \gamma L_0 v^2/c^2$$
$$L = \gamma L_0\,(1 - v^2/c^2)$$
$$\text{or: } L = L_0/\gamma$$

The result: the rest frame measured a shorter length (L) for the ruler than the ruler's proper length L_0.

The Lorentz transformations also manifest Einstein's second postulate, the *constancy of the speed of light.* To see this we first need to introduce the equation for the addition of velocities between observers in relative motion. Suppose, as before, observer O is at rest at the origin, and observer O′ moves along the x-axis at velocity u_1. Now let a second observer, O″, move along the x-axis at a velocity u_2 with respect to O′. What is the velocity u_3 of O″ with respect to O?

According to classical physics via the Galilean transformations (and common sense!), velocities add, and the relative velocity u_3 would be:

$$u_3 = u_1 + u_2$$

But according to SR (using the Lorentz transformations) the relative velocity u_3 is:

$$u_3 = (u_1 + u_2)/(1 + u_1\,u_2/c^2)$$

For everyday life, u_1 and u_2 are minute compared to c, and their product, $u_1 u_2/c^2$, essentially vanishes, returning us to the Galilean result.

Now we can return to our question about the constancy of the speed of light. If we replace observer O″ with a photon emitted by a flashlight that O′ is carrying so that $u_2 = c$, we can easily calculate that u_3 is in fact c:

$$u_3 = (u_1 + c)/(1 + u_1 c/c^2)$$
$$= (u_1 + c)/(1 + u_1/c)$$
$$= (u_1 + c)/(c + u_1)/c$$
$$= c$$

and Einstein's second postulate, the constancy of c, is sustained.

An immediate consequence of the constancy of c is that the light cone structure is another relativistic invariant, like the hyperbola of one-second ticks and the proper time, τ. As can be seen in diagram 4.10, two observers in relative motion will draw the same worldlines for photons emitted at the point where the observers' worldlines cross. Like the invariance of the hyperbola of one-second events, the invariance of the light cone will allow us to move to a single generalized spacetime diagram that includes the perspectives of both observers in relative motion (see below).

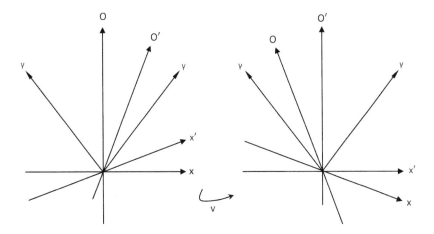

Diagram 4.10. The Invariance of the Light Cone
(Shown Here as the Worldlines of Two Photons γ)
between Observers O and O' in Uniform Relative Motion

A further consequence of the relativistic rule for the addition of velocities is that we can never start at velocities less than light speed and accelerate to velocities equal to or greater than light speed. The speed of light can be approached but never exceeded and is therefore the maximum speed by which phenomena can influence each other, that is, the maximum speed for the transmission of causes in nature.

Finally, the Lorentz transformations ensure *causal invariance.* On the one hand, events that are temporally and thus causally ordered in relation to an event O are temporal and causal invariants for all observers independent of their relative motion with respect to O. On the other hand, the Lorentz transformations ensure that events that are temporally ambiguous and acausally related to O remain so for all observers in relative motion with respect to O.

We can understand this claim more easily by employing a three-dimensional spacetime diagram that includes two spatial dimensions x, y and time t, along with the light cone centered on the event O (diag. 4.11).

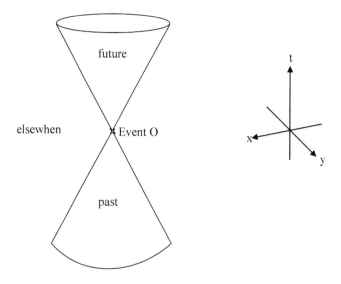

Diagram 4.11. Light Cone Structure for Event O in the x-y-t Format

Because the speed of light is a maximum for physical phenomena, the worldlines for all observers A, B, . . . , moving through event O, regardless of their relative speed, will lie within the light cone at O. The light cone divides spacetime into three separate regions. We will make claims both about the temporal and about the causal relations between events in the three regions:

1. Events that lie in the future[22] of event O are contained in the upper region of the light cone called the "future light cone." Because all events that can be influenced by O lie within this same region, it is also called the "causal future" of O.

2. Events that lie in the past of event O are contained in the lower region of the light cone called the "past light cone." Because all events that can influence O lie within this same region, it is also called the "causal past" of O.

3. Events that can neither influence O nor be influenced by O lie within what is called the "acausal elsewhen" of O[23]. Because these events lie outside the causal past and the causal future of O, their temporal relations to O are ambiguous and depend, as we shall see, on the motion of an observer whose worldline passes through O.

Taking these claims into consideration, we are now prepared to discover another crucial set of invariants yielded by SR. First, the Lorentz transformations guarantee that the events that lie in the causal future of event O, according to one observer, will also lie in the causal future of event O, according to all other observers regardless of their relative motion. A similar fact holds for events that lie in the causal past for event O. Moreover, events that lie in the elsewhen of event O according to one observer will also lie in the elsewhen of event O according to all other observers regardless of their relative motion. This holds not only for observers whose worldlines cross at event O; it holds for all observers in spacetime. We can summarize this by saying that the causal future, the causal past, and the acausal elsewhen of event O are invariant.

Second, the Lorentz transformations preserve the order of events along every observer's worldline but not the order between an event and the other events in its elsewhen. For simplicity we return to the x-t format. Note, however, that as in the x-t format of our previous diagrams, the "connectedness" of the elsewhen region is obscured and spacetime appears to be made up of four separate regions instead of three, as seen here in diagram 4.12.

Thus, when using the x-t format, we should remember that the elsewhen regions to the left and the right of the origin are actually connected, as diagram 4.13 below makes clear.

Diagram 4.13 shows the varieties of causal-preserving and acausal relations that occur in spacetime. We start with two observers, A and B, whose worldlines cross at the event O, and a third observer, C, whose worldline lies to the right of event O. Because all observers move at relative speeds that are less than the speed of light, the worldlines for these

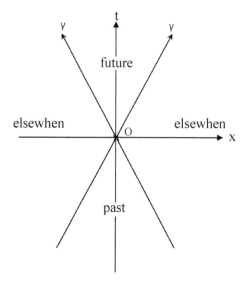

Diagram 4.12. Light Cone Structure at Event O in the x-t Format:
The Elsewhen Falsely Appears as Two Disconnected Regions
to the Left and Right of Event O

and all other observers that cross at O will lie within the light cone at
O. Consider observer A whose worldline includes three events ordered
temporally and thus causally: α, β, and δ. That is, event α occurs before
event β and thus could influence it, and similarly for event γ. Because
these events are temporally and thus causally ordered for A, all other
observers will agree that they are temporally and thus causally ordered
for A. One way to see this relation is to let A send photons from these
events, as shown in diagram 4.13. These photons will strike observers B
and C in the same order along their worldlines, and they will therefore
agree that the events are causally ordered for A as α, β, and δ. On the other
hand, events lying in the elsewhen of O are not causally related to O; they
can neither influence nor be influenced by O (diag. 4.13). This means
that their temporal relations with respect to O are ambiguous.

Consider events δ and ε along C's worldline, where C lies some
distance to the right of the origin O. This time we use a second way to
show the temporal and causal relations; we construct axes of simultaneity for observers A and B at the event O as indicated by the dashed
lines in diagram 4.13. Clearly observers A and B will construct differ-

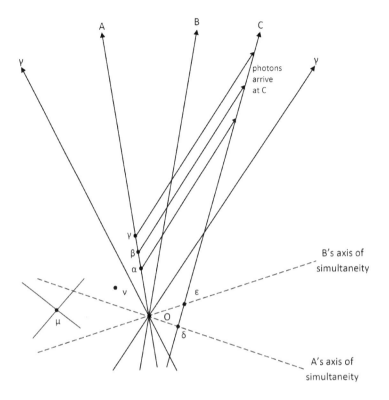

Diagram 4.13. Varieties of Causal-Preserving and Acausal Relations
between Events in Spacetime

ing axes of simultaneity due to their relative velocity. As these axes
suggest, A considers δ to be simultaneous with O while B considers ε
to be simultaneous with O. This in turn means that A understands β to
be in his acausal future while the opposite is true for B. Nevertheless
A and B agree that δ precedes ε along C's worldline, and that δ can
causally influence ε. Now consider a second set of events, μ and ν, such
that ν lies in the elsewhen of μ, and thus vice versa (note that the small
light cone associated with μ makes this clear). Once again, A and B
might disagree about which of these events lies in the acausal future or
the acausal past of A and B. Nevertheless, they once again agree on the
causal relations between μ and ν, namely that they cannot causally influence each other.

7. The Pole-in-the-Barn Paradox

Summary

Finally, spacetime paradoxes arise from the Lorentz transformations and represent variations on the themes of time dilation, length contraction, the relativity of the present (the axis of simultaneity), causal invariance, and so on. As these paradoxes are treated in detail in most of the texts mentioned in the notes above,[24] I present only one here. It nicely illustrates many of these paradoxes and, more important, it will serve us well in the upcoming discussion of the competing interpretations of SR, including my own construction of a new flowing time interpretation. It is called "the pole-in-the-barn paradox,"[25] and it goes like this: A pole 10 meters long is carried by a runner at enormous speeds toward a barn 5 meters wide and with left- and right-hand doors that can open and close. Will the pole make it through the barn undamaged? And how can both observers (one in the barn, the other carrying the pole) agree with the answer to this question if the pole is twice the length of the barn and if the doors' motions are synchronized?

I first present this paradox in the standard way by using two separate spacetime diagrams, one for the pole carrier and the other for the observer in the barn (section a). The resolution of the paradox lies in the playoff between length contraction and the relativity of synchronization (section b). I will then present the paradox in terms of what I am calling a "generalized spacetime diagram" (section c). The striking advantage of a generalized diagram is that it does not privilege one or the other observer as the "rest" observer, as the two separate spacetime diagrams do. Instead, it subsumes all of the relevant physics from the perspective of both observers into one diagram from which we can extract all the details lodged in the two separate diagrams. Such a single generalized diagram is possible precisely because the hyperbola of one-second events and the light cone are spacetime invariants—their structures are independent of any specific coordinate system, and therefore they can be drawn in a single diagram without the need for the usual x-t coordinate axes. And the "hidden treasure" in the generalized diagram (section d) is that we can understand how the worldview of each observer and the "story" each tells about the events that transpire are constructible out of the pieces of the worldview and the "story" of the other observer.

Discussion

a. Presentation of the paradox. A runner carrying a pole dashes at a ve-locity v nearly half the speed of light toward a barn with doors on either end that can open and close. The pole is 10 meters long, the barn is 5 meters wide, and their relative velocity[26] is such that $\gamma = 2$. According to the barn, the doors on the left and right are synchronized to open and close such that the left door closes at precisely the moment that the right door opens. Will the pole fit inside the barn undamaged (fig. 4.1)?

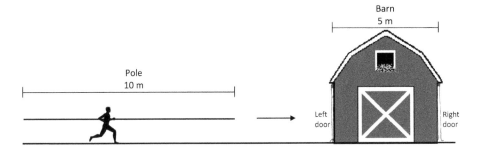

Figure 4.1. A Runner Carrying a 10-Meter Pole Races toward
a Barn 5 Meters Wide

Yes: From the *barn's* perspective, the moving pole has been Lorentz contracted. Since $\gamma = 2$, the pole is only 5 meters long and it will just fit inside the barn at the moment when both doors are closed. The result: *no damage to the pole* (fig. 4.2).

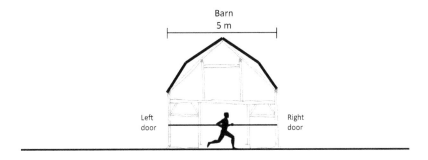

Figure 4.2. The Barn's Version: The Pole Has Been Contracted
to 5 Meters and Just Fits inside the Barn

But here is the paradox: from the *pole's* perspective the pole is at rest and its length is 10 meters, but the barn, racing toward the pole, has been Lorentz contracted. Since $\gamma = 2$, the barn is only 2.5 meters wide. It seems that the pole's right end will be smashed against the barn's right door just before the door opens and that its left end will be crushed by the barn's left door just as it closes (fig. 4.3). The result: *the pole is wrecked!*

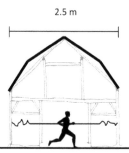

2.5 m

Figure 4.3. The Pole's Expected Version: The Barn, Lorentz
Contracted to 2.5 Meters, Has Smashed the Pole

Both versions cannot be true: either the pole makes it through the barn intact or it is smashed. So which version is correct?

b. Resolution of the paradox. The resolution of the paradox comes from the fact that from the pole's perspective the opening and closing of the barn's doors are not synchronized. Instead, the time between these events is just long enough to let the entire pole go through the barn undamaged (fig. 4.4).

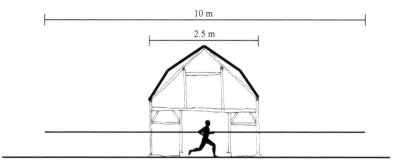

10 m

2.5 m

Figure 4.4. The Pole's Actual Version: Both Barn Doors Are Open

Now the two perspectives agree on what happens physically to the pole—it is not damaged—but they disagree about the timing of the barn doors opening and closing and about whether the pole or the barn is Lorentz contracted. The take-away point is this: what really counts, what physically matters, is that the pole is undamaged according to both the pole's and the barn's worldviews. What fades out of the spotlight is the timing of the events—whether the opening and closing of the barn doors is synchronized—and the distances involved—whether the length of the pole or the length of the barn is contracted. Let us explore this point more carefully by examining moment by moment what happens in the barn's account and then in the pole's account. After this, we will see how to combine these separate and apparently contradictory accounts into a single overall and unified account.

We start with two spacetime diagrams, one from the barn's perspective, the other from the pole's perspective. Figure 4.5 is drawn from the barn's perspective. Here the worldlines of the barn's left end, L, and right end, R, are 5 meters apart. The 10-meter pole, with left end L' and right end R', races to the right. Wavy lines along a barn door's worldline indicate that the door is closed. Figure 4.5 depicts five "snapshots," or worldviews, of what happens from the barn's perspective, each at a distinct moment of time labeled (a) through (e) and representing the barn's x-axis (i.e., its axis of simultaneity, or global present) moving forward in time. In (a) the pole hurries toward the stationary barn whose left door L is open and right door R closed. In (b) the pole is partway inside the barn. A third snapshot (c) shows the pole entirely inside the barn, with both left and right doors closed. In (d) the pole is partially out of the barn whose left door L is closed but whose right door R is open. Finally in (e) the pole is entirely out of the barn. Again the left barn door L is closed and the right door R open.

The key moment is clearly (c). Here the left barn door L closes just as the left end of the pole L' is inside the barn (event A). *At the same moment,* the right barn door R opens as the right end of the pole R' just exits the barn (event B). *Thus, according to the barn, the moving pole is length contracted and fits between the two barn doors, and events A and B are simultaneous.*

Now from the pole's perspective, figure 4.6 shows the barn moving rapidly to the left with the pole at rest. The pole's left and right ends, L'

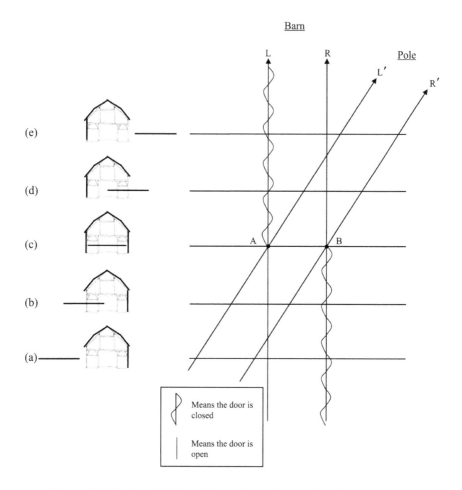

Figure 4.5. The Barn's View: A Spacetime Diagram Showing a Temporal Sequence of Events, (a) through (e), from the Barn's Perspective

and R', are 10 meters apart while the barn is length contracted to a width of 2.5 meters. The barn's right door R opens (event B) long before its left door L closes (event A). Thus, while the pole does not fit inside the barn, it does not matter because the events A and B are not simultaneous. Instead event B happens long before event A. Again five representative snapshots, or worldviews, are depicted but this time from the pole's perspective showing its x'-axis (or global present) moving forward in time. In (a′) the barn, one quarter the width of the pole, races toward the stationary pole from the right. The second snapshot (b′) shows that the barn's left door, L′, is still open, allowing it to contain part of the pole even though its right door, R′, is closed but about to open. In (c′) both barn doors are simultaneously open, so that even though the pole is much longer than the barn it extends out through the two open doors and no damage is done to it. In (d′) the barn's left door, L′, just closes as the pole's left end L is just inside the barn (event A), but the right door, R, remains open and the right end R′ of the pole extends well beyond the open right door, R, of the barn (event C). According to (e′), the barn is completely past the pole with its left door still closed and its right door still open.

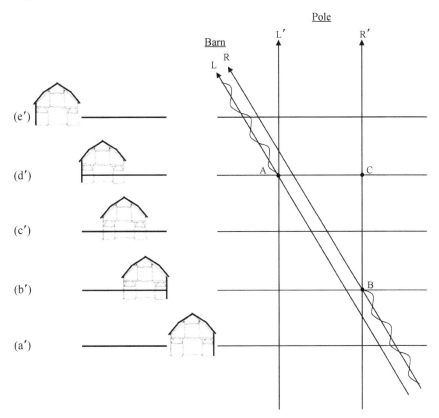

Figure 4.6. The Pole's View: A Spacetime Diagram Showing a Temporal Sequence of Events, (a′) through (e′), from the Pole's Perspective

Now let us compare the scene at the moment of (c') with those of (b') and (d'). Here the pole extends to the left and right beyond the tiny, length-contracted barn, but both of the barn doors are open. The barn's right door has already opened (event B now in the past) while its left door has not yet closed (event A now in the future). Thus, according to the pole, the moving barn is length contracted but the pole survives being damaged because events A and B are *not* simultaneous as they were from the barn's perspective. To reiterate the "take-home" point of above, what physically matters is that the pole is undamaged according to both the pole's and the barn's worldviews. What fades out of the spotlight is the timing of the events—whether the opening and closing of the barn doors is synchronized—and the distances involved—whether the length of the pole or the length of the barn is contracted.

Is there any way to highlight the key feature, the central drama and narrative point, of both these accounts—that *both the pole and the barn agree that the pole is undamaged*—and to let all the "staging" and its props—differing coordinate systems, different timing of the opening and closing of the barn doors, different distances involved—recede entirely from the picture? There is!

c. Combining the two spacetime diagrams into a single, generalized spacetime diagram. What is truly exciting is that all of the physics of the presentation and resolution of the pole-in-the-barn paradox can be represented by a single, generalized four-dimensional spacetime diagram. This diagram incorporates the perspectives of both the barn and the pole and accounts for why they have such differing versions of "what really happened." More important, it delivers on what was asked for above: in it the staging and props have disappeared and all that remains is an austere figure, the interlacing worldlines of the pole and the barn, and the crucial events that define their interaction, namely events A and B.

Figure 4.7a is just such a generalized spacetime diagram for the pole-in-the-barn paradox. Note the almost serene sparseness of the elements: there are no t and x axes for each observer, only the juxtaposition of worldlines as the paths actual observers take in the four-geometry. Events A, B, and C are the events defined above in both figures 4.5 and 4.6:

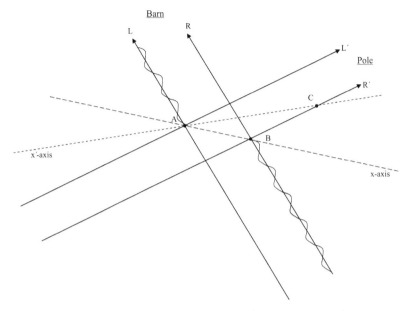

Figure 4.7a. Generalized Spacetime Diagram Incorporating
Previously Distinct Sequences of Worldviews from the Perspectives
of the Barn (Fig. 4.5) and the Pole (Fig. 4.6)

In figure 4.7b I have added the perspectives of both the pole and the
barn to show how they arise from the minimum number of the elements
in figure 4.7a. So, when we take A and B as simultaneous we obtain the
barn's narrative at its moment (c), with the pole neatly fitting within its
doors that are both momentarily closed. When we take events A and C
as simultaneous we obtain the pole's narrative at its moment (d′), with
the pole extending well beyond the open, right door of the barn.

In this single, unified diagram we see displayed the essential physics of both of these points of view. The reason why the accounts of the
barn and the pole differ is due entirely to the fact that they are based on
different slices of time, a different construal of what events in spacetime
are synchronized. Yet both accounts agree with the physics of "what
happened." This agreement emerges smoothly from the unity of the overall scenario of the diagram. What they both agree with is that the pole is
undamaged by its encounter with the barn. The rest is, if not superfluous,
of much lesser importance and its intrinsic differences of little concern.

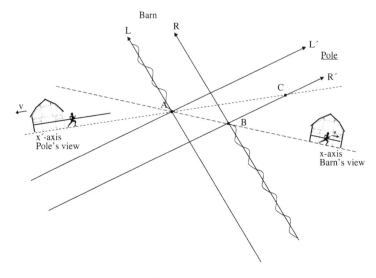

Figure 4.7b. Generalized Spacetime Diagram Incorporating
Previously Distinct Sequences of Worldviews from the Perspectives
of the Barn (Fig. 4.5) and the Pole (Fig. 4.6)

little concern.

d. The hidden treasure: the present of each as constructed from the history of the other. Once we have obtained this level of simplification regarding the essential physics of the paradox we can make an elegant and surprising move: we can construct the barn's view at a specific moment in time by "sewing together" pieces of the consecutive presents from the pole's perspective, and vice versa.[27]

Let us first chose that crucial moment in the barn's version where the pole just fits entirely within the barn and both doors are momentarily closed, indicated by worldview (c) in figure 4.5 above. To do so we return to figure 4.6 and add the barn's axis of simultaneity, or x-axis, at that precise moment, as shown in figure 4.8a.

We then extract tiny portions of the pole's worldviews from its axes of simultaneity that coincide with the x-axis of the barn for worldview (c). Again, in principle this requires an infinity of infinitesimal extractions (extractions of infinitesimal width along the x-axis) each drawn from the pole's worldview for the event determined by where the x-axis and x'-axis cross each other. Here we extract these pieces from (b'), (c'),

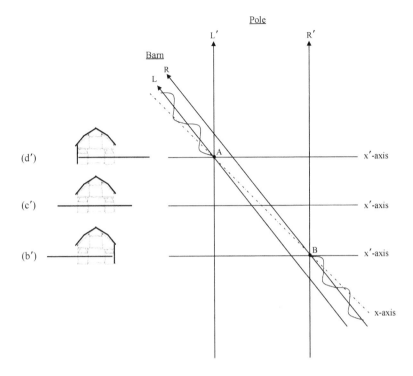

Figure 4.8a. Constructing the Barn's Worldview at Moment (c) in Figure 4.5 by First Adding the Barn's Axis of Simultaneity (Dashed x-Axis) to the Pole's View in Figure 4.6

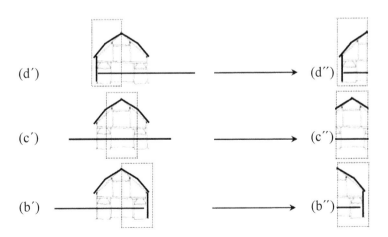

Figure 4.8b. Tiny Extractions Are Then Taken of the Temporal Sequence (b′), (c′), and (d′) of the Pole's Views of the World for the Events along the Barn's x-Axis

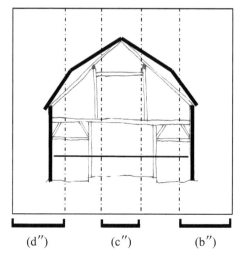

Barn's view
(c) = (d″) + (c″) + (b″)
showing the pole
fully inside the barn

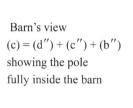

(d″)   (c″)   (b″)

Figure 4.8c. The Barn's View of the World at the Time of (c):
Composite of Extractions (b″), (c″), and (d″)

and (d′) to illustrate the process. The extractions result in (b″), (c″), and (d″) as shown in figure 4.8b.

Finally, we sew together the extractions represented here by (b″), (c″), and (d″) to form the barn's view of the world at the time of (c). This view is now seen as a composite of tiny portions of the pole's temporally successive worldviews b′, c′, and d′ as in figure 4.8c.

The resolution of the paradox, represented in terms of the generalized spacetime diagram (fig. 4.7a), thus opens onto a quite unexpected and profound insight: the present, from one observer's perspective, is generated by "sewing together" pieces of the consecutive presents from the other observer's perspective.[28] One might even say that this is a beautiful example of the general relation of "the many and the one": what is for me one momentary global view of the world is a seamless integration of infinitely many infinitesimally tiny local pieces of your temporally successive global momentary views of the world, bound together as a unitive worldview. Conversely, each of your momentary worldviews, taken globally as a whole, is similarly constructed from an endless sampling of tiny pieces of my sequential global views of the world.

Two final notes about the philosophical meaning of the paradox: First, contrary to the way SR is sometimes mistakenly used as a justification for a "relativistic" and even nihilistic view of truth, SR does not allow for an indefinite number of distinct perspectives on reality. For example, there is no way to "sew together" pieces of the consecutive views of the present from either the barn's or the pole's perspective such that the barn doors remain closed or remain open or break the pole or the pole goes around the barn instead of through it, etc. In essence, SR incorporates two distinct, factual accounts of reality (that of the pole and of the barn) into a single spacetime perspective, while ruling out of bounds an endless, perhaps infinite, series of possible misaccounts of reality. I summarize this important philosophical argument with the catchphrase "relativity is not relativism."

Second and closely related, we can interpret the resolution of the paradox as pointing to what is often called in metaphysics the problematic of "the many and the one." This relation between the relativistic paradox and the underlying metaphysics is based on the physics shared between both observers, such as the fact that the pole is not destroyed by the barn doors as mentioned above. The "many and the one" relation does not hold for all the other possible worldviews that are inconsistent with the physics, such as the barn doors remaining closed, etc. In this satisfying sense, the unique relation between the two worldviews, that of the pole and the barn, which are available to the "many and the one" description suggested above, points to the true, underlying physics, while the many relations between the pole's view and that of the barn, such as the barn doors remaining closed, are not available to a "many and one" relationship. Relativistic physics, while expanding the notion of "truth" to include a certain latitude in our knowledge and interpretation of the world (i.e., the legitimacy of the views of the pole and the barn), does not expand this view inordinately. Once again, "relativity is not relativism."

8. Spacetime as a Four-Dimensional Geometry

Summary

We have come a long way in our introductory exploration of Einstein's SR. Only three topics remain. The first two topics, the transition from space and time to spacetime and the recognition of the traditional global present as an anthropocentric illusion, will be treated here. The third, the challenge of offering a flowing time interpretation of spacetime—an interpretation which might well seem like an oxymoron given the final arguments for spacetime as a geometry and for the illusory character of the global present that follow here—will be deferred to chapter 5. This is primarily because this philosophical interpretation will draw in part on new insights which we will obtain below from the reconstruction of Pannenberg's theology. But more on that later. With thanks again for the reader's forbearance, I return to the last two topics of this chapter. Here my introduction to SR would not be complete without a clear presentation of the reasons why the spacetime interpretation is now almost universally accepted.

Discussion

Einstein's original work on SR was done, of course, in the worldview of classical physics—at least insofar as he assumed a separate coordinate system (or frame of reference) for each inertial observer in uniform relative motion. This is the traditional picture of a three-dimensional space as separate from a one-dimensional time. Nevertheless, in the two years following Einstein's publication of SR, the mathematician Hermann Minkowski (1864–1909) produced a series of papers and lectures that pointed in the direction most physicists now take: that space and time are not separate worlds but form a single, integrated four-dimensional geometry called "spacetime." As Minkowski declared in a now famous statement:

> The views of space and time which I wish to lay before you have sprung from the soil of experimental physics, and therein lies their strength. They are radical. Henceforth space by itself, and time by itself, are doomed to fade away into mere shadows, and only a kind of union of the two will preserve an independent reality.[29]

Minkowski's suggestion that we think of spacetime as a geometry can be understood by reflecting on the significance of time dilation and the way the Lorentz transformations preserve the hyperbola of one-second events for two observers, O and O', in relative motion (see diag. 4.8

above). As we saw before, the individual coordinates of events A . . . F in the (x′, t′) frame of O′ differ from the coordinates for these same events in the (x, t) frame of O. Nevertheless, when we shift mathematically between the (x′, t′) coordinates of observer O′ and the (x, t) coordinates of observer O by using the Lorentz transformations (equation 2), the events recorded by both observers lie along the same hyperbola of one-second events. To be specific, the hyperbolas are "the same" (or more precisely, invariant) because both are characterized by the same value of the proper time, τ:

$$\tau^2 = t^2 - x^2/c^2 = t'^2 - x'^2/c^2$$

In short, the Lorentz transformations ensure that the hyperbola of one-second events is invariant.

Now in a crucial step we shift our focus from the invariance of the hyperbola as such and focus instead on a specific path from the origin to a single event along the hyperbola. Let us chose event C for a clock at rest for observer O as shown in (a) of diagram 4.14. Here the path from the origin to event C lies entirely on the time axis of observer O and the event has coordinates (τ, 0) where τ is the proper time. For observer O′, however, moving to the right with a velocity v according to observer O, event C lies along a different path than for observer O. Event C now occurs at a spatial distance x′ from the origin and at a time t′ so that its coordinates are (t′, x′). Nevertheless when we compare these very different paths from the origin to C with the fact that the value of τ^2 for the first path is identical to $t'^2 - x'^2/c^2$ for the second path, the invariance previously understood as pertaining to the hyperbola of one-second events can be seen in a new light.

The invariance now suggests that the two different paths described by O and O′ are best understood as a single path in spacetime, and the differences between them arise entirely from the differences in the perspectives of O and O′. Moreover these differences in perspective are due entirely to their relative motion. In essence, the path should best be understood as lying in a single, unified "spacetime" as described by Minkowski, and not in the two different "space plus time" frameworks that we have been using for the observers O and O′. Moving explicitly into the standard interpretation of SR (although this will be the very concept whose *additional* "block universe" interpretation I will challenge later[30]), we should say that the path from the origin to event C represents a "distance" in

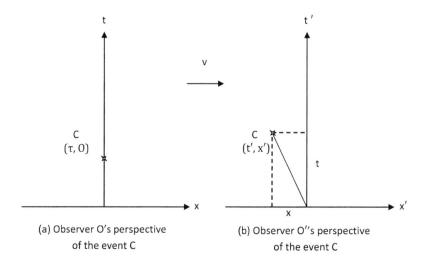

(a) Observer O's perspective
of the event C

(b) Observer O''s perspective
of the event C

Diagram 4.14. The Same Event C from the Perspective of Observers O
and O' in Uniform Relative Motion with Velocity v

the four-dimensional geometry of spacetime, a distance whose "size," τ, is invariant even though the coordinates that describe this path vary according to observers O and O' in relative motion.

To complete the geometrical interpretation of spacetime we now interpret equation 1 as a "metric," that is, the rule for measuring a distance in a given geometry. Equation 1 is analogous to the Pythagorean metric. If a point P with coordinates (x, y, z) lies at a distance r from the origin in ordinary three-dimensional Euclidean space, the size of the distance r can be calculated using the Pythagorean metric, namely $r^2 = x^2 + y^2 + z^2$. When we include y and z, the spacetime metric becomes

$$\tau^2 = t^2 - (x^2 + y^2 + z^2)/c^2 \tag{3}$$

Note that the spacetime metric has a minus sign where only the positive sign occurs in the Euclidean case. In a very important sense, this difference—the minus sign—is the most crucial feature of Einstein's theory. It underlies all the phenomena we have discussed, starting with time dilation, and it leads to the paradoxes of SR.[31] Since the invariance of the metric is guaranteed by the Lorentz transformations,[32] it is often called the Lorentz metric. Finally, since the term "distance" would make it a foregone conclusion that the path in spacetime is strictly spatial, and thus undermine the option to reinterpret it, an option we shall develop in chapter 5, I use a different but equally standard term: I will call it a "spacetime interval."

9. The Illusion of the Global Present in the Loss of the Ambiguous Elsewhen

Summary

Finally, we are prepared to understand that the idea of a global present, as we seem to experience in everyday life and as it is represented formally in classical physics, is actually an anthropocentric illusion. It arises from the enormous speed of light compared with ordinary speeds in nature, and it represents the conflation of three features of spacetime at each event O: the light cone structure for event O, the elsewhen region within the light cone structure, and the infinity of axes of simultaneity correlated with observers in relative motion that move through the event O.

Discussion

Consider the elsewhen region for the event O in diagram 4.15. If we imagine the speed of light, c, to increase indefinitely, the photon worldlines will rotate away from the t-axis toward the x-axis, flattening the elsewhen region in the process (diag. 4.15, a, b, and c, consecutively). If the speed of light, c, were infinite, the photon worldlines would be congruent with the x-axis and with each other, and the elsewhen would have vanished (diag. 4.15, d). The causal future of event O would have expanded to fill the entire upper half of the diagram, and the causal past would have expanded to fill the entire lower half of the diagram. We call the congruence of photon worldlines and the x-axis "the present."

Now consider the same process of imagining the speed of light to increase to infinity in the case of *two* events O and O', each with their associated light cones (diag. 4.16, a, b, and c). If we have chosen these events correctly, the present for event O is congruent with the present for event O', and they share the same causal future and causal past. Imagine this process repeated indefinitely for all correctly chosen events in spacetime

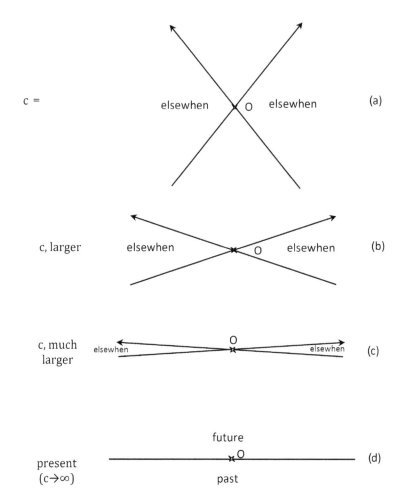

Diagram 4.15. Congruence of Photon Worldlines and the x-Axis

Note: The elsewhen region for event O shrinks and vanishes as the speed of light goes to infinity and the light cone flattens to the x-axis.

along the global present of O and O'. The result is the global present that we take for granted in everyday life and in classical physics (diag. 4.16, d). The elsewhen region for events O and O' shrinks and vanishes as the speed of light goes to infinity, the respective light cones flatten into the x-axis and the x'-axis, and the congruence of the x-axis and x'-axis forms the shared present for O and O'. Clearly the so-called global present of our everyday experience is an anthropocentric and mythological[33] illu-

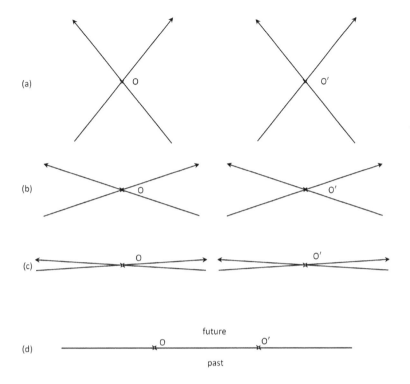

Diagram 4.16. Congruence of Photon Worldlines and the x-Axis
with Two Events

Note: The same process as in diagram 4.15 but with two events O, O'.

sion that arises from the relatively slow velocities of ordinary experience compared with c. It is what we take to be the actual world of everyday life, and while this is fine from a practical point of view it is not in fact the way the physical world actually is.

When the two suppressed spatial dimensions (y and z) are added into the picture, the global present moves from its appearance as a line in the previous diagrams into a volume in the four-dimensional space-time. This infinite volume is our concept of "everything in the world right now." From the preceding description of how that volume is formed, it should be relatively easy to understand why physicists prefer to call it a three-dimensional "hyper-surface" in four-dimensional spacetime.

This concludes the brief overview of Einstein's special theory of relativity.

B. REFLECTIONS ON PANNENBERG'S COMMENTS
ON SPECIAL AND GENERAL RELATIVITY

As we saw in chapter 1, Pannenberg begins his discussion of time and
space in special relativity with a basic and unarguable point: while *absolute* simultaneity is challenged by special relativity, it does not eliminate simultaneity altogether. Instead, it defines simultaneity "relative to
the standpoint of the observer," and in doing so relativizes spatial measurements as well.[34] So far, well and good! Next, Pannenberg claims
that time is more basic than space. He first gives a strictly formal reason
for this: time is "constitutive" to the idea of space because the concept of
simultaneity, which is fundamentally a temporal idea, allows us to identify things in space as different and not just as successive. Again, well and
good! Pannenberg also notes that this constitutive relation lends "philosophical plausibility" to special relativity's idea of spacetime "as a multidimensional continuum" (i.e., a spacetime geometry).[35]

Pannenberg then argues for "the reduction of space to time."[36]
This idea seems to be based on two claims. The first is grounded in the
basic distinction Pannenberg makes between eternity and omnipresence: put simply, eternity is the presence of things in the world to God
and omnipresence is God's presence to things in the world. The second
is the claim that God's eternity in relation to the world takes precedence
over God's omnipresence to the world. In fact, for Pannenberg, God's
omnipresence to the world is *based on* God's eternity. I will agree with
and adopt both of these claims in the constructive project below. But I
am hesitant to agree with Pannenberg when he argues that the concept
of the presence of God in space presupposes the "reduction of space to
time." I am reluctant to go this far. It is true, as we shall see in chapter 5,
that some scholars argue for a "temporalizing of space" in an effort to
avoid the "spatializing of time" (i.e., the "block universe" interpretation of spacetime). Instead, my hope in chapter 5 will be to show that by
constructing a "flowing time" interpretation of spacetime as opposed
to the block universe interpretation, I will give fresh support to the idea
of the "temporalization of space" without going as far as to reduce space
to time.[37]

In his discussion of general relativity (GR), Pannenberg discusses
the relation between scientific concepts of space and theological con-

cepts of omnipresence. He notes that Einstein followed Leibniz in adopting a geometrical/relational view of space in contrast to Newton's receptacle view. While Pannenberg supports this move he recognizes that it does not lead to the unity of space that he seeks theologically, because a geometrical space is still a divisible space. Pannenberg turns instead to God's omnipresence to the world as the basis for the indivisible unity of creaturely space.

Pannenberg also argues that while the space of creation might be unlimited and thus potentially infinite, it is not actually infinite. Only God's omnipresence to all creation is infinite. Here I find Pannenberg deferring to the traditional understanding of infinity as the *aperion,* which I discussed in chapter 3. This understanding of infinity plays a useful role in the doctrine of creation *ex nihilo* for distinguishing sharply between God as infinite and creation as finite.[38] But my suggestion in chapter 3 was that modern mathematics, particularly due to the work of Georg Cantor, offers a *tertium quid* in the form of the infinite hierarchy of the transfinites: those infinite sets that share some properties with finite sets but that differ radically from them in other properties.[39] This means that the universe might indeed be actually, and not just potentially, infinite in the sense of Cantor's transfinites, and yet we can still radically distinguish between the universe as created—and thus at most *trans*finite—and God as Absolute Infinity, to use Cantor's concept rooted in both his mathematics and his theology. So, while I value Pannenberg's understanding of the pivotal role of a relational, versus a container, concept of space in GR, I would offer Cantor's threefold classification as more helpful for the God/world distinction than Pannenberg's traditional twofold classification.

C. A COVARIANT CORRELATION OF ETERNITY AND OMNIPRESENCE IN LIGHT OF SPECIAL RELATIVITY

The primary purpose of this section is to offer a way to correlate the two divine attributes that were the focus of my analysis of Pannenberg's theology in chapters 1 and 2—eternity and omnipresence—in light of the spacetime interpretation of SR. The correlation should reflect the fact that SR, as I presented it above, is almost universally interpreted geometrically in terms of "spacetime": a four-dimensional manifold with a Lorentz metric. Such a view contrasts and replaces the "three-dimensional space plus one-dimensional time" view of nature in ordinary experience and classical physics. This change in interpretations is suggested by the symbol "3 + 1 → 4."

But the "3 + 1" dimensional framework pervades systematic theology, reflected keenly in the separate treatment of eternity and omnipresence that is standard practice and that Pannenberg, too, follows. Somehow we must find a plausible way to reformulate theology in light of the "3 + 1 → 4" dimensional shift to spacetime. I refer to this goal as a "covariant correlation of eternity and omnipresence." That is, I suggest that we can correlate the divine attributes of eternity and omnipresence in a *covariant* way by first identifying particular events in time for a given observer with particular events in space for that observer in an *invariant* way. These spacelike events then constitute the global present for that observer.

With this, the eternal duration that holds together this particular set of timelike events along the observer's worldline will be the basis for God's omnipresence to the particular spacelike events constituting the observer's global axis of simultaneity, or global present. It is these events to which God is simultaneously present for that observer. But this correlation is Lorentz invariant: the relation between a given observer's worldline and her construction of a global present is preserved by the Lorentz transformations between her and all other observers in uniform relative motion to her. In essence they will all agree on which events, for a given observer, constitute her global present at a particular moment along her worldline. We may therefore conclude theologically that God's eternity *for each observer* will be uniquely and invariantly correlated with God's omnipresence *to that observer in her global present as it moves forward in time.*

This covariant correlation of eternity and omnipresence reflects the spacetime interpretation of SR without falling into the "block universe" interpretation of spacetime, as I argue in chapter 5. The immediate implication of this result is that God's relation to the world is not to be understood in the traditional way, namely a general and single global presence to the world at a universal moment in time. Instead, God's relation to the world, including personal existence, historical experience, and natural processes, is always given to the particular global present, associated with each particular observer and defined by her particular worldline through the present event P. It is to *this world* and for *this observer* that God is omnipresent—and this is true for *all observers*. Moreover, God's particular omnipresence to each observer reflects, and is based on, God's gift of particular eternal duration for the specific events along that observer's worldline. The key point is that *we are led to this expanded theological understanding of eternity and omnipresence by SR!*

Three things follow from this. First, I return to the way we extracted the observer-invariant physics from the apparent contradictions of the "pole/barn" paradox. It was through a comparison of the multiple views of the relations between events in space and time, as suggested by the pole/barn paradox, that we can glimpse what are the invariant underlying physical processes. In particular we constructed the barn's point of view at a unique moment in (its) time from an endless succession of infinitesimal pieces of the pole's point of view in flowing time sewed together in just the right order. And conversely we constructed the pole's point of view from that of the multiple global presents of the barn, suitably chosen.

The upshot, theologically, is that compared to our ordinary experience of the world, the unimaginably more complex interweaving of these multiple histories and worlds, as revealed in the simple pole/barn paradox, greatly enlarges our understanding of the divine eternity and omnipresence. Perhaps we should put this in the plural. SR leads us to discern the divine eternities and omnipresences in endless interwoven complexity as they move in, through, and under creation, interrelating different "lengths" and "times" at the physical level in nature in ways that reveal what is much more important than them: the reality of what is true for us all in nature, personal experience and history, and independently of its multiple interpretations.

Second, in section 3 below I will invite the reader to explore the theological implications of the discovery of the multiple elsewhen regions associated with different events in spacetime and the light cones for these events that form their boundaries. We will note in particular the implications generated by the collapse of the light cones in the limit that the speed of light, c, goes to infinity. In this limit the light cones conflate into a single global present characteristic of the worldview of classical physics and ordinary experience: I will stress the fact that the stark simplicity of the classical global present masks and obscures the endless beauty and astonishing complexity of the region of the elsewhen with its countless relativistic axes of simultaneity. But because of SR we are able to glimpse something of the utter complexity of God's particular omnipresences to the vast diversity of observers and their differing views of the world and God's particular eternities to their personal histories and flowing time experiences of the world.

Third, in section 4 below, I claim that it is God's omnipresence to events in space that gives to these distinct and separate events a differentiated unity, one that preserves their spatial distinctions but overcomes their spatial separations. We will see how this relates to Pannenberg's interpretation of the philosophical debate between Newton and Leibniz about a container versus a relational concept of space and Clarke's appeal to a space prior to all divisions. I will suggest that theology thus has something of value to offer to this debate by following Pannenberg in pointing to the differentiated unity of the Trinitarian nature of God.

1. The Basic Argument

In its simplest formulation the process for correlating eternity and omnipresence proceeds as follows:

a) We start with a generalized spacetime diagram depicting an observer O and her worldline A (fig. 4.9). A set of timelike events {a, b, c . . .} lie along her worldline and mark off her proper time τ. Worldline A is then the t-axis for observer O.

b) Next we chose one of these events, say event a, to be the present moment. SR then defines the unique global present, or axis of simultaneity, for observer O at event a. We can construct this axis, that is, the "x-axis," in any of the usual ways (e.g., using photons for clock synchronization; using equal angles between worldlines A and the x-axis in relation to the null lines; etc.).

Note the Lorentz invariance of the global present for the observer O at event a. All observers in relative motion to O will agree on the axis of simultaneity for observer O at event a (fig. 4.9):

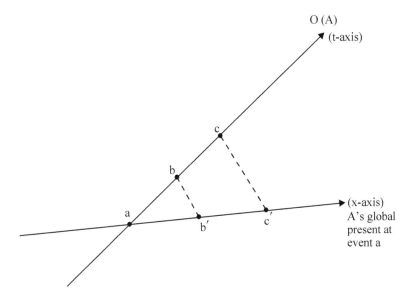

Figure 4.9. Photon Worldlines Correlate Temporal Events b, c
with Spacelike Events b′, c′ along the Global Present of Observer A
at event a. Eternity in Relation to {a, b, c} Covariantly Correlates
with Omnipresence in Relation to {a, b′, c′}

c) We then identify the set of spacelike events that lie along A's
global present at event a and for which photons emitted from these spacelike events toward the worldline A arrive at timelike events b, c. We will
designate the spacelike events with primes: spacelike event b′ correlates
to timelike event b and spacelike event c′ correlates to timelike event c.

d) Following Pannenberg we can think of the eternal temporal duration T_A of observer A as including timelike events a, b, c. This means that
the timelike events a, b, and c are co-present in the divine eternity, held together by God as distinct but not separate. We can then define S_A as the
"spatial co-presence" between the spacelike events a, b′, c′, and this spatial co-presence is the result of God's omnipresence to them. Finally we
can relate T_A and S_A as the specific events for observer O at event a that
are the events related to God's eternity and omnipresence for that observer at that event a. In essence, the divine eternity in relation to observer
O at event a is the basis of God's omnipresence to observer O at event a,

and conversely the divine omnipresence to observer O at event a depends on God's eternity in relation to observer O at event a. This means *we have correlated eternity and omnipresence with respect to the specific temporal event a of observer O in a way that is Lorentz invariant.*

Note the details of the Lorentz invariance of this correlation: according to SR, the causal relations between timelike events {a, b, c} is Lorentz invariant. All other observers B, C, etc., in relative motion to A will agree with their causal ordering. The identification of the events a, b′, c′, as forming the global present to event a for observer A is also a Lorentz invariant. All observers will agree that these events form the global present for observer A at event "a." Finally, the photon worldlines between events b′ and b, c′ and c, are also Lorentz invariant. All other observers B, C, etc., will agree that these photon worldlines uniquely connect events b′ and b, c′ and c. Thus all observers will agree with the specific correlation of divine eternity for observer A with divine omnipresence for observer A; this correlation is covariant. In short, the Lorentz invariance between observer O′s t-axis and x-axis is the structure provided by SR for a covariant spacetime correlation of God's eternity and omnipresence *for observer O at event a.*

e) Finally, we obtain a flowing time view of the eternity-omnipresence correlation (for defense of a flowing time interpretation of special relativity, see chapter 5). As the observer moves forward in time, a continuous succession of spacelike events are picked out as forming the temporal sequence of global presents for her. These successive global presents are also Lorentz invariant, in that all observers will agree that, for observer O, they constitute the changing set of global present events. Thus they would agree, in turn, that they constitute the particular events gifted successively in time with divine omnipresence relative to observer O.

2. The Covariant Sewing Together of Eternity and Omnipresence in Light of the Pole-in-the-Barn Paradox

Additional theological insights for a covariant correlation of eternity and omnipresence for different observers in relative motion can be gained by reconsidering the resolution of the pole-in-the-barn paradox within the theological discussion of the divine attributes. The key to the resolution was the recognition that we can reconstruct the barn's version of the world, that is, the barn's global present at a specific moment in time, by taking tiny segments of the pole's view from each of the consecutive moments in the pole's story of what happened. In particular, each segment was taken from the event where the barn's x-axis intercepted the pole's x′-axis as the latter moved forward in time (fig. 4.8a above).

To explore this more carefully we start with a single view of the world at a single moment in time *from the barn's perspective.* Let us chose the moment when the pole is entirely within the barn and both of the barn's doors are shut (fig. 4.5 above, time indicated by (c), with events denoting the left and right ends of the pole just inside the barn, denoted A and B). This moment in the barn's time is reconstructed by sewing together an endless set of vanishingly small views of the world (i.e., views for each of which $\Delta x \to 0$) at those specific moments in time defined by the Lorentz transformations and lying along the pole's axis of simultaneity as it moved forward in time. Recall that these moments were chosen as the barn's x-axis (defined when the pole is entirely inside the barn) coincides with specific events along the pole's x′-axis as it moves forward in time (fig. 4.8a). Each tiny view of the world is a present event of extremely limited spatial extension (fig. 4.8b). These tiny views are then sewn together to form the barn's view of the pole entirely inside it (fig. 4.8c, which corresponds to fig. 4.5, time indicated by c). In this way the entire spatially extended view of the world from the barn's perspective at a specific moment of time (i.e., its global present) is composed of "present moments" fully real to themselves even though they occur at different moments in time *from the pole's perspective.* We can follow the same covariant process and reconstruct a given moment in the pole's view of the world from a sequence of tiny views of the world from the barn's story generated successively in time. Let us once again start with event A, when the pole's left end just touches the left door of the barn (fig. 4.6, time indicated by d′). According to the pole's view of the world its right end, event C, sticks well beyond the right, but open, barn door. I leave it to the reader to construct a figure for the pole's view that shows how to carry out the process as we did for the barn in figure 4.8.

In short, SR tells us that the *barn's global view* of the world at a given moment in time is a composite of an endless set of infinitesimally narrow views of the world taken one by one from temporally successive global views of the world from *the pole's perspective* and sewn together

following the precise instructions contained in the Lorentz transformations. Again, the converse is true for the *pole's global view* at a given
moment in time. Finally *both* views of the world, the barn's and the
pole's, can be constructed out of a *single* generalized spacetime diagram
(fig. 4.7a), where the key events A, B, and C (which I will call the ABC
triangle) are indicated and where both views are inserted for ease of reference (fig. 4.7b). And again, while the views differ remarkably about spatial and temporal issues (i.e., Does the pole fit entirely within the barn?
Are its doors simultaneously closed? etc.), they agree on the underlying
physics represented by the ABC triangle: an unbroken pole that travels
through the barn.

We are now ready to import this lesson from SR into the theological
discussion of the divine attributes. Frankly, the discussion of eternity is,
in fact, relatively unaffected by the shift to a spacetime perspective, but
the meaning of God's omnipresence to all events in light of SR clearly
requires our attention. Again we start with Pannenberg's insights: eternity
is the presence of things in the world *to God* and omnipresence is God's
presence *to things* in the world. In this way our experiences in time of the
world are bound together in what I call God's eternal co-presence, and
each of our views of a flowing-time global presence in time is the locus
of God's omnipresence to the world.

To put this more accurately we can delineate two claims: First, all
events along the worldline of my life have their own distinct events labeled past and future (i.e., the ppf structure of each present event), where
the terms "past" and "future" are understood as temporal relations to
each present event. Moreover, each present event has a unique, Lorentz-
invariant global present, and this global present moves forward in time
from the past into the future. Now all the events along my worldline are
taken up and participate in the divine eternity, where eternity is characterized as a differentiated temporal unity or duration as co-presence.
It is this participation in God's eternity that gives unity to the otherwise
seemingly separated events of my life. Second, for each moment along
my worldline God is present to all the particular events distributed in
the elsewhen of that moment, that is, my global present. This in turn
holds for each present moment along my worldline with its own global
present. Each such global present is my "view of the world" at a particular moment in life, and God's presence to all of these particular

moments of my life is God's global omnipresence for me as it evolves in time.

In this way I believe we have been able to restructure our concept of the *form* of God's presence, entailed by the divine *omni*presence, to the world now understood in a Lorentz invariant, and not a classical, way. We sew together tiny, sequential pieces of the time-evolving global present to which God is omnipresent from the perspective of the pole observer, and thus compose a single global present moment in time from the perspective of the barn's observer. *That* world according to the barn, with the pole just nested within the barn's doors before it flashes out of the left side as the left door opens (fig. 4.5, moment c), is the world to which God is *omni*present for the barn at the moment c. *But it is not the only such world.* God is also *omnipresent to the pole's world* just as the right barn door is about to open and the pole's right end is just about to exit that open door (fig. 4.6, moment b'). Moreover God is present to the infinity of worlds for observers moving at various speeds through event O according to either the observer of the pole's or the barn's viewpoints.

In addition, we can capture both the pole's and the barn's worldviews in a single, generalized spacetime diagram (fig. 4.7), as we have seen already in detail. The reason for recapping this here is that the diagram leads to the underlying theological insight about God's multiple omnipresences to each of the observers in uniform relative motion in a Lorentz invariant way (see triangle ABC in fig. 4.7 above). We know that, while the measurements of such things as simultaneity and length contraction differ radically between these observers, they will entirely agree on what counts physically: for example, the pole passes unbroken through the barn. I now take this to be what is theologically relevant and view as of lesser theological importance the individual descriptions of the simultaneous events of the world by the differing observers—those moving with the pole or with the barn. And again to avoid any charge of "relativism" I stress that there are *many* "accounts" of what happened that are rejected by the Lorentz transformations of SR (i.e., the pole was broken by the barn doors, the pole missed the barn, etc.).

In sum, we can relate each instance of God's omnipresence to each event in God's eternity by the Lorentz-invariant correlation with those events along both the barn's and along the pole's worldlines. Eternity for the barn is duration across those successive events in time that correlate with God's successive global omnipresences for the barn. But each of these successive events in time for the barn correlates with a tapestry of individual events sequentially located along the pole's global present as it moves forward in time. Omnipresence at a moment for the barn is constituted by sequential partial divine omnipresences for the pole, and vice versa. In this way, consistent with importing SR into theology when properly interpreted philosophically, the divine eternity and the divine omnipresence are intimately correlated in light of the relativistic correlation of time and space.

3. The Elsewhen of Relativity and the Interwoven Tapestry of God's Eternity and Omnipresence in Creation

I want to close this chapter with some further insights on relativity and the divine attributes related to the intrinsic complexity of the region of spacetime called the elsewhen, and then, in section 4 below, with a suggestion about the unity of space based on God's omnipresence. Recall our previous discussion of the elsewhen in section A.9 above. There we considered an event O and its light cone (diag. 4.15). Again the elsewhen of O is that spacetime region bounded by the upper and lower portions of the light cone. It includes all the axes of simultaneity (i.e., x-axes or global presents) for all observers passing through event O.[40] Now we imagine that the speed of light, c, increases to infinity. The result is that the light cone flattens (diag. 4.15b and c) until it collapses into a single line (diag. 4.15d). At the same time the region of the elsewhen, and the infinity of x-axes within it, decreases in size until it vanishes. The crucial point is that all the axes of simultaneity in the elsewhen for event O are gradually conflated within the collapsing light cone until all that remains of the light cone, the elsewhen, and all the axes of simultaneity in it, is a single axis of simultaneity, which we call the classical global present.

Previously we extended the argument for two events O and O′ with their respective light cones and elsewhens (diag. 4.16). Here, for properly chosen events O and O′, the result is that the x-axes in the elsewhen of O and x′-axes in the elsewhen of O′ become congruent with the elsewhen regions of O and O′ and the light cones that form their respective boundaries. Altogether they form the single, shared, global present for O and O′.

The uptake of the preceding is this: the global present of classical physics and ordinary experience does not actually exist in nature, although it is a reasonable approximation for speeds much less than that of light. In its place, in principle, is the complex phenomena of light cones, elsewhens, and multiple axes of simultaneity conflated to appear as a unique global present because of the enormous size of the speed of light compared with ordinary velocities. The assumption of a shared global present that universally demarcates the single, global future from the single, global past, which we take to be self-evident and given in everyday life, is actually a mirage that we need to abandon like the edge of the world, with its waterfall, and the ancient geocentric cosmology.

I can make this concrete by returning to the pole/barn paradox (fig. 4.7). As we imagine the speed of light increasing to infinity, the x-axes of the pole and the barn move toward each other and finally overlap perfectly, forming the unique global present of ordinary experience. In this limit, events B and C, which represent two alternative versions of what is present to event A, converge into the same event. Both are now simultaneous with A and lie in the unique global present in relation to event A. Here the length of the pole and the length of the barn are identical. Moreover, the timing of the barn's doors in closing and opening is the same from the viewpoint of the pole and of the barn. All the paradoxes of SR are gone, and the underlying physics—the pole passing through the barn undamaged, etc.—would match the specific narratives of the observers moving with the pole and with the barn—stories about the sequence of the closing of the barn doors or whether the pole fits within the barn. In short, classical physics is *unable to distinguish* between what really counts about the world—such as the pole being undamaged in passing through the barn—and what its narratives will render about the timing of the barn's doors or whether the pole fits inside the barn. Relativistic physics, on the other hand, *is able to make this distinction* about what really counts about the world.[41]

Relativity, in turn, can inspire theology to celebrate an infinite complexity and beauty of the relation between eternity and omnipresence, one simply not available within the classical and everyday-experiential view of the world. Assuming the interpretative lens of ordinary experience and classical physics, theology took as given the existence of a single global present each moment of which offered a "snapshot" of the world in three dimensions as separate from time and as changing in time. In such a world it was natural to treat the divine eternity as separate and distinct conceptually from the divine omnipresence. We now see the world differently, thanks to the interpretative lens of relativity.

From this much more complex perspective, we can take up Pannenberg's point about events as present to the divine eternity and God's omnipresence as present to events while modifying it in light of SR. The result, I suggest, is the following: the events along *each* worldline are eternally present to God just as God is endlessly present, that is, omnipresent, to the events of *this particular* worldline's global present as it moves forward in time. But as we switch to other worldlines, the events eternally present to God and God's omnipresence to them shift—and they shift in accordance with the Lorentz transformations. Nevertheless what is preserved in the shift is the Lorentz invariance of the relation between each worldline's events as eternally present to God and the particular set of events to which God is omnipresent for that observer. Another way to say this is that while in principle we see the world from our own perspective, one described alternatively by observer O *or* O', and while these views will always include paradoxes, relativity gives us the possibility of discovering a unitive view of the world (i.e., spacetime) that goes beyond paradoxes and expresses the underlying and unambiguous physical processes. And it is in relation to these processes that God as Creator acts, even while God's action might be clothed in the paradoxical descriptions given by observers in relative motion.

This in turn leads to a portrayal of the eternity and omnipresence of God as endlessly interwoven within the interchangeable stories of this world, all framed in a way that covariantly preserves the true, Lorentz-invariant aspects of the natural world that underlies all our experiences, including those of God's eternity and omnipresence. Through the gift of relativity we now understand that God is multiply related to the world in ways we would not previously have imagined: every event of every worldline is taken into the co-presence of the divine eternity, and God is omnipresent to all these events in their legitimate but relativized simultaneity. Moreover, every worldview of events in spacetime is itself a tapestry woven together from an endless source of infinitesimal worldview slices woven following the Lorentz transformations of SR, and this, in turn, is true in converse.

4. The Divine Gift of a Differentiated Unity to
Otherwise Separated Spacelike Events

Finally I want to revisit the ideas explored in section C.1 above, but with a different goal in mind. There I suggested that Pannenberg's understanding of duration, as it relates time in creation to eternity can be correlated with God's omnipresence to a specific spacelike global present in a relativistically correct way. In essence, I described a way to correlate the divine attributes of eternity and omnipresence by identifying particular events in time for a given observer with particular events in space for that observer in a relativistically invariant way. These spacelike events then constitute the global present for that observer, and in turn the events to which God is omnipresent.

We return to Pannenberg's insight regarding eternity and time, namely that events in time are present *to God in eternity* while *God is omnipresent* to things in the world. Here I begin again with the claim that it is the co-present duration of the divine eternity that gives to our distinct and separate events in time a form of differentiated unity, one that preserves their distinction but overcomes their separation. Now I want to extend this claim to suggest that it is God's omnipresence to events in space that overcomes their distinct and separate spatial characteristics and gives them a differentiated spatial unity. Thus, while the particular spacelike events that compose the axis of simultaneity of a given observer are relative to that observer, the unity of these spacelike events is not grounded in nature. Instead, it is bequeathed theologically by God's omnipresence to them. Moreover the form of God's omnipresence to them is invariantly correlated with the form of God's eternity to events for that observer.

This reformulation of the relation between eternity and omnipresence in light of SR offers an elegant response to the philosophical problem of the spatial unity of the global present for each observer. To see this, first recall Pannenberg's discussion of the Newton-Clarke/Leibniz debates (chapter 1, section B.2). According to Pannenberg, Newton and Leibniz radically disagreed over the concept of space. Newton thought of space in the Aristotelian sense of it being a container of matter while Leibniz thought of space as the set of relations between matter. Clarke, however, appealed to a third concept of space, one that is infinite, unitary, and undivided, in his defense of Newton against Leibniz. But Clarke's concept of space was not that of Newton because the latter, while infinite and unaffected by matter, is not unitary and undivided. Note that this is evident from the fact that it includes the Euclidean metric based on the Pythagorean theorem for distances in space. Clarke, however, imagined a space prior to all divisions or relations, and this space would be that of the divine omnipresence, or as Pannenberg put it, the divine omnipresence as "the effect of God's infinity in his relationship with the world of his creatures."[42] Pannenberg deemed Newton to be unsuccessful in his search for an adequate concept of space because "he did not develop his thought in terms of trinitarian theology."[43] It is here in the question raised by the debates between Newton, Leibniz, and Clarke over the philosophy of space that I propose that the theological concept of the differentiated unity of the divine attributes offers a highly promising suggestion. Moreover, it reflects an underlying basis in Pannenberg's concept of the differentiated unity of the Trinitarian nature of God.

We start with the fact that with Einstein our concept of space and time has shifted, following the symbol "3 + 1 → 4," from space-plus-time to spacetime. It has shifted further with Einstein's adoption of a Leibnizian, and not an Aristotelian, concept of spacetime. Nevertheless, Einstein's work leaves unresolved the underlying philosophical problem of the unity of spacetime. In response my proposal is that theology can offer the needed insight to resolve this philosophical problem. My beginning point is Pannenberg's claim that the divine eternity receives and unites distinct and separate timelike events in the world into the co-presence of eternity. I then extend this claim by suggesting that the divine omnipresence unites distinct and separate spacelike events in creation by God's ubiquitous presence to them. Thus, while it is God's eternity that gathers up and unites separate events in time while preserving their distinctions, it is, in my view, God's omnipresence to and in the world that gives to the world, fragmented into individual spacelike solitary events, the underlying differentiated unity that Clarke sought unsuccessfully. And if this holds true, it is surely a promising discovery about the gifts theology has to offer as it engages creatively with the natural sciences. And it in turn is particularly indebted to the theological writings of Wolfhart Pannenberg.

PART TWO

TRP → SRP

The Theological Reconstruction of Pannenberg's
Views in Light of Mathematics, Physics, and
Cosmology as Offering Suggestions for New Research
Programs in the Philosophy of Time and in Physics

A New Flowing Time Interpretation of Special Relativity Based on Pannenberg's Eternal Co-presence and the Covariant Theological Correlation of Eternity and Omnipresence

The Special Theory of Relativity constitutes a serious challenge to any tensed view that maintains that the future is not real.

—Michael Tooley, *Time, Tense, and Causation*

If the speed of light had been much smaller, reality would presumably have seemed very different to us and we would never have fallen into the error of assuming that, at each moment of our experience, the whole universe divides into events that have not yet happened, and those that have.

—Chris J. Isham and John C. Polkinghorne,
"The Debate Over the Block Universe"

In this chapter I lay out and defend a new flowing time interpretation of special relativity (SR) based on my interpretation of Pannenberg's concept of eternity, namely what I am calling eternal co-presence (chapter 2) and on what I refer to as the covariant correlation of eternity and omnipresence (chapter 4). If successful, this new interpretation will be a worked example of the way in which theology can be a conceptual resource for research in other nontheological disciplines—here, the problem of time in the philosophy of physics. In the following chapter I will, in a similar way, offer suggestions for research directions in theoretical physics.

A. DEBATES OVER THE INTERPRETATION OF SR: FLOWING TIME OR THE BLOCK UNIVERSE?

Should SR be taken as supporting a philosophy of being or a philosophy of becoming? In this section I present a brief summary of the many detailed arguments over this question.

1. The Block Universe of Taylor and Wheeler

I begin with the problem facing defenders of flowing time: the argument for the block universe. I have chosen the way in which Edwin F. Taylor and John Archibald Wheeler[1] make the case for the block universe because their approach is easily understandable and because it has been widely used in courses on physics.

Before presenting their case, I want to set the stage by resuming the discussion about viewing spacetime as a four-dimensional geometry[2] that we found in chapter 4. There, in section A.8, I noted the striking similarity between the Euclidean metric by which distances are measured in three-dimensional space and the Lorentz metric by which proper time is measured in spacetime. If x, y, and z represent the three spatial dimensions, and t the dimension of time, then distances d in Euclidean space between two points, one lying at the origin and the other at coordinates x, y, z, are given by the Pythagorean formula:

$$d = (x^2 + y^2 + z^2)^{1/2} \tag{1}$$

Similarly, in spacetime the proper time τ between two timelike events, one lying at the origin and the other at coordinates x, y, z, t, is given by the Lorentz metric:[3]

$$\tau = (t^2 - (x^2 + y^2 + z^2))^{1/2} \tag{2}$$

Of course the Lorentz metric (2) includes a difference in sign between the way time and spatial coordinates are included, while the Euclidean expression (1) has no minus sign. Nevertheless an analogy is often suggested between proper time τ—the invariant interval as expressed in (2) between two events viewed by observers in relative motion in spacetime—and distance d—the invariant length of an object such as a ruler rotating in Euclidean space as expressed in (1). The analogy is this: just as the length of the ruler is an invariant of the Euclidean metric (1), so too the interval between two events is an invariant of the Lorentz metric (2). The analogy is then pressed further to suggest that, just as the length of the ruler is a real property of the ruler in three-dimensional Euclidean space, so too proper time is a real property of the interval in four-dimensional Lorentzian spacetime.[4] Finally, and here is the coup d'état: like a ruler, all of whose parts are simultaneously present and real in space, so all the events that make up the interval are simultaneously present and real in spacetime. But these events are timelike. Therefore all timelike events are simultaneously present and real, yielding the block universe.

By using an analogy that is similar to the rotating ruler, Taylor and Wheeler argue that spacetime is a four-dimensional geometry and, by implication, all events in spacetime are equally present and real. They call their analogy the Parable of the Surveyors. Here two different kinds of surveyors measure the distance from the center of town to a distant point. Daytime surveyors use directions such as north and east to make their measurements while nighttime surveyors use directions based on the North Star to make theirs. Obviously the coordinates that result will differ between daytime and nighttime surveyors. Nevertheless, in this delightful story "a student . . . with openmindedness" shows up who recognizes that the square root of the sum of the squares of the daytime surveyors' measurements equals that of the nighttime surveyors' measurements. This clever student has discovered "the principle of the invariance of distance."[5]

Now comes the crucial move: this discovery is generalized to the spacetime interval given in SR. As we know, calculations of the interval between events in space and time are equal regardless of our coordinate system. This in turn suggests that the spacetime interval between these events is in fact a distance. And since a distance in ordinary space presupposes that all events along it are present at the same time, so too we can presuppose that the events in spacetime are equally present. The upshot: contrary to our ubiquitous experience of time as flowing, the "block universe" interpretation of spacetime is one in which all events, past, present, and future, are simply "present."

2. Costa de Beauregard vs. Čapek

In back-to-back chapters published in J. T. Fraser's monumental 1966 collection, *The Voices of Time,*[6] Olivier Costa de Beauregard and Milič Čapek support philosophies of being and becoming, respectively, as they debate the correct interpretation of SR. We will look to Costa de Beauregard first, and then to Čapek.[7]

Costa de Beauregard's basic move is to claim that covariance in physics is intimately related to objectivity in nature. When the mathematical definitions of the properties of a physical system are covariant, that is, when the definitions are independent of the particular coordinate system that expresses them, then these properties are objective features of that system. Because SR provides such covariance, "physics is . . . related to the intrinsic geometry of space-time."[8] From here, Costa de Beauregard argues against the objective status of the classical present defined as a global division between past and future. He acknowledges that the global present was covariant in Newton's mechanics, but in SR it is no longer covariant; instead it is defined relative to the observer. In its place we have the covariant light cone whose trifold division of spacetime at each event represents the Lorentz metric. As Costa de Beauregard contends, the basis for the difference between classical physics and SR lies precisely in the Lorentz metric and its light cone structure, and not in the invocation of a spacetime arena per se.

In addition, as Costa de Beauregard notes, we may understand why classical physics contained a covariant present by considering the elsewhen, or "elsewhere,"[9] region of spacetime for an event O. If we imagine the speed of light, c, increasing indefinitely, the elsewhen shrinks in volume. In the limit c → ∞, the elsewhen "disappears altogether and spacetime consists only of the future and the past, separated by the present instant." Thus, in the limit c → ∞, we regain the classical worldview of an objective global present.[10] I discussed this point in chapter 4 (diag. 4.15), relabeled here as diagram 5.1:

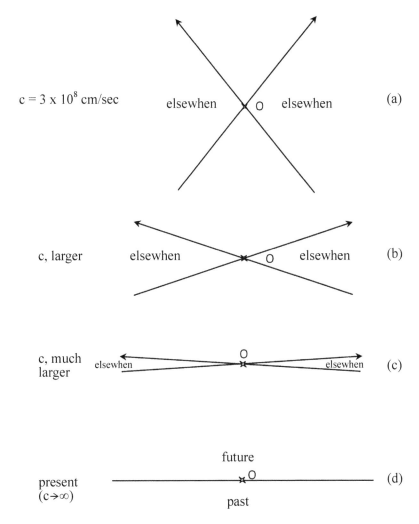

$c = 3 \times 10^8$ cm/sec   elsewhen   O   elsewhen   (a)

c, larger   elsewhen   O   elsewhen   (b)

c, much   elsewhen   O   elsewhen   (c)
larger

future

present   O   (d)
(c→∞)   past

Diagram 5.1. The Elsewhen Region for Event O Shrinks and Vanishes as the Speed of Light Goes to Infinity and the Light Cone Flattens into the x-Axis

But according to SR the speed of light is finite. Thus "there can no longer be any objective and essential . . . division of spacetime" via a global present that divides all of spacetime into events that are past and events that are future.[11]

The loss of a global present and, in its place, the trifold division of spacetime at each event also suggests that matter, which is extended in space, must be seen as extended in time as well.[12] "This is why first Minkowski, then Einstein, Weyl, Fantappiè, Feynman, and many others have imagined space-time and its material contents as spread out in four dimensions." In a winsome understatement, Costa de Beauregard writes that "there is inherent in this small fact a small philosophical revolution." He concludes with the claim that "relativity is a theory in which everything is 'written' and where change is only relative to the perceptual mode of living beings."[13]

Čapek challenges this conclusion, insisting instead on a philosophy of becoming and temporal flow, referring specifically to the paper by Costa de Beauregard that we have just discussed. Čapek starts with the oft-made claim that the relativization of simultaneity undermines the objective temporal order because events that appear simultaneous to one observer may appear as successive to another observer. He acknowledges that this claim is "the most frequent and superficially most plausible argument" in favor of a philosophy of being. Nevertheless the relinquishing of Newtonian time and its global, objective present no more requires us to give up temporality than does the undermining of Euclidean geometry by the Lorentz transformations lead inevitably to the loss of geometry. What we need is a more nuanced understanding of the impact of the light cone structure for our appreciation of objective temporality.[14]

In pursuit of this goal, Čapek first agrees with Costa de Beauregard that, given SR, the classical stratification of spacetime by a global present into a global past and future is no longer possible. But what, he asks, does this actually entail? Certainly for events in the acausal elsewhen of an observer "the simultaneity and succession of . . . causally unrelated events is fully and without qualification relative." This means that two events in the acausal elsewhen may be seen as simultaneous by one observer and as successive by another. But if we consider events that lie in the causal past or future of an observer and that are therefore causally ordered from the perspective of that observer, such events can never be viewed either as simultaneous or in reverse causal order from the perspective of another observer. Instead, "the succession of causally related events is preserved in *all* frames of reference" because causal succession is a Lorentz invariant.[15]

Čapek concludes that the irreversibility of events along an observer's worldline, and the causal order that constitutes it, is a "topological invariant" even while the succession of acausal events in the elsewhen is "completely relativized." Hence we should not regard the relativization of simultaneity as meaning that the "now" has no objective status whatsoever and that becoming is merely a subjective response to what is in fact an objective world of being. Instead, Čapek argues that the "now," though confined to the "here," is absolute in a crucial sense: it precedes its own causal future, and this fact is true for all observers.[16] This means that events in the causal future of the "now" can never be seen as lying in the causal past of that same "now" for any observer in relative motion. Čapek summarizes his entire argument by stating that "the relativistic union of space with time is far more appropriately characterized as a *dynamization of space* rather than a spatialization of time."[17]

3. Isham vs. Polkinghorne

A recent example of the debates over the interpretation of SR was published in 1993 in the first volume of the series of volumes that resulted from the two-decade long research collaboration between the Vatican Observatory and the Center for Theology and the Natural Sciences. In a chapter written jointly by Chris Isham and John Polkinghorne, Isham defends the block universe interpretation and Polkinghorne the flowing time interpretation of SR.[18] By the term "block universe," Isham means that "all spacetime points have an equal ontological status." Hence "no fundamental meaning is to be ascribed to the concepts 'past,' 'present,' and 'future.'"[19] Flowing time, and with it an openness to the future, is purely "a construct of the human mind that has no reference to reality" at least as it is understood by modern physics. Polkinghorne, on the other hand, defends flowing time as a barrier that continuously travels into the open future, leaving behind an unchangeable past.

Isham and Polkinghorne do agree on several key points: 1) The light cone associated with every spacetime point E divides spacetime, in relation to that point, into three causal regions based on the finite speed of light: the causal future of E, the causal past of E, and the acausal else-when of E. "This causal structure is the single most important feature of spacetime. . . . The crucial point is that, although it is meaningful to talk about the past and future of any event, there is no unique way of identi-fying those events in the 'elsewhere' of an event E that can be regarded as contemporaneous with E and with each other. . . . Consequently no meaning can be ascribed to the notion of *the* future or past."[20] 2) We should distinguish between a kinematic and a dynamic theory. On the one hand, a kinematic theory, such as SR, allows us to compare the de-scriptions of the motion of a system given by observers in relative mo-tion through a mathematical transformation between these descriptions, such as the Galilean or Lorentz transformation. On the other hand, a dynamical theory, such as Newtonian mechanics, includes kinematics in its more general explanation of the causes of motion of the system (e.g., contact force) and the kinds of motion that result from these causes (e.g., accelerated motion). 3) This means that we can use Galilean kine-matics in a deterministic dynamics (e.g., classical mechanics) and in an indeterministic dynamics (e.g., nonrelativistic quantum mechanics). We can also use Lorentzian kinematics in a deterministic dynamics (e.g., Maxwell's electromagnetism) and in an indeterministic dynamics (e.g., quantum electrodynamics).[21]

The upshot of points 2 and 3 is that SR, as a kinematic theory, does not determine whether a particular form of dynamics is the correct physi-cal theory. Instead, SR is compatible with both determinism and indeter-minism and thus with either an "open" future filled with genuine physi-cal novelty or with a "closed" future in which everything is determined in advance. Still, Isham and Polkingham disagree over which view of SR is correct: a closed future that already exists (Isham) or an open future that does not yet exist (Polkinghorne).

According to Isham, who supports a block universe worldview, the "essential ingredient of relativistic physics [is] the eternal reality of the spacetime manifold." The subjective experience of "becoming" and "un-certainty" do not affect the view of "the true reality of the entire space-time as a single mathematical entity." Philosophically, supporters of the block universe are "unashamed Platonists." The main scientific issue, however, that divides them from the supporters of flowing time is language about "the future" or "the past" because this requires "[a unique global] 'passing now' which is incompatible with both special and general relativity." Clearly the light cone for each event along a given world-line divides spacetime into causal past and causal future, but such light cones are tied to particular events and are not global structures defining the traditional notion of a global present. "The main objection is to the idea that the 'nows' of different events can be related in some way" as "the future" suggests. If such a special slicing of spacetime into a global past and future really exists, "how is it determined, and where is the evidence for it?"[22]

Isham also addresses a set of theological concerns about the flowing time interpretation of SR. He does not object, on theological grounds, "to positing a special sequence of spatial reference frames for God [to experience physical reality]." For Isham, the real problem with this idea arises if God is said to act on the world through this special sequence of spatial frames. This action "would be equivalent to introducing a preferred slicing of spacetime." And how can God do so and still preserve relativistic physics? Moreover, is it possible to construct a model of such an interaction? "It is not easy to add any external influence on the physical world and maintain the full fabric of relativity."[23]

Finally, Isham levels an important challenge to supporters of flowing time: "Can the debate be taken beyond the level of conceptual issues to the point where a genuine theoretical model is developed?" Do opponents of the block universe "seek merely to *reinterpret* the existing theories of physics, or do they make the much stronger claim that their metaphysical views can be sustained only by *changing* the theories?" Isham continues,

> If the current laws of physics really are inadequate, should this not reveal itself by a failure in some well-defined extension of the current domain of applicability of these laws . . . ? This challenge is not intended to be facetious — the question is genuine. It would be a major achievement if conceptual worries about the nature of time could be transformed into a real change in the theories of physics.[24]

According to Polkinghorne, the underlying determinism of classical mechanics allows for a block universe view, and this view has been carried over into SR. Still, the rates at which clocks tick is a "secondary consideration," and the relativity of simultaneity is merely a "retro- spective construction." The global present is not directly knowable by a specific observer.[25] In addition, the intrinsic indeterminism seen in much of contemporary physics, including quantum mechanics and chaotic dy- namics, encourages a flowing time view of nature. In the end, if science fails to provide a robust basis for the unalterability of the past and the "not yet" of the future, "the philosopher will reply, 'so much the worse for scientific theories!'"[26]

Theologically, Polkinghorne distinguishes between God's omni- presence to the world, God's knowledge of the world, and God's action in the world. God can be thought of as present to a particular observer, experiencing each spacetime event for that observer *as and when it happens.*" According to Polkinghorne, God's experience will reflect the causal order of the observer's experience. Thus God does not need to know "all that happens at once." God's action, in turn, will "respect those self-limitations he has imposed as expressions of his will" including those given by SR. Issues that might seem to arise from God's action in rela- tion to acausal, spatially separated events can be resolved easily: God will not act to introduce irrationality into the world. The key here is Polking- horne's commitment to a "temporal pole to God's experience" of the world. A timeless view of divine eternity would be "destructive" to a reli- gion such as Christianity. "Salvation history would have to give way to a theology of timeless Gnostic truth. . . . God knows all that can be known, but in a world of true becoming the future is not yet there to be known."[27]

4. Craig's Neo-Lorentzian Flowing Time Interpretation of SR

In his debate with Isham, Polkinghorne gives us a conventional version of the flowing time interpretation of SR. There is a lesser known version that, its supporters claim, avoids many of the technical challenges to the conventional one. This is the so-called neo-Lorentzian interpretation of SR. Its key feature is that it permits a unique, global present that divides the past from the future in a way that is consistent with SR. But it does so by viewing such phenomena as time dilation and length contraction as real, physical effects in nature with their own dynamic causes, and not just as the kinematic results of the Lorentz transformations. This in turn leads to the problem of explaining the cause of these phenomena. Further problems arise from the neo-Lorentzian interpretation because it requires us to set aside the routine assumption in SR that the speed of light in the vacuum is the same in all directions. Instead, it assumes that while the time it takes light to travel round-trip between two observers is the same whether the observers lie along the x, the y, or the z-axis, the time it takes light to travel one way between them can differ. In essence, light might propagate faster along the positive x-axis than along the negative x-axis; it could go faster going left to right than right to left, just as long as the total time it takes to travel back and forth along the x-axis is the same as the total time to travel back and forth along either the y- or z-axes. By introducing this generalized form of SR, however, the neo-Lorentzian interpretation allows for a unique global axis of simultaneity, the precise feature that SR in its standard form denies to nature. This in turn allows for a flowing time interpretation in which the unique axis of simultaneity, the global present, defines a unique global future and past as we find in classical physics.[28]

The neo-Lorentzian interpretation of SR is a view that few physicists would consider as preferable to the conventional arguments for flowing time. As we proceed let me reiterate the radical step taken by neo-Lorentzians: according to them we should move back behind Minkowski's 1908 spacetime interpretation of SR and return to Einstein's original 1905 conceptual framework, namely that of classical physics with its three-dimensional space and linearly flowing time. Here the existence of a universal physical present moving forward in time and dividing all events into a single past and a single future is taken for granted.

William Lane Craig is a strong supporter of the neo-Lorentzian view of SR. In 2000–2001, Craig published an extraordinary tetralogy of scholarly texts that offer a massive survey and a detailed critical analysis of the philosophical debate over dynamic versus static concepts of time with particular attention to SR.[29] His semi-popular publication *Time and Eternity* gives a readable account of this debate.[30] Throughout these works, Craig has consistently defended an A-theoretic, dynamic view of time in general and, in specific, a neo-Lorentzian interpretation of SR. His arguments are persuasive but by no means incontrovertible.[31] I present several of Craig's key arguments drawn from *Time and the Metaphysics of Relativity* and from *Time and Eternity*.

Table 5.1. Three Interpretations of SR: Two Dynamic and One Static

Two views of time	Three interpretations of SR	Who holds them
A-theories of dynamic, flowing time	Relativity interpretation	Einstein's original view
	Neo-Lorentzian interpretation	Some current scholars (e.g., Craig, Tooley)
B-theories of static time	Spacetime interpretation	Minkowski's view: Block universe (Einstein's later view)

Before presenting a brief overview of his work I need to make some initial comments on Craig's terminology, his adoption of presentism, and his strategy for defending it in light of SR. Regarding terminology, Craig calls the conventional flowing time view of SR the "relativity interpretation." He claims that this was Einstein's original view of SR, and he distinguishes it from the neo-Lorentzian interpretation, an A-theoretic flowing time view of SR that he defends. He contrasts both the relativity interpretation and the neo-Lorentzian interpretation to what he calls Minkowski's B-theoretical, or block universe, interpretation of SR, the view that Einstein eventually adopted. Craig refers to this view as the "spacetime interpretation" of SR (table 5.1).[32]

Next, it is important to note that Craig supports presentism. As we saw in chapter 2, presentism is a specific version of the dynamic view of flowing time according to which "present entities are the only temporal entities which exist."[33] Finally, a comment on Craig's strategy for supporting presentism is needed. We often find Craig defending the spacetime interpretation (Minkowski's block universe) over the relativity interpretation (Einstein's original flowing time interpretation). This may seem surprising given that Craig is an ardent supporter of flowing time. His strategy becomes clear, however, when we find him claiming that the neo-Lorentzian version of flowing time is preferable to the spacetime, block universe interpretation of SR and, by implication, the relativity interpretation of SR. So, although the static, spacetime interpretation at first seems to be gaining ground over the dynamic, relativity interpretation of flowing time, a flowing time view is eventually the real winner according to Craig—but in the neo-Lorentzian form.

Let us look briefly at a sample of Craig's argument. In chapter 5 of *Time and the Metaphysics of Relativity,* Craig engages a crucial question: If only present events are real, how do we define which events are present to each other given the relativity of simultaneity in SR?[34] According to Craig, there are three ways one might respond to this question, but Craig also claims that none of them really work. The first option is to regard the entire elsewhen of an event P as real, but this leads to an ontological contradiction. Consider events S and T in the elsewhen of P such that T is in the causal future of S. On the one hand, because they both exist in the elsewhen of P, S and T are simultaneous. On the other hand, since T is in the causal future of S they cannot be simultaneous (fig. 5.1):

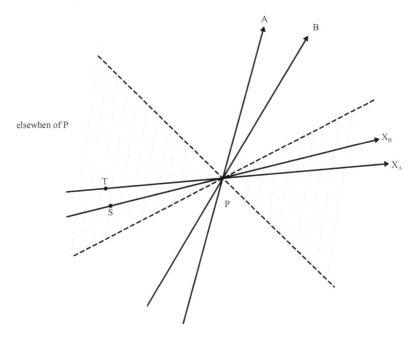

Figure 5.1. How Can T and S Both Be Real If T Is in the Causal Future of S?

A second option is to "shrink the present down to a single spacetime point," the "I-now." Lawrence Sklar calls this move the "solipsism of the present moment," and it entails seemingly unacceptable implications.[35] For example, if no spacetime point, the "I-now," is located in

London, then the sentence "It is raining in London" is never true. The third option is to make an arbitrary choice: pick one frame of reference and let its axis of simultaneity define simultaneity and thus reality. The problem of course is that SR offers no basis for which one to choose, so that any choice would be ad hoc, and according to Craig a claim about what exists should not be based on an ad hoc choice.

When we turn with Craig to his *Time and Eternity,* we find closely related issues for defenders of the relativity interpretation of flowing time in SR: simultaneity, fragmentation, and real effects without causes.[36] We start with the problem of simultaneity. According to the presentist account, entities only exist in the present moment. Thus two observers, A and B, can only be said to coexist if two events, one along each of their worldlines, exist simultaneously. The problem arises for the relativity interpretation of SR when A and B are in relative motion. Here, A may consider an event P along A's worldline as simultaneous to an event Q along B's worldline, but the converse is not true: being in relative motion, A and B have different axes of simultaneity, and B will consider event R, not P, as simultaneous to Q (fig. 5.2). Because of this disagreement over simultaneity, A and B cannot be said to coexist if we adopt the flowing time/relativity interpretation of SR.

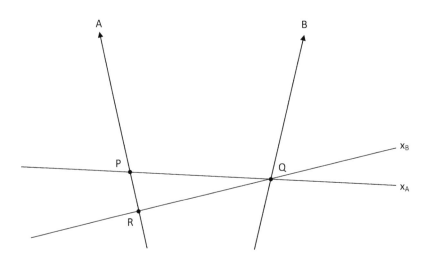

Figure 5.2. A Considers Events P and Q as Simultaneous— But Not So for B!

Second is the problem of fragmentation. According to the relativity interpretation of SR, spacelike events that are simultaneous for one observer will be in the acausal future or acausal past for other observers in relative motion. S and T in figure 5.3 are simultaneous with P for B. For A, however, S lies in the acausal past while T lies in the acausal future, while for C the reverse is true: S lies in the acausal future while T lies in the acausal past. This means that the world of events that B believes are simultaneous and therefore real is not the world of events that A and C believe are simultaneous and therefore real. As Craig stresses, "the relativity interpretation results in a fantastic fragmentation of reality . . . . Reality is relative to reference frames. One can change one's reality just by changing one's relative motion."

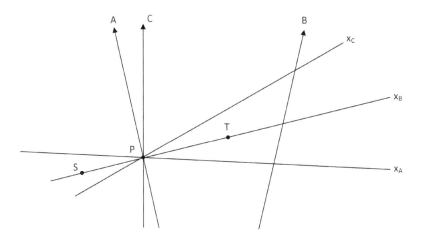

Figure 5.3. Events S and T Are Simultaneous with Event P for B but A and C Assign Them to Differing Acausal Pasts and Futures

Third, we have the issue of real effects without causes. This problem is actually quite different from the first two. According to the relativity interpretation of SR, time dilation and Lorentz contraction are real, physical effects, and thus they require a causal explanation. But this interpretation "neither provides nor permits any causal explanation" of these effects.[37]

Craig now turns to the static, spacetime interpretation of SR, that is, the block universe. He claims that this interpretation, unlike the relativity interpretation, can easily address all of these problems. For example, according to the block universe all events in spacetime are equally real. Hence the first problem, that of simultaneity, never arises. Again, according to the spacetime, block universe interpretation, reality is not fragmented but unified as an objectively existing, four-dimensional world. Finally, in the spacetime interpretation, time dilation and length contraction arise from our taking different three-dimensional slices through the all-existing four-dimensional Minkowskian geometry. Time dilation and length contraction are mere appearances or aspects of the block universe, not physical effects requiring a physical cause. From these reasons, Craig concludes that "the space-time interpretation of SR is superior to the relativity interpretation. But if that is the case, then, as Einstein came to believe, tense and temporal becoming are illusions of human consciousness. Reality is tenseless, and the static theory of time is correct."

Craig, however, appropriates a third interpretation of SR that he claims is "equivalent to the Einsteinian and Minkowskian interpretations and is fully compatible with a dynamic theory of time," namely the neo-Lorentzian interpretation.[38] The beauty of the neo-Lorentzian interpretation of SR is that, like the relativity interpretation, it is an A-theoretic type, dynamic, and tensed theory of time, but, unlike the relativity interpretation, the neo-Lorentzian interpretation includes a physically meaningful axis of absolute simultaneity. In Craig's view, this solves the problems that undercut the relativity version of flowing time seen above. Coexistence now holds for everything that exists on the true axis of absolute simultaneity. Observers moving in relation to this axis will disagree on the simultaneity of distant events, but this is an epistemic disagreement, not an ontological one, as it was in the relativity interpretation. Relativistic phenomena, such as time dilation and length contraction, are real phenomena for observers moving relative to the axis of absolute simultaneity and produced by physical causes in nature. According to Craig, the neo-Lorentzian interpretation is "at least empirically equivalent to the rival views." Moreover, Craig writes that the neo-Lorentzian interpretation of SR is "wholly compatible with the reality of tense and temporal becoming since these are characteristics of absolute time." For these and other reasons he sees it as "no less plausible" than the spacetime view that is "superior" to the relativity view, and he adopts it as his preferred interpretation of SR.[39]

B. A NEW FLOWING TIME INTERPRETATION
OF SPECIAL RELATIVITY BASED ON
PANNENBERG'S ETERNAL CO-PRESENCE AND
THE COVARIANT THEOLOGICAL CORRELATION
OF ETERNITY AND OMNIPRESENCE

We now continue a two-part move that began in chapter 4 and reaches completion in the remainder of this chapter. After presenting this interpretation I offer responses to the positions taken by the scholars we have just discussed. In chapter 4, section C I outlined a Lorentz-covariant reconstruction of the two divine attributes taken from my analysis of Pannenberg's theology: eternity and omnipresence. We first identified particular events in time for a given observer with particular events in space for that same observer in a Lorentz-invariant way, events that then constitute the flowing global present for that observer. Next the eternal duration that holds together this particular set of timelike events along the observer's worldline becomes the basis for God's omnipresence to the particular spacelike events constituting the observer's global axis of simultaneity. It is these events to which God is simultaneously present for that observer at each moment along her worldline. But this correlation is Lorentz invariant: the relation between a given observer's worldline and her construction of a global present is preserved by the Lorentz transformations between her and all other observers in uniform relative motion to her. In essence, they will all agree on which events, for a given observer, constitute her global present at a particular moment along her worldline. We therefore concluded theologically that God's eternity for each observer will be invariantly correlated with God's omnipresence to that observer in her global present. This covariant correlation of eternity and omnipresence reflects the spacetime interpretation of SR without falling into the block universe interpretation of the spacetime interpretation of SR.

Still, as I mentioned in chapter 4, using SR in the process of such theological reconstruction will require a defense of a flowing time interpretation of SR in order to avoid inadvertently introducing a static, block universe interpretation into the theological program. This is my present task. Moreover, such a flowing time interpretation differs from all three of the options discussed above: it is obviously not a block universe interpretation of spacetime such as Taylor and Wheeler propose; it is not the conventional flowing time interpretation of SR that typifies the theology/science literature and is represented by Čapek and Polkinghorne; nor is it a neo-Lorentzian flowing time interpretation that returns us to the "3 + 1 dimensional" (space plus time) approach to SR as is defended in detail by Craig. Instead, it will be a truly four-dimensional, spacetime interpretation of SR, but it nevertheless will be a flowing time interpretation. To put this more carefully, it will be a *flowing time interpretation of the spacetime interpretation of SR.*

In chapter 4, and reiterated above, I suggested how the new flowing time interpretation would be based on the correlation of eternity and omnipresence in a Lorentz-invariant way. Here, then, I will seek to demonstrate that when Pannenberg's theology is reformulated in light of science (represented by the symbol SRP → TRP) as it was in chapter 4, such a reformulated theology can lead to new insights (represented by the symbol TRP → SRP) both here in chapter 5 in the philosophy of science and, as we will explore in chapter 6, in theoretical physics.

1. Delivering on a Flowing Time Interpretation through a Relational and Inhomogeneous Spacetime Ontology

In this new flowing time interpretation the key move is to argue that all the events along the observer's worldline do not have the same temporal ontology: they are not all "present" moments. The latter, of course, would make a flowing time interpretation impossible. To make this move work, I once again draw on the relational and inhomogeneous ontology developed in chapter 2 as providing a path forward here—but now in full light of SR. Thus, in order to develop a flowing time interpretation of spacetime, I will need to defend what I call a relational and inhomogeneous *spacetime* ontology.

Recall that in chapter 2 I developed a relational and inhomogeneous ontology for theological reasons, namely to elucidate one of the main concepts drawn from Pannenberg's discussion of the divine eternity: copresence. There, such an ontology provided a way to interpret Pannenberg's twofold claim: First, the *distinction* between the unique past and the unique future of each present moment (or what I call time's ppf structure) is preserved in eternity, so that each present moment retains its identity to the extent that its identity is expressed by these distinctions. Second, the *separation* between present moments, which we experience as the loss of the present when it is irretrievably past and the inaccessibility of the present when it is still future, is overcome in eternity. The co-present duration of the divine eternity holds these multiple present moments, with their distinct pasts and futures, in a unity that preserves their temporal distinction while overcoming the separation perceived in nature, represented in the laws of classical physics, and experienced in our lives.

The important factor here is that this co-present duration was based, in part, on a relational and inhomogeneous ontology for time. A relational ontology was the key to understanding these tensed designations—past and future—not as properties of a given event but as relations between events. In this way, we avoided the usual paradoxes in the philosophy of time, such as McTaggart's paradox. An inhomogeneous ontology was essential in giving these temporal relations differing ontologies—the ontologies of being past (actually real and determinate) and future (potentially real and indeterminate), which in chapter 2 we symbolized as follows:

p, the past event, is real, actual and determinate but unavailable: $R^{A, D, U}$
f, the future event, is potentially real and indeterminate: $R^{P, I}$

The combination of a relational and inhomogeneous ontology offered a new way forward to interpret time as dynamic/flowing in regard to the general philosophical debates.

The Key Ontological Move

Here I assume that a relational and inhomogeneous ontology is coherent with and supports Pannenberg's theological concepts of eternity and omnipresence when they are reformulated in light of SR, as we saw in chapter 4. I now propose we *reverse* the claim and explore whether such an ontology might also be coherent with and supportive of creaturely flowing time—not only subjective time but also objective time in nature. That is, if we adopt Pannenberg's theology of eternity and omnipresence as reconstructed in light of the spacetime interpretation of SR and based on a relational and inhomogeneous ontology, then the robustness of Pannenberg's work in theology will lead us to expect that a relational and inhomogeneous ontology must be preferable, compared to its competitors, when we return to the context of SR and its ontological grounding. In this way, Pannenberg's theology, when reconstructed in light of science, can provide fruitful implications for research in the philosophy of science, specifically the philosophical interpretation of SR.

Our first task is to show how we can assign a relational and inhomogeneous ontology to the four-dimensional spacetime interpretation of SR, thus giving to spacetime a flowing time interpretation. We start with the generalized spacetime diagram we saw in chapter 4 (fig. 4.7b), relabeled here as figure 5.4:

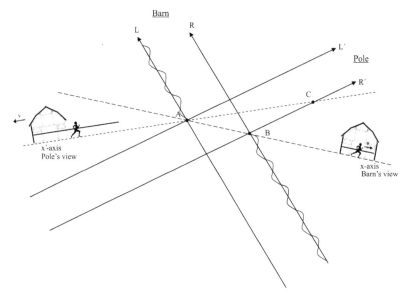

Figure 5.4. Generalized Spacetime Diagram Incorporating Sequences of Worldviews from the Perspectives of the Barn and the Pole

Recall that events A and B are simultaneous from the barn's perspective when the pole is contained within the barn, while events A and C are simultaneous from the pole's perspective when the pole extends out beyond the barn's open right door. From a flowing time perspective we seem to have a contradiction: How can both events B and C be simultaneous with event A when event C is in the causal future of event B—a fact on which both pole and barn agree since causality is Lorentz invariant? The only way out would seem to be to give all three events A, B, and C the same ontology, namely that all three are simultaneously present. This, of course, is the block universe, the interpretation I am trying to avoid. But can we?

My response to this key question is to view the events A, B, and C in terms of the inhomogeneous and relational ontology just discussed. It is evident from figure 5.4 that there are in fact two *different* temporal relations between events B and C:

> p:BrC (B is past in relation to C as present)

and

> f:CrB (C is future in relation to B as present)

Because p:BrC and f:CrB are different temporal relations we can assign them different temporal ontologies without contradiction:

> $p:BrC \longleftrightarrow R^{A, D, U}$
> $f:CrB \longleftrightarrow R^{P, I}$

This assignment of different temporal ontologies to past and future temporal relations is precisely what I mean by an "inhomogeneous temporal ontology." Since the context here is the spacetime interpretation of SR, we have arrived at what I will call a "relational and inhomogeneous *spacetime* ontology," one that avoids McTaggart's paradox even in light of SR. When both A and B are construed as present and therefore as simultaneous, C is construed as future in relation to B (f:CrB). Similarly, when A and C are construed as present and therefore as simultaneous, B is construed as past in relation to B (p:BrC). And,

> $p:BrC \longleftrightarrow R^{A, D, U} \neq f:CrB \longleftrightarrow R^{P, I}$

In this way a relational and inhomogeneous temporal ontology provides a consistent flowing time interpretation of SR—or, more precisely, of the spacetime interpretation of SR.

The bottom line: remarkably, the resolution of the temporal paradox that I proposed in the context of classical physics (see chapter 2) offers a similar resolution of the temporal paradox that arises in the

context of SR. It would appear that although there are fundamental differences between the worldviews of classical physics and SR, the use
of temporal relations to support flowing time in the context of classical
physics works to support flowing time in the context of SR, too. If true,
this is an unexpected and pleasing result.[40]

2. Responses to Scholars Previously Discussed

Response to Taylor and Wheeler

We have seen that Taylor and Wheeler argue for an analogy between
surveyors, on the one hand, who discover that the distance between two
points in space is invariant under Euclidean transformations, and scientists, on the other hand, who discover that the timelike interval between
events in spacetime is invariant under Lorentz transformations. If the interval in spacetime is like a distance in space, and since all points along
a distance in space are equally real, then by analogy all events which we
think of as past and future in spacetime are in fact equally real, and we
arrive at the block universe.

My response is that one can fully preserve Minkowski's spacetime
interpretation of SR without accepting the inference to the block universe. The discussion just above about the pole-in-the-barn paradox
shows that the Lorentz invariance of the timelike spacetime interval between B and C need *not* force us to accept the purported analogy between distances in space and the spacetime interval. The coordinates of a
spacetime interval do indeed yield the invariant τ, and invariants like τ
are indeed objective features of nature, but that does not require that τ
refers to an objective feature of nature *with a uniform ontology* such as
we have for a distance in space. Instead, if I deploy a relational and inhomogeneous spacetime ontology, we arrive at a genuine flowing time interpretation of Minkowskian spacetime, in which the spacetime interval
is invariant. In essence, Taylor and Wheeler's argument, as I read it, begs
the question. To argue for an analogy between a distance in space and
an interval in spacetime one needs to assume that a distance and an interval share a common ontology. But that assumption is what Taylor and
Wheeler seem to come to as a conclusion, one based on the invariance
of distances and intervals. For my part, since the invariance of distances

and intervals is possible for a relational and inhomogeneous ontology as well as for a homogeneous ontology such as the block universe, I am free to assume that the ontologies of space and spacetime are not equivalent without challenging the fact that they both share certain kinds of invariant measures, namely distance and interval respectively.

Response to Costa de Beauregard, Čapek, Isham, and Polkinghorne

Despite the radical differences on static and dynamic views of time, there are several areas of agreement between Čapek, Costa de Beauregard, Polkinghorne, and Isham, and these are places with which I too agree.[41] We are in agreement on the Lorentz invariance of causality between timelike events; the temporal relativity of acausal, spacelike events; and the crucial role of the light cone in distinguishing SR from classical physics. Finally, and this is more philosophically charged, we hold in common the assumption that causal invariance is indicative of what is objective in nature.

    All this notwithstanding, I disagree with Costa de Beauregard and Isham in the particulars of their support of the block universe interpretation of SR. For example, I do not agree with Costa de Beauregard that "space-time and its material contents [are] spread out in four dimensions." This is a "perdurance" view of objects as extended in spacetime. Instead I adopt an "endurance" view in which objects with respect to a given inertial observer occupy an axis of simultaneity defined by that observer at a given moment, and that their properties can change in time.[42] The subtlety, however, is that observers in relative motion will understand enduring objects and their properties in different, and seemingly paradoxical, ways, as the pole-in-the-barn paradox shows. In response, my interpretation of this paradox is as follows: It is certainly true that there are differing ways to reconstruct the world, for example, a moment in the barn's framework is knit together from a series of moments in the pole's framework and vice versa. Nevertheless, each such reconstruction is causally invariant, for example all observers in relative motion would agree that the barn "got it right" from its perspective and similarly for the pole. In this clearly limited epistemic sense each reconstruction is objectively true.[43] Thus, rather than see matter as "spread out" in spacetime I propose viewing matter and its causal relations as multiply-interpretable from differing spacetime perspectives. Using tensed relations with an in-homogeneous ontology I claim that we can do so without falling into a B-theory of tenseless time.

I do not agree with Isham that the "essential ingredient of relativistic physics [is] the eternal reality of the spacetime manifold." I take the idea of "the eternal reality of the spacetime manifold" to be a valid metaphysical option, but one not forced on us by physics.[44] If one is already an "unashamed Platonist," as Isham admits he is, then of course the "eternal reality of the spacetime manifold" is a natural conclusion, but this does not make it the "essential ingredient" of physics. This to me seems clear for a variety of reasons. If spacetime includes at least one "essential" singularity such as $t = 0$ in standard big bang cosmology, it is not at all obvious what it would mean to assert that spacetime is "eternal." It might make more sense to claim this for spacetimes ranging from Hoyle's steady state cosmology, which intentionally lack a $t = 0$, to multiverse string theories with landscapes of 10^{500} universes.[45] Still, even the multiverse is not self-evidently "eternal." Indeed, from a theological point of view, the existence of an eternal universe is contingent, requiring the creative act of God *ex nihilo*.[46] Moreover its eternity, endless time, is not the kind of eternity Pannenberg ascribes to the divine eternity as the source and fullness of time, as we saw in chapter 1.

I also do not agree that a unique global present is "required" by flowing time theories, as I hope to have shown above. I do agree with Isham, however, that this is the commonly held view, and I further agree that supporters of a conventional view of flowing time in the context of SR have not given sufficient attention to the technical details required for its robust support. I also agree with Isham that "it would be a major achievement if conceptual worries about the nature of time could be transformed into a real change in the theories of physics." To a certain extent what Craig (and Tooley) have done can be construed in this way.[47] And in this sense I can applaud their support of a neo-Lorentzian interpretation of SR even though I do not follow them on this path, the reasons for which I will discuss below. Instead, I take up this challenge in a rather different way, as will be seen in chapter 6, where my goal is to show the fruitfulness of theology for new directions in research physics but not to suggest changes in SR. Regarding Isham's theological concern, I agree with him that one can readily posit a special sequence of spatial reference frames in terms of which God can experience the world. Indeed that is part of what the proposal of this book is about. The challenge, as Isham points out, is to elaborate a theory of divine action that takes these spatial reference frames into account without violating relativity.[48]

Although I agree with them on supporting a flowing time interpretation of SR, I have several disagreements with Čapek and Polkinghorne. I believe that Čapek, in emphasizing the causal importance of the "here-now," underemphasizes the challenge raised by the relative axes of simultaneity and overemphasizes the importance of the single event, the "here-now." Regarding Polkinghorne, I disagree with his choice of what I term an "ontology II growing universe," according to which the future simply does not exist. I prefer an "ontology III growing universe," according to which the future exists but only as potentially real, waiting to be actualized as the present. Clearly, however, science alone can force this choice one way or the other. More important, I am not certain whether Polkinghorne's choice about the future is entirely consistent with his striking commitment to an eschatology in which the future is the realm of God's new action already present proleptically, to use Pannenberg's term, in the Easter Resurrection of Jesus.[49]

Response to Craig

I am grateful to Craig for his extensive treatment of both historical and contemporary materials regarding the conflicting interpretations of flowing time and, more recently, flowing time in light of SR, and for his persistent and creative engagement with these interpretations. Nevertheless, I have two critical responses to Craig. The first is theological and the second concerns his adoption of the neo-Lorentzian interpretation of SR.

1) I do not hold the view of God's temporality that Craig seems to hold. According to Craig, God exists "causally, but not temporally, prior to the Big Bang."[50] Craig also claims that once God creates the world God's temporality changes; God now lives and acts in time. "God must therefore be timeless without the universe and temporal with the universe. . . . On such a view, there seem to be two phases of God's life, a timeless phase and a temporal phase."[51] I would modify these statements to claim that God's eternity is not timeless "prior" (as it were) to the big bang. Rather the divine eternity is fully temporal regardless of the existence of the big bang universe (i.e., we could say that it is fully temporal "prior" to the big bang if we want to speak this way), and it is the true

source of creaturely temporality and the basis of creaturely temporal unity (co-presence) both now and as transformed and consummated in the New Creation. Moreover, Craig's radical distinction in God's temporality—timeless without the universe and temporal with it—I find both theologically and philosophically problematic, particularly from the Trinitarian perspective on the divine eternity that I have followed in this book. Theologically, God's act of creating the world *ex nihilo* does not change God's eternal nature; instead it expresses God's boundless and free love as an act of divine self-giving. Philosophically, God's eternal nature does not change with the changing character of God's temporal creation. Finally, while I agree with Craig that we need a tensed interpretation of SR for *theological* reasons, I do not believe this requires us to adopt a neo-Lorentzian interpretation. Instead, as I proposed above, a relational and inhomogeneous *spacetime* ontology can offer a better flowing time interpretation of the spacetime of SR.

2) I am also critical of Craig's reasons for adopting a neo-Lorentzian viewpoint. As I outlined in section A of this chapter, Craig claims that the "relativity" interpretation (Einstein's original view) forces us into a series of problems that the spacetime (block universe) solves but that are better addressed by the neo-Lorentzian interpretation. Let me summarize these problems and offer my response from the perspective of a relational and inhomogeneous spacetime interpretation of SR.

According to Craig, the first set of problems elicits three responses: either ontologizing the entire elsewhen, limiting the ontology to the "I-now," or choosing a unique axis of absolute simultaneity. The first response is the block universe which Craig and I want to avoid. The second response is "temporal solipsism," which Craig and I also want to avoid. The third response is ruled out by SR, but Craig's concern is that this means we cannot speak about the co-existence of two observers in relative motion. Consequently he modifies SR into a neo-Lorentzian format. My proposal instead is that we do not need to identify a *unique* axis of absolute simultaneity to talk about their co-existence as long as we do so without requiring that "co-existence" must mean what it did in classical physics. SR allows for an axis of simultaneity for *each* observer, and my claim is that these axes provide *all* we really need to speak about co-existence. Recall how the pole-in-the-barn "paradox" led to the conclusion that there were no *physical consequences* of the fact that there are differing views of simultaneity between the pole's frame of reference and that of the barn. Recall, too, that both views included all the relevant physical invariants, and it is these physical invariants that are the key to what is objective in nature, not the complementary narratives offered by observers in the barn and carrying the pole. This means that both views provide an objective interpretation of what "co-existence" means in nature, even if these interpretations are different.

The second problem involves a multiplicity of axes of simultaneity. It in turn leads to Craig's concern about the fragmentation of reality: namely that acausal events can be both simultaneous and acausally ordered. My response is that SR does not imply a fragmentation of reality but, instead, the discovery of a multiplicity of diverse objective perspectives on reality. Through this discovery, the subtlety of reality is enhanced, not undermined, particularly because other perspectives on reality, such as the pole missing the barn, the pole crashing into the barn, the barn's doors never opening, etc., are rejected entirely (i.e., "relativity ≠ relativism").

The third problem, according to Craig, is that both the relativity interpretation and the neo-Lorentzian interpretation view phenomena such as time dilation as real effects in nature, but only the neo-Lorentzian provides real causes for them. My response is that the relational and inhomogeneous spacetime interpretation of SR understands these phenomena to be perspectival. They do not need a causal explanation. Rather the key to reality, as I argued above, is found in the relativistic invariants.

In short, then, we need not go to the extreme of adopting a neo-Lorentzian interpretation of SR to preserve a flowing time account of the world. Instead, we can offer a new flowing time account by assuming a relational and inhomogeneous spacetime interpretation of SR as proposed in this book. It is a spacetime view as adopted by most physicists and philosophers of science, but it is amenable to a flowing time interpretation of spacetime, and flowing time is supported by most scholars in theology and science and presupposed by most theologians.

3. Widening the Search for a Physical Global Present: Why General Relativity and Big Bang Cosmology Do Not Help

Perhaps we should follow the lead of some scholars in theology and science[52] and seek to extend the search for a physically significant global present by moving from SR to Einstein's relativistic theory of gravity, the general theory of relativity (GR), and its application to the universe in big bang cosmology. We shall see that the four-dimensional space-time framework of SR, with its relativizing of the global present, seems to be suppressed under the intuitively familiar "space + time" picture of the expanding universe offered by the standard big bang models. Hence we are tempted to think of relativistic spacetime as a three-dimensional spatial surface expanding in time. It is then this three-dimensional surface that is often taken as providing a new basis for the "global present" that is missing in SR and thus for the locus of God's experience of, and action in, the world. Before challenging this move I want to review briefly the grounds for the standard big bang models in GR. We shall see immediately that the problem of the ambiguous global present, grounded in SR, is ubiquitous to GR itself.

In the introduction to this volume I described GR as providing a relativistically correct approach to the physics of gravity, that is, a theory of gravity consistent with SR while going beyond it. One way to see this consistency is in terms of what is called the equivalence principle. According to the equivalence principle, while large regions of spacetime may be curved in highly complex ways, spacetime is flat when looking at small regions that are far from massive objects. A convenient shorthand for this claim is that "spacetime is globally curved and locally flat." The spacetime of GR is locally the spacetime of SR. Hence, as I indicated, even in GR, spacetime comes with the attendant problems regarding a unique present that we discussed above in light of SR.

But even if the problem of the ambiguous global present of SR is ubiquitous to GR, can the specific big bang cosmological models give us an *unambiguous* global present? We saw in the introduction that the fundamental equations of GR can be written in the following compact form:

$$R_{\mu\nu} - \tfrac{1}{2}Rg_{\mu\nu} = 8\pi\,T_{\mu\nu} \qquad\qquad (1)$$

The left-hand side of equation (1) describes the curvature of spacetime, while the right-hand side represents the distribution of mass-energy-stress in spacetime. Most solutions of equation 1 mix spatial and temporal coordinates in complex ways that go far beyond the way spatial and temporal coordinates relate to each other in such SR paradoxes as the pole-in-the-barn. In doing so, they challenge any simple picture of physical reality.[53] Nevertheless there is one set of solutions that can be easily visualized: those of standard big bang cosmology.[54]

This cosmology arose in the first half of the twentieth century. As early as the 1920s astronomical observations increasingly suggested that the distribution of clusters of galaxies in the universe was homogeneous and isotropic, that is, the same as seen from all points in space and in all directions, and that these clusters of galaxies were moving away from us at a speed v proportional to their distance d, giving rise to Hubble's Law, v = Hd, where H is the Hubble constant. The theoretical framework for representing this astronomical data is known as the Friedmann-Lemaître-Robertson-Walker (FLRW) equations. Because of the enormous simplification brought about by the homogeneous and isotropic distribution of the galactic clusters, these equations allow us to factor the description of the universe into a linear combination of a temporal component and a three-dimensional spatial component that expands or contracts in time.[55] Such a description leads to the standard representations of the big bang universe as portrayed in figure 5.5:

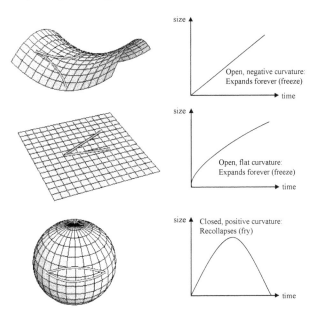

Figure 5.5. Three Models in Standard Big Bang Cosmology: "Freeze" (Open with Negative or Flat Curvature; Expands Forever) and "Fry" (Closed with Positive Curvature; Recollapses)

Note: The vertical axis is the "radius" of the universe and the horizontal axis is cosmic time.

Source: Images adapted from NASA, "End of the Universe," *Wikipedia,* March 20, 2006, http://en.wikipedia.org/wiki/File:End_of_universe.jpg.

Because of the astronomical data on galaxy distribution and the resulting simplification in the FLRW equations it is routine to speak of the "expanding universe" and to mean something like a classical "space + time" universe with a well-defined size and shape at each moment of time t. This picture of the universe easily lulls us into forgetting that it is made possible by the observational evidence for the symmetric distributions of clusters of galaxies.

This in turn tempts us to claim that the shape of the universe at a given moment in time is the unique global present that was lost in SR.[56] Viewed in this way, we can once again frame the concept of God's temporal relation to the universe, including the time and eternity question, as we usually do intuitively: God experiences, acts in, and interacts with the world in the present moment. The world in each present moment *is* the three-dimensional universe pictured in figure 5.5 and the present moment is the age of the universe measured starting at the beginning of time, t = 0, approximately 13.7 billion years ago.[57]

While this might work for the standard big bang models, it is problematic for several reasons. First, it is quite possible that our universe does not fit the standard big bang models. A number of scientists are working on non-FLRW cosmologies.[58] Many of these cosmologies lack the kind of symmetry found in standard big bang cosmology; as a consequence they may not include a unique global present. Second, as I indicated above, GR includes SR in a profound way: over small distances and times the curved spacetime of GR reduces to the flat spacetime of SR. This means that, even in big bang cosmologies, the challenge raised by SR to a "global present" is unavoidable.

Third, we must face an "in principle" question: What should we take as a more apt characteristic of the universe? There really are two options: (1) the empirical distribution of matter and the big bang models that fit this distribution, or (2) the theoretical framework of spacetime as incorporated from SR into GR. According to option 1 the empirical state of the universe points to the ontology of nature, and the "global present" offered by the standard big bang models solves the problem of the "present" posed by SR. According to option 2 the fundamental laws of physics point to the ontology of nature, and because this ontology is always locally that of SR, in which there is no unique "global present," the problem of the "present" is not solved by appealing to cosmology—big bang or otherwise.

Option 1 clearly favors the significance of the empirical character of the universe as a clue to its underlying reality over that of the fundamental laws of physics. Hence the three-dimensional surfaces found in standard big bang cosmology are what we think of as the present moments in which God acts in, perceives, and/or experiences the universe. Option 2 stresses the importance of the fundamental laws of nature as pointing to the underlying ontological structure of the universe. For this option, the empirical character of the universe is interesting phenomenologically, but it is inadequate to reveal the true, ontological character of the universe and thus respond to the question of how God relates to it "in time."

It seems clear to me that Pannenberg would favor the importance of the fundamental laws of nature (option 2) over happenstance phenomenological factors such as the symmetries in the geometry of the universe based on contingent, empirical data. If this is so, the big bang models of GR may seem to provide what SR cannot—a "global present"—but in fact they fail to do so for the reasons I have just outlined. Because of this I return to the previous discussion of SR and the goal of offering a flowing time interpretation of the spacetime interpretation of SR. Such an interpretation would be user friendly not only to theology in general, which presupposes flowing time, but to Pannenberg's doctrine of the divine attributes, which I believe requires it, as I discussed in part one of this volume. More important to the agenda of part two of this volume, such an interpretation might well be of interest to philosophers of science, particularly those specializing in the philosophy of physics and its subfield, the philosophy of time in physics.

C. THE REAL LESSONS OF SR:
"RELATIVITY ≠ RELATIVISM,"
THE "SIMULTANEITY RICHNESS" OF
THE ELSEWHEN, AND THE "AUSTERE PAUCITY"
OF THE ANTHROPOCENTRIC CLASSICAL PRESENT

I now take up a series of points that are akin to our discussion at the end of chapter 4, except here I address them as they arise in the area of philosophy of science and not in theology. They involve the fact that SR does not lead to physically significant disagreements between the views of observers in relative motion, and thus that SR does not lead to a form of epistemic relativism. They also underscore the discovery of the temporal richness of the elsewhen and, in turn, the paucity of the classical global present, a discovery that I am led to celebrate for its rich view of nature.

Flowing time theorists assume the need for a single, physically meaningful definition of simultaneity—a global present—in order to de-marcate the future from the past. Atemporalists view the lack of a unique global simultaneity, according to SR, as a key argument against flowing time. But is this the best way to understand the impact of SR on the mean-ing of simultaneity and thus the plausibility of flowing time? Granted that SR shows us simultaneity is observer-dependent—but so what? I believe the problem is not that, within SR, we cannot choose a single, observer-independent axis of simultaneity that corresponds to the clas-sical present. The problem is actually about why we believe we need to do so. Suppose the implications of SR are correct: the region of the else-when surrounding a given event P is actually composed of an uncountably infinite number of equally valid axes of simultaneity. Does construing "the world" using any of these axes lead to physical problems? My point is that it does not.

To see this more clearly, let us return to events A, B, and C in fig-ure 5.4 above. Here no two events along any axis of simultaneity (say events A and B along one axis or A and C along another) are causally re-lated: being in the elsewhen, neither can physically affect the other. Al-ternatively, events that are causally related according to one observer (say, events B and C) are causally related in the same way according to all observers regardless of their relative motion. We know both of these points to be true because causality is Lorentz invariant. Finally, as we saw with the pole-in-the-barn paradox, there are two kinds of narratives about "what happened" reflecting the pole's view and the barn's view. More-over, each of these two narratives at a given instant for one observer can be seen as constructed out of the other observer's narrative by sewing to-gether tiny fragments of "what happened" along the moving axis of si-multaneity of the other observer. So what we get is an intimate relation-ship between narratives, each of which has a physically legitimate, but different, view of "the whole" and "what happened." However, what is crucial to add here is that there are dozens of purported construals of "the whole" and "what happened" that are disallowed by SR: the pole missed the barn, the barn's doors sliced the pole into pieces, the pole poked through the barn doors, one of the barn's doors failed to close or to open, etc. To repeat, the "relativity" of SR in this sense means that there are "two truths" about the world that carry equal veracity and that there are "many falsehoods" about the world that SR undermines. In short, the "relativity" of SR does not lead to epistemic relativism: "relativity ≠ relativism." If we lay aside that concern and if we grant that a flowing time interpretation of SR via a relational and inhomogeneous ontology is possible, then flowing time theorists should no longer be concerned about the infinity of equally valid axes of simultaneity in the elsewhen.

To the contrary, I propose that we "invert the question." In my view the best way to understand SR is not to ask how close it can be forced to fit into a classical picture of the world, with its unique, global present, in order to rescue a conventional flowing time interpretation. Instead, the way to understand SR is to start with SR as a fundamental physical theory (which it is!) and with the spacetime interpretation of the world that it offers. Then we can ask how the classical picture of the world as an anthropocentric "limit case" arises for observers moving with relative velocities that are minuscule in comparison with the speed of light.[59]

To illustrate this point let us return to the elsewhen as a region in spacetime surrounding a given event P and bounded by its light cone. If we imagine the speed of light, c, as increasing until it becomes infinite, we discover, as we saw previously in diagram 5.1, that the light cone flattens out, leaving more and more events that were once in the elsewhen now in the causal past or the causal future of P. In the limit $c \to \infty$, the light cone itself "becomes" the classical global present for P that divides P's entire world into a global past and a global future.

Now let us apply this to the spacetime diagram for the pole-in-the-barn paradox (fig. 5.4). As the light cone flattens, the choice we are required to make about which is the correct axis of simultaneity is undermined by the flattened light cone. For example, in the limit $c \to \infty$ there are no longer two distinct timelike events, B and C, each of which could be considered as simultaneous with event A. Instead there is only one event P somewhere along the worldline of the pole's right end, between events B and C, which is uniquely simultaneous with A, and B and C are past and future to A, respectively. Similarly the issues raised by the fact that events B and C, being timelike, are causally related, and the impression that this fact favors the block universe interpretation, are rendered mute as the null cone collapses to the classical present.

In short, classical physics provided ordinary human experience with a theoretical justification for what we now know to be a fictitious, un-physical, and anthropocentric view of nature—that it has an absolute global present. In classical physics we could value what a global present offered: a simpler construal of "the world" than SR supports, one that is deeply imbedded in our ordinary, intuitive experience of the world. If we start by taking this classical construal as true, we are led to ask why SR involves such profound and counterintuitive paradoxes, and from there it is a slippery slope to the block universe as a "solution," albeit hard to swallow, to these paradoxes. Instead if we start with the elsewhen of SR as an objective fact of nature, with its plethora of equally valid axes of simultaneity such as the pole's view and the barn's view, we are free to discover how this plethora leads to Lorentz-invariant views of the world. Moreover, these invariant views admit a flowing time interpretation when we adopt a relational and inhomogeneous temporal ontology.

The "simultaneity richness of the 'now'" bequeathed to us by Ein-stein's astonishing discovery, SR, is preferable, in my view, over the "aus-tere paucity of the sole present" given to us by personal experience and by classical physics. Frankly, SR is the correct physical theory, and it offers a vastly reliable understanding of the natural world no matter how counterintuitive this understanding might seem. Hence, I urge us not to fall back into the worldview offered by classical physics even if it seems to confirm our ordinary experience of space and time. Instead I believe that the "simultaneity richness" of the elsewhen compared to the "aus-tere paucity" of the classical view of a purported unique global present leads to a more satisfying understanding of nature, one to be celebrated and embraced rather than explained away. I also believe that a relational and inhomogeneous ontology of spacetime is a better candidate for inter-preting SR than its competitors. Finally, I hope to have shown the value of a relational and inhomogeneous interpretation of SR, reflecting the reconstructed theology of Pannenberg, in pointing us in the right direc-tion for novel *philosophical* research on time and space in physics and cosmology.

CHAPTER 6

Duration, Co-presence, and Prolepsis

*Insights for New Research Directions
in Physics and Cosmology*

A. INTRODUCTION: RESURRECTION AS TRANSFORMATION AND THE PRECONDITIONS FOR THE POSSIBILITY OF DURATION, CO-PRESENCE, AND PROLEPSIS IN CREATION

In the introduction to this volume I indicated that I would follow eight paths for the mutual interaction between theology and science. Five are paths for reconstructing Christian theology in light of science as expressed by Barbour's now-standard phrase, "a theology of nature." These were followed throughout chapters 2–4 of part one. I also offered three paths in which a theology, so reconstructed, might provide insights into research in science and in the philosophy of science. These paths are taken up in part two. The paths into philosophy were pursued in chapter 5 that began part two, where I developed a new flowing time interpretation of special relativity in light of key themes in Pannenberg's theology, which, in part one of this volume, were reconstructed in light of science.

In this concluding chapter of part two I look for ways in which these key themes in Pannenberg's reconstructed theology might illuminate the search for criteria of choice among existing scientific theories as well as the search to construct new scientific theories. The style of this chapter is highly schematic and unapologetically speculative, since by engaging directly in TRP → SRP (Theological Research Program → Scientific Research Program) we are entering unexplored and most likely contentious territory. I will try to strike a balance between an exposition of individual scientific topics that is sufficiently general and detailed to make it readable for the nonscientist and a succinct itemization of a diversity of scientific topics that will convey a hint of the vastness of this new interdisciplinary research. I acknowledge at the outset that some of the research described in this chapter awaits further study and eventual confirmation or disconfirmation by the scientific community. Still, the fact that it is here serves as a "proof of concept" that theology can offer creative suggestions to interested scientific researchers while respecting and endorsing the methodological naturalism that underlines and shapes the natural sciences.[1]

In the introduction I also laid out what constitutes the overarching conceptual structure of this chapter: that "resurrection as transformation" means that some of the preconditions for the possibility of the New Creation that God is bringing about, starting at Easter and ending in the eschatological future, must already be present now. This is because such preconditions constitute one element of continuity between the world as we know it now through the natural sciences and the New Creation into which it is both already being transformed (i.e., "realized eschatology") and into which it will be transformed (i.e., "apocalyptic eschatology"). Of course, there undoubtedly are considerably more elements of discontinuity between the eschatological New Creation and the present world than there are elements of continuity.[2] Nevertheless, here I make a key move and presuppose that these preconditions and elements of continuity, even if they play a minor role compared to the preconditions and elements of discontinuity, include the themes that I have drawn, interpreted, and reconstructed in light of science from Pannenberg: duration, co-presence, and prolepsis.[3] To recapitulate:

Duration. In my reading of Pannenberg, time is not merely point-like, the terminus of an infinite divisibility of a one-dimensional temporal continuum. Instead, there is duration, or temporal thickness, not only in our conscious experience of memory and anticipation within the lived present but also in nature, including its fundamental physical processes as studied by physics and cosmology. Pannenberg grounds duration, in life and nature, in the temporal structure of eternity. Hence it is eternity as a divine attribute that provides the unity of our everyday temporal experience, even if we only experience this unity briefly in the fleeting events of our daily lives.

Co-presence. Pannenberg's concept of the unity of eternity is more subtle than the two "garden-variety" options so often found in philosophical and theological literature. Eternity is neither timelessness: the conflation of all moments of time into a single timeless "now" in which all temporal thickness is lost. Nor is eternity endless ordinary time: a continuing succession of separate temporal moments, each of which exists only for an instant as the "present" and then is gone forever. Instead, the divine eternity is one of duration, but a duration that includes co-presence: it is a differentiated unity that holds together all present events in the history of the universe both now and in the eschatological New Creation as proleptically anticipated in the present age. In this way the divine eternity "bridges" the temporality of the present creation and that of the New Creation, even while it finds its complete expression in the New Creation. Within this eternity each event retains its unique memories and anticipations held together within the enveloping temporal thickness of that event. I call this time's ppf structure, in which each event as present has its own relations to events as past and as future.[4] All events, in turn, each with their own ppf structure, are held together "simultaneously," without conflation and without separation in the differentiated unity of the duration found in eternity. And in perhaps his greatest *ansatz* in this topic, Pannenberg makes the bold claim that this understanding of eternity is possible only because God is Trinity: the divine eternity as a differentiated unity is the eternity of the differentiated unity of the Trinitarian God.

In addition, I have suggested that for Pannenberg, the relation of time and eternity can be seen through the following claim: that the distinction between events in time will be sustained in eternity while the separation between events in time will be overcome in eternity. By this I mean that the distinctive character of every event as present, reflected in its unique ppf structure, will be preserved in the unity of the divine eternity. This argues against eternity being a mere conflation of all events into a single unstructured present. I also mean that the separation of every event as present from all other events as present—reflected in the fact that each event can only be experienced as present once, and for each event its future relations to events are never available as present nor are its past relations to events available again as present—is overcome in eternity, in which all events are equally available to be savored, forgiven, and re-experienced endlessly. The relation between time as we know it and the divine eternity is true not only in the "now" of our creaturely lives, although experienced only partially and by way of anticipation, but even more so as a relation between this lived "now" and the endless life we will have eschatologically in the New Creation, when we shall experience God's eternity immediately. As Pannenberg writes, the distinction between God and creation will remain in the New Creation but the distinction between holy and secular will be overcome.

Prolepsis. I also argued that Pannenberg's view of prolepsis includes two distinct concepts. First, I suggested that prolepsis includes the immediate "causality of the future": in every present moment in life and in nature the factors predisposing, but not predetermining, the character of that present moment include not only the immediate efficient causality of the past, as represented by forces and interactions in physics and as reflected in our ordinary experience of memory and anticipation. These factors also include what I call the immediate causality of the future in which the future is manifested (or as Pannenberg phrases it, "appears") in the present as an additional contributing causal factor in making the present concretely what it is. Second, I suggested that prolepsis includes a strikingly new topological view of the relation between creation and the New Creation. Such a topology would allow the eschatological future to "reach back" and be revealed in the event of the resurrection of Jesus. This "reaching back" would not be entirely within the secular topology of the universe as described by GR or by any other scientific theory. Instead, it would be the topology only fully describable at a theological level, a topology that connects the present universe with the New Creation, even if it had some minimal implications for nature that science might explore. Let me parse out this claim in a bit more detail. On the one hand, since both creation and New Creation are part of a single divine act of creation *ex nihilo* this topological connection would be part of God's single creative act and thus in some way it would be a feature of nature that science could, in principle, study. Yet on the other hand, this connection is not so much a part of the present creation—it does not come from the ordinary future of nature—but more properly it is a manifestation of the proleptic act originating in the New Creation even as the New Creation emerges by God's act beginning at Easter out of the present creation. In this sense it would be a topological connection only fully recognizable by theology even if there are hints of it which science might uncover.

In this chapter I will look at preconditions in nature, as we know nature through contemporary physics and cosmology, which might be suggestive of the three distinctive elements in Pannenberg's theology: duration, co-presence, and prolepsis. This is in keeping with the guidelines I laid out in the introduction relating to the role of theology both as suggesting new research programs in science and as offering criteria of theory choice between competing scientific research programs. I refer to the scientific research programs to which these theological elements lead as "SRPs," and I explore each SRP, as generated by either duration (thus SRP 1), co-presence (SRP 2), or prolepsis (SRP 3), in sections B–D below.

But before exploring them we need to step back and ask how one choses the SRPs? Frankly, there are numerous possible candidates in physics and cosmology reflecting the complex transition from classical physics and cosmology to the discoveries and theories of the twentieth and twenty-first centuries. These include special and general relativity, quantum mechanics, quantum field theories, the Standard Model of fundamental particles, string and superstring theories, big bang cosmology, inflationary cosmology, eternal inflation, the multiverse, and so on. All of these theories involve the three elements of temporality we have seen to be central to Pannenberg's theology.[5] Here I will chose a small assortment as a first go, and pursue the wider range of further possibilities in future writings.

B. DURATION: SRP 1, THE SEARCH FOR DURATION IN PHYSICS

In my view, Pannenberg claims that duration exists in the natural world as well as in human consciousness. Is there evidence that a theological view such as this can play a fertile role in choosing between competing

research programs in science (following guideline 8); or in suggesting fruitful directions for new research programs in science (following
guideline 9)?[6]

One way to approach this question is to look to philosophy as a me-
diator between theology and science. Here, I turn to the philosophy of
Alfred North Whitehead and its influence on two scientific research programs: David Bohm's approach to quantum mechanics and Ilya Prigogine's formulation of nonlinear thermodynamics. I then look to specific
theories in science as suggestive of what one might pursue if one assumed a theology such as my interpretation of Pannenberg offers. To do
so, I briefly explore recent developments in string/M-theory, which can
be taken as pointing to temporal duration in nature.

1. Whitehead, Bohm, and Prigogine

Time as becoming and time as duration are in one way or another, and despite enormous technical differences, crucial central concepts in modern
philosophy, as even a cursory reading of Hegel, Heidegger, Samuel Alexander, and Henri Bergson demonstrates. Yet it is in the metaphysics of Alfred North Whitehead, as well as that of Charles Hartshorne, that becoming and duration play a conceptual role with revolutionary consequences.
It is for this, and for many other reasons, that process philosophy has
figured so prominently in the theology and science literature, particularly
in the writings of Ian Barbour, Charles Birch, Philip Clayton, and John
Haught. Although I will focus on Whitehead's philosophy and on two representative research programs in science that, arguably, reflect and in-
stantiate a representative part of his metaphysics within science, it should
be kept in mind that on many key points Pannenberg separates himself
sharply from Whitehead's metaphysics.[7] In the long run, these points of
difference between Pannenberg and Whitehead might lead to differences
in the kinds of scientific research, compared with those cited here, to
which Pannenberg's understanding of temporal duration might direct our
attention. Nevertheless it is still valuable to consider Whitehead and his
relation to current scientific research under the general concept of "duration" that he and Pannenberg so thoroughly share.

One of the most striking features of Whitehead's immensely complex metaphysics, as developed largely in *Process and Reality,* is his cen-

tral concept of an "actual occasion" as the fundamental element of reality. Unique to Whitehead's concept of temporal becoming and duration is the idea that an actual occasion incorporates two modes of temporality.[8] Whitehead describes these two modes, and it is worth pausing to examine them briefly here.

Whitehead begins with a remarkable statement about "the flux of things": "Without doubt, if we are to go back to that ultimate, integral experience, unwarped by the sophistications of theory, that experience whose elucidation is the final aim of philosophy, the flux of things is one ultimate generalization around which we must weave our philosophical system."[9] Here he has replaced the generic idea, rooted in Heraclitus, that "all things flow" with the specific concept of "flux." Essentially this is because Whitehead will replace the classical concept of substance as that which persists in and flows through time with the novel concept of a series of momentary, atomistic ("epochal") actual occasions. Nevertheless, Whitehead recognizes and affirms the Parmenidean yearning for permanence in this world of change, a yearning whose sole fulfillment lies in eternity.

How are these two concepts of temporality—flux and permanence—to be brought together? Or to reframe this question within the context of this volume: What is Whitehead's proposed solution to the problem of "time and eternity"? Strikingly, Whitehead turns to religious experience to address this question and here he finds inspiration in the couplet of a now famous hymn. He even places its two classic verses center stage in his erudite exposition of the metaphysical concept of an actual occasion at the start of chapter 6 of *Process and Reality*:

> Abide with me;
> Fast falls the eventide.[10]

In a profound and stunning comment, Whitehead then tells us that in these two lines "we find formulated the complete problem of metaphysics. Those philosophers who start with the first line have given us 'substance'; and those who start with the second have developed the metaphysics of 'flux.' But, in truth, the two lines cannot be torn apart in this way."[11] But how, then, should substance and flux be combined? Whitehead offers a brilliant solution to this problem, one that melds the two notions—flux and permanence—into a single metaphysical structure. It combines what he calls "concrescence" and "transition" and locates them within the fundamental elements of reality, what Whitehead calls "actual occasions." By "concrescence" he means the coming to be of an actual occasion even as it is perishing. By "transition" he means the way in which the past universe of actual occasions are "prehended" (i.e., very roughly, "experienced") during concrescence and result to a certain extent in the conformity of the new occasion to the past (i.e., very roughly, efficient causality). For Whitehead, it is through this conformity that we are led to attribute to a long series of occasions the appearance of an enduring substance governed by previous efficient causes. And it is through the inherent novelty of each actual occasion that its specific form arises during concrescence. Finally God's subjective lure draws each occasion forward toward its greatest possible realization while respecting the inherent novelty and prehended causality of the occasion. "Concrescence moves towards its final cause, which is its subjective aim; transition is the vehicle of the efficient cause, which is the immortal past."[12] Thus Whitehead offers a metaphysics that involves a form of duration as the extended character of an actual occasion.[13] Out of this metaphysical concept of duration, he constructs a temporal concept of duration that provides a basis for viewing flowing time and genuine becoming at every level of complexity in nature—including fundamental physical processes. This returns us to a central question of this section of our chapter: Could Whitehead's metaphysics, as reflecting in part Pannenberg's concept of duration, inspire new research directions in physics and cosmology that understand time as duration?

This type of question—the efficacy of conceiving of time as irreducible duration instead of instantaneous present—was "in the air" in the late 1920s, the time of Whitehead's transition from physics to philosophy. The question of the infinite divisibility of space, time, and motion trace back, of course, to the ancient Greeks where they received their classic expression in the famous paradoxes of Zeno. As I have noted already, the independent development of the infinitesimal calculus by Newton and Leibniz in the seventeenth century included their idea of infinite divisibility: both point-like space and time and continuous motion in space through time (i.e., locomotion), as well as their assumptions about infinite divisibility, were embedded in the mathematics of the calculus. It was the calculus, in turn, that helped make physics so incredibly successful both at predictions and at inspiring new, culture-changing technologies.

In December 1900, however, a revolution in our classical conception of nature, grounded in Newton, was launched when Max Planck proposed his radical idea that electromagnetic energy can only be emitted in discrete, finite amounts (i.e., that energy is "quantized"), earning him the title "founder" of quantum mechanics. By 1913, Niels Bohr had developed his "solar system" model of the atom in which electrons orbit the atomic nucleus in circular orbits. But these orbits are highly nonclassical: they are "quantized," meaning that they only come in discrete sizes (no intermediate orbits are allowed), and when electrons absorb radiation they "jump" instantaneously from a smaller to a larger orbit without moving continuously between orbits. Within a very short period, scientists such as Henri Poincaré, J. J. Thomson, and—notably—scientist-philosopher Whitehead began to speculate that the continuum model of space, time, and motion might break down at the atomic level. In the 1930s, the terms "chronon" and "hodon" were proposed to designate this "atomic" structure of time and space, respectively. While the sizes of these irreducible temporal and spatial "atoms" are tiny—roughly 10^{-22} seconds and 10^{-13} centimeters respectively—this fact does not, in principle, render "the difference between the classical continuous space and time and its modern atomistic counterparts less radical."[14]

Still, there are significant conceptual problems associated with atomic time and space. One is how to put them into the relativistic framework of spacetime, in which they can be combined into a single four-dimensional "atomic" structure. Another, and perhaps much more significant, problem lies in the very concept of atomic versus point-like space and time. Take the idea of the chronon, for example. If a chronon has an irreducibly finite duration in time, say of 10^{-22} seconds, this implicitly assumes that the chronon is a duration or extension in time with a boundary composed of two instants in time, instants that begin and end with its duration and are separated by 10^{-22} seconds. Thus, as Milič Čapek argues, "the concept of chronon seems to imply its own boundaries; and as these boundaries are instantaneous in nature, the concept of instant is surreptitiously introduced by the very theory which purports to eliminate it."[15] It is in the context of these considerations and others that Whitehead constructed his metaphysical system that we touched on above. Published in 1929, *Process and Reality* offers a radically novel treatment of duration as ontologically prior to space and time (i.e., the "extensive continuum"), reflecting Whitehead's intense involvement with physics and mathematics prior to the near completion of nonrelativistic quantum mechanics circa 1930, as well as his contributions to the history of western philosophy.

Jumping ahead, now, by six decades, we focus on the year 1984 when process philosopher David Griffin convened a landmark conference at the Center for Process Studies in Claremont, California. Its purpose was to explore the relation between process philosophy and the research programs of David Bohm and Ilya Prigogine, both of whom were participants at the conference. Griffin's fascinating publication resulting from the conference, *Physics and the Ultimate Significance of Time,*[16] includes an historical background to the philosophical dimensions of Bohm's and Prigogine's research and the constructive relationship between Bohm, Prigogine, and process philosophy. Griffin's preface to the volume is a masterful assessment of the entire project. Moreover, the research programs begun by Bohm and Prigogine continue to this day to stimulate ongoing research in physics. It will therefore be interesting to explore these programs, even if briefly, as a response to the question of the positive influence of a philosophy of temporal becoming, shared in part by Pannenberg and Whitehead, on scientific research.

David Bohm is well known for his development of a nonlocal form of the "hidden variables" interpretation of quantum mechanics in the early 1950s.[17] Perhaps his most important move was to introduce what he called the "quantum potential," U:

$$U = -(\hbar^2/2m)(\nabla^2 R/R)$$

Here, $\hbar$ is Planck's constant h divided by 2π, and R is part of the mathematical expression for the wavefunction ψ, which is governed by Schrödinger's equation. It is the irreducibly holistic (i.e., nonlocal) character of the quantum potential U that rules out any interpretation of the time evolution of the wavefunction as due entirely to local, though hidden, causes (i.e., "local hidden variables"). Instead, the quantum potential arises from the entire environment—in principle, the universe. More recently, George Greenstein and Arthur G. Zajonc have given a particularly bold interpretation of Bohm's nonlocality, one that "goes beyond simple nonlocality, and calls upon us to see the world as an undivided whole. Even in a mechanical world of parts, the interactions between the parts could, in principle, be nonlocal but still mechanical. Not so in the quantum universe."[18]

In the decades that followed his initial work, Bohm sought to construct a metaphysics that would undergird his interpretation. As we will see it includes an unusual form of temporal duration. In what he called "the implicate order," Bohm viewed the world at its most fundamental level as a perpetual enfolding and unfolding of an intrinsic multidimensional manifold in which the whole of reality is implicit in each part.[19] The implicate order, as this manifold, extends over all space and time. The result of the enfolding and unfolding of this manifold is a series of momentary and isolated events in space and time that sequentially give the appearance of subsisting elementary particles moving through space in time. Because the implicate order enfolds all of reality within itself, these momentary events it gives rise to, while seemingly isolated in space and time, implicitly contain the universal global feature of "wholeness." Here indeed are grounds for a concept of duration in physics.[20] It is this global wholeness that represents, to me, Bohm's unique take on what can be called temporal duration, or "thickness." If one were to start with Pannenberg's views on time and eternity, one might be led to explore research in physics as stemming, at least in part, from Bohm's pioneering achievements.

And Bohm's approach to quantum mechanics (QM) does in fact continue to generate new directions in scientific research. For example, in 1999 Hua Wu and D. W. L. Sprung studied the application of Bohm's interpretation of QM to the problem of trajectories in chaotic quantum systems.[21] As Wu and Sprung stress, Bohm offers a classical interpretation of the motion of particles, one in which a particle simultaneously possesses precise position, x, and momentum, p. Such precision is forbidden by the Heisenberg uncertainty principle in QM: the product of the uncertainty in the position of the particle, Δx, and the uncertainty in its momentum, Δp, must be greater than or equal to $\hbar/2$, where h is Planck's constant and $\hbar = h/2\pi$. Nevertheless, it is possible in Bohm's approach because of its underlying classical view of particles following well-defined trajectories even while the "forces" affecting these trajectories include the highly nonclassical quantum potential.

More recently B. J. Hiley and R. E. Callaghan[22] have studied "delayed-choice experiments" involving quantum interference. The idea of a delayed-choice experiment was first introduced by J. A. Wheeler in 1978.[23] In these experiments, our choice of how to measure a beam of light in an interferometer seems to influence whether the light acts in a wave-like way or in a particle-like way *previous* to when the choice is made. Wheeler's conclusion is that "no phenomenon is a phenomenon until it is an observed phenomenon." Moreover, startling is his claim that "the past has no existence except as it is recorded in the present."[24] Wheeler's interpretation of the delayed-choice experiments might be seen as reflecting a form of what I call "retroactive causality" in discussing Pannenberg's understanding of time. But according to Hiley and Callaghan, the Bohmian interpretation allows us to understand the delayed-choice experiments entirely in terms of particles with well-defined trajectories and without claiming that "the past only (comes) into being by action in the present."[25]

In 2005 Roderick Sutherland published a "causally symmetric" or "time symmetric" formulation of Bohm's interpretation. (We will explore additional time symmetric formulations of QM as well as of electromagnetism and gravity in some detail in section D below.) In such a formulation, physical influences are included that move backward in time from the future along with the usual physical influences, that is, efficient causes, which move forward in time from the past. Such retroactive causality allows the future to play a causal role on the present state of the quantum system along with the usual contribution of past, efficient causality. In Sutherland's reformulation, Bohm's theory is consistent with special relativity: it is Lorentz invariant, it does not require a preferred reference frame (i.e., unique global axis of simultaneity), and it explains the nonlocality of Bell's theorem without its being in violation of special relativity.[26]

For decades Henry Stapp has found inspiration from Bohm's work for his own approach to QM, one which is related to process philosophy. Stapp's special focus is on the bearing of QM to consciousness and the mind-brain problem.[27] In addition, in 1991 Hiley and F. David Peat edited a crucial publication honoring Bohm's work and exploring a variety of implications for physics and philosophy.[28] Two more items should be noted: first, the University of Innsbruck, Austria, has a division on Bohmian mechanics,[29] and second, there is an ongoing multinational research group connected with the University of Munich, Germany, studying Bohmian mechanics. (Their website includes an extensive reference to books, preprints, events, and debates.)[30] Finally, a textbook on Bohmian mechanics is now available.[31]

Ilya Prigogine received the 1977 Nobel Prize in chemistry for his research on non-equilibrium thermodynamics, particularly his contributions to the theory of dissipative structures. These remarkable structures are "self-organizing." As background, in classical thermodynamics, the famous second law states that the entropy of systems closed off from their environment will necessarily increase until reaching a maximum. Here entropy can be thought of as a measure of the disorder of a system or, roughly the equivalent, as a measure of a system's loss of available energy.[32] But Prigogine shifted his research to the study of open, nonlinear systems, which exchange energy and matter with their environment and which are far from their equilibrium state. Such systems can undergo spontaneous and rapid transitions into new and more complex states. This means that in these open systems, new levels of order and complexity, and thus lower entropy, spontaneously occur as entropy is exchanged with the environment containing them. Of course the total entropy of the enveloping system (i.e., the open system and its environment) increases because it is governed by the second law. Nevertheless the system within the environment can become more complex and ordered.[33]

Note that the thermodynamics of both closed and open systems entails the possibility of a fundamental direction to time based on the second law. The opposite is true in the dynamics of classical mechanics in which the processes of nature are reversible in time.[34] This led Prigogine to prioritize thermodynamics over dynamics and to insist on the fundamental role of temporal irreversibility in nature as found in thermodynamics.[35] This groundbreaking research in thermodynamics led Prigogine to a richer concept of time as "internal" to natural processes and not just as something external to them and measured by simple physical clocks.

The following comparison can help make clear Prigogine's idea of internal time. Some objects seem to have a well-defined age, while others have what can be called an average age embodying a series of irreversible transformations over time. For example, the ruins of Pompeii are basically identical to what they were when Vesuvius erupted in 79 CE and "froze" Pompeii in time. In this sense we can say that if one visits Pompeii today (let's say 2012 CE), it has a well-defined internal age, namely 1933 years. On the other hand, it is easy to find buildings in Rome that have been remodeled many times over many decades and centuries. They might well contain both twentieth-century ornaments and eighteenth-century furniture, but they also might contain marble columns and brick walls dating back over two millennium. The effect of these successive reconstructions and additions together constitute the building's present structure and its age. In this sense, unlike the ruins of Pompeii, the buildings in Rome often do not have a unique, well-defined age. Instead, their age or "internal time" is an average over the ages of their reconstructions and additions in their irreversible development in time up to the present.

Prigogine mathematically formulates this intuitive idea of internal, irreversible time in terms of what is called a Baker transformation. This transformation operates on a thermodynamic system's phase space—the space of the system's canonical variables such as position x and momentum p.[36] When a Baker transformation is first applied to a region in phase space it splits the region in several small regions. When reiteratively applied to these smaller regions, more and more successively smaller regions are created. Eventually every region in phase space will be a composite of previous regions, and the trajectories of the system as it evolves in time will diverge from each previous region into new composite regions imitating the way the physical system spontaneously fluctuates into new states of increasing complexity and order. This series of new composite states represents Prigogine's idea of an internal time with an irreversible and historical character.

Research has continued over the past three-plus decades that takes up Prigogine's ideas in a variety of directions. In 1988 a major review article was published by Peter Coveney in *Nature* surveying work done on entropy, irreversibility, and dynamics based on Prigogine's work.[37] R. M. Keihn has extended Prigogine's work to a generalized theory of self-organization for non-equilibrium thermodynamic systems of four topological dimensions in comparison with closed non-equilibrium systems of three dimensions. He describes their application to a variety of synergetic systems ranging from mechanical systems to biological and even economic and political systems.[38] Giorgio Sonnino has studied thermo-dynamic field theory, aimed at determining the nonlinear flux-force relations for thermodynamic systems far from equilibrium using non-Riemannian geometry.[39] And today, dozens of scientists work at the Center for Complex Quantum Systems (formerly the Ilya Prigogine Center for Studies in Statistical Mechanics and Complex Systems and co-founded by Prigogine) at the University of Austin, Texas.[40] Their research involves chaos theory, the arrow of time, and the many-body problem. As a whole, these research programs reflect various aspects of Pannenberg's under-standing of the concept of temporal duration. Such programs, in turn, might provide avenues for further scientific research stimulated in part by Pannenberg's theology of time and eternity.

2. String Theory and Duration in Nature

I turn now to recent developments in string theory. With Prigogine's re-search we inquired about the possibility of duration in nature at the or-dinary level of experience as described by thermodynamics. With Bohm's work we probed the atomic and subatomic levels of nature as under-stood by quantum mechanics. Now we move far below the subatomic realm via string theory. Here, beginning in the 1960s, physicists sought to represent fundamental particles not as point-like objects but as one-dimensional loops, that is, "strings." And while strings are almost un-imaginably small—some twenty orders of magnitude smaller than an atomic nucleus—they are nevertheless finite in size and cannot be re-duced to point-like objects. String theory, in turn, opens up yet another theoretical possibility that physical objects may have irreducible *tem-poral* extension just as they have intrinsic *physical* extension, even at the almost unimaginably small scales involved here. And although string theory continues to be highly controversial, its proponents, including Mi-chael Green, Stephen Hawking, and Leonard Susskind, claim that string theory helps address what might well be the central conundrum of twenty-first-century physics.[41] This conundrum is how to achieve the unification of the three fundamental forces, electromagnetism, weak and strong nu-clear forces,[42] with the fourth fundamental force, gravity.

First, let me say a brief word of background to string theory. Quan-tum mechanics was formulated in the first three decades of the twentieth century by such luminary physicists as Niels Bohr, Albert Einstein, Irwin Schrödinger, and Werner Heisenberg. It is famous for the Schrödinger wave equation, the Heisenberg uncertainty principle, the Bohr model of the atom, and such wholly nonclassical phenomena as superposition, entanglement, and nonlocality.[43] From the 1930s to the1960s, physicists such as Paul Dirac, Freeman Dyson, and Richard Feynman reformulated QM in a way that is consistent with special relativity, resulting in relativistic QM. They then focused on the interaction between matter and the electromagnetic field, producing the first quantum field theory, quantum electrodynamics (QED).

By the 1970s, physicists had developed quantum field theories for the weak and the strong nuclear forces. A crucial step was taken when Sheldon Glashow, Abdus Salam, and Steven Weinberg merged the weak nuclear force with electromagnetism to produce electroweak field theory.[44] The weak and strong nuclear forces and electromagnetism, and such elementary particles as electrons, neutrinos, and quarks, form the basic ingredients of today's "standard model" of particle physics.[45]

But what about gravity? In 1916 Einstein published his general theory of relativity (GR), a theory of gravity that is consistent with SR. Having successfully passed several crucial tests, most physicists now consider general relativity to be a fundamental cornerstone of contemporary physics comparable to QM and the standard model. Nevertheless, the repeated attempts to unify general relativity and quantum field theory have been unsuccessful—hence the conundrum mentioned above. The central reason is that general relativity assumes that the universe can be represented by a curved spacetime manifold. Such a manifold is locally flat even while it is globally curved: when viewed at closer and closer ranges its curvature vanishes and the flat spacetime described by special relativity emerges (see chapter 5, section B.3). QM challenges this view. The Heisenberg uncertainty principle implies that everything, even empty space, undergoes constant and violent fluctuations whose intensity increases at smaller and smaller scales. Here, then, is the conundrum. As Brian Greene writes, "the notion of a smooth spatial geometry, the central principle of general relativity, is destroyed by the violent fluctuations of the quantum world on short distance scales."[46]

It is precisely here where string theory offers a way forward. As we saw above, the basic premise of string theory is that fundamental particles of the standard model are in fact tiny strings: oscillating one-dimensional loops. Their vibrations produce the masses and charges of the fundamental particles. Again, their size, roughly the Planck length (10^{-33} cm), is 10^{20} times smaller than the nucleus of an atom. Nevertheless they are finite, in stark contrast to the routine point-like concept of electrons, neutrinos, and quarks. Given this fundamental premise, the door is opened to reconciling gravity with the three forces in the standard model. In essence, when we study interactions between fundamental particles using string theory, the finite size of the strings precludes our probing the interaction below the Planck length. The result is that the intense fluctuations arising from the quantum treatment of spacetime can be ignored. In addition, the graviton, the particle that transmits the force of gravity, can be included in string theory.

Starting from this, a plethora of complex developments over the past four decades have convinced many theorists that string theory offers a fruitful path for unifying gravity and the three nongravitational forces. This unification starts with the classical symmetries in nature and their associated conservation laws[47] and combines them into what is called supersymmetry, based on the relation between the (integer and half-integer) quantum spins of fundamental particles. It continues with the realization that strings as one-dimensional loops in space can be generalized into two-dimensional membranes and higher-dimensional objects known as branes. Physicists have also discovered that while there are five alternative ways of formulating string theory, these formulations can actually be placed within a single framework called "M-theory." The striking implication, however, is that the universe is eleven-dimensional. Four of the eleven dimensions are the usual three dimensions of space and one of time, but the universe also includes seven spatial dimensions that are so compacted that they cannot be observed even with our most powerful instruments.[48]

How might such developments in string theory bear on our concern for duration as an irreducible temporal extension in nature? One possibility is that strings might actually be extended minutely *in time as well as in space.* When I raised this possibility to string theorist Gerald Cleaver his response was that string theory "strongly supports the idea of quantization of time, and not just space, with the Planck time, 10^{-44} seconds, [as] the fundamental block of time."[49] Cleaver offers two reasons for this idea. "One argument," Cleaver notes, is based on the idea of information exchange across a string (of Planck length). While we portray the fundamental particle as a string (membrane) with finite length (area), the state of the entire string is transformed as a whole when a string is involved in an interaction. We can't view this information as just traveling around the string at the speed of light. We understand this either as instantaneous transmission of information across the string or as occurring within a fundamental block of time. I believe the latter presents the better picture.[50]

A related argument comes from physicist S. Roy, who begins with a fundamental question raised by quantum gravity and string theory: Is nature continuous or discrete at or below the Planck scale? If, as Roy believes, it is in fact discrete, then

> there are basically two attitudes towards this discreteness. . . . One starts from the continuum concept and then tries to detect or create modes of "discrete behavior" on finer scales. . . . [The other tries] to describe how macroscopic space-time . . . emerges from a more fundamental concept like a fluctuating cellular network around the Planck scale. . . . It is generally believed that no physical laws which are valid in [ordinary] continuum space-time will be valid beyond or around the Planck scale. . . . Some scientists suspect space or time should be considered as emergent properties.[51]

Roy then connects quantum gravity and string theory with such discrete behavior at the Planck scale and with the idea of the emergence of physical time at larger scales. He also offers a fascinating discussion of several compelling philosophical implications of this research.[52]

The second reason Cleaver gives for why duration might show up as a precondition at the subatomic level is the modification of Heisenberg's uncertainty principle in string theory. According to Cleaver,

> the modification is an additional term which gives a lower bound to the smallest possible uncertainty in distance (Planck scale), and hence a corresponding minimum time unit. . . . Related arguments imply that it is impossible for the measurement of any closed distance to be smaller than a Planck length. Thus, under a Lorentz

transformation, the shortest corresponding time block is a Planck time unit.[53]

Finally, let us shift from the subatomic to the cosmological level. Suppose we assume that the universe itself, and not just the particles/strings in it, is characterized by the eleven dimensions of string/M-theory—four of them being spacetime and the remaining seven being highly compacted. Could the universe have additional temporal dimensions along with ordinary time, dimensions that in some ways represent the preconditions for what Pannenberg calls duration and that are compacted in a way similar to the extra seven curled-up spatial dimensions in M-theory?

Although highly speculative, this idea offers further suggestions for scientific research. For example, Itzhak Bars at the University of Southern California is exploring what he calls "2T" or "two time" physics that is related in several ways to string/M-theory. Bars starts with the premise that our "3 + 1" dimensional view of the world (which he refers to as "1T") should be extended with the addition of one temporal and one spatial dimension (thus "2T"). We can think of the 1T world (3 + 1 dimensions) as a shadow of the real 2T world (4 + 2 dimensions). The 2T world also includes curled up dimensions following the insights of string theory.[54]

In sum, research from the Planck length to subatomic physics to the physics of the everyday world to cosmology offers a variety of hints that nature might already include something like the physical preconditions for the possibility of temporal duration. It will be fascinating to see how this scientific research develops in the future.

C. CO-PRESENCE: SRP 2, THE SEARCH FOR NON-SEPARABILITY IN TIME

In section B, we explored the possible implications of my interpretation of Pannenberg's generic concept of duration as temporal thickness for research in physics, touching on the metaphysics of Whitehead and the theories of Bohm, Prigogine, and string theory. Now we turn to Pannenberg's unique contribution to the theological understanding of duration: duration as having an internal structure that I call co-presence, a structure whose ultimate ground is the differentiated unity of the eternity of the Trinitarian God. This idea of temporal co-presence arises within the general theme of "time *in* eternity." Here co-presence stands for the internal structure of duration both in time, as a present phenomenon in human experience and in nature, and in eternity, as the foundation for the Trinity's eternal creation of the temporal character of the world. Co-presence also points to the relation between time in this present creation, in which events as present are distinct by their ppf structure but separated into the unavailable future and irretrievable past, and time in the eschatological New Creation in which events retain their distinctive character while their separation is overcome by the complex structure, or co-presence, of the divine duration gifted to the New Creation by God its Creator. In the introduction to this volume I sharpened this idea of the relation between creation and New Creation by focusing specifically on the interpretation of the bodily Resurrection as "transformation." This interpretation implies that there will be elements both of continuity and discontinuity between the present creation and the New Creation. We now turn to one of the central questions of part two: Can Pannenberg's theological concept of duration as co-presence offer creative directions for scientific research under the rubric of "elements of continuity"? We will begin to explore this question here.

1. The Search for QM-like Non-separability in Time

In chapter 2, section B.3, I introduced the idea of quantum non-separability as an ontological interpretation of quantum entanglement, that is, the correlations between spacelike quantum events that defy a classical interpretation. As we saw there, spacelike quantum correlations between two particles could be thought to arise because of a superluminal interaction between them (which might be seen as violating special relativity). I chose instead to adopt the interpretation that quantum correlations between two particles are due to the fact that these particles are part of a single, spacelike, entangled quantum system with a non-separable ontology.[55]

So, is it possible to think analogously of *non-separability in time?* Will two particles, once bound in a single system and then released, continue to show quantum correlations that defy a classical explanation when these correlations are measured along a *timelike* trajectory between the particles, that is, when one particle lies in the causal future of the other *even when no causal interactions are involved?* And, if so, do these correlations point again to an underlying ontological non-separability between these two particles?

For example, suppose two timelike events show the kind of correlations noted by Bell's theorem. Normally we would assume that the correlations resulted from a routine physical interaction between them, with the earlier event affecting the later event. But is it possible that the correlations arise, in part, from the non-separable ontology of the two particles and due to the fact that the two particles form in fact a single "object"? Could this mean that the correlations are *not merely* the result of the first event determining the outcome of the second event by efficient causality, as one would routinely assume? Instead could the correlations arise, along with the efficacy of efficient causality, from an underlying ontology that gives rise to what I will call temporal non-separability?

Frankly, none of the references I have explored so far directly consider the possibility of temporal non-separability, although retrocausality is certainly discussed, as we will see below (section D.1). I leave this as an open question for future research.

2. The Search for Non-Hausdorff-like Non-separability in Time

As we have seen already at the Planck scale and with the idea of the emergence of physical time at larger scales, a non-Hausdorff manifold[56] might offer an illustration of what I am calling temporal co-presence as the internal structure of duration both in time, as a present phenomena in human experience and in nature, and in eternity, as the basis for the differentiated unity given to time by God both now and most fully in the New Creation. Following the guidelines set out in the introduction to this volume (see guidelines 6a and 6b, appendix to the introduction, section E.1), we can sharpen this idea by focusing specifically on the interpretation of the bodily Resurrection as "transformation." This interpretation implies that there will be elements both of continuity and discontinuity between the temporality of the present creation and the eternity of the New Creation. On the one hand, if temporal co-presence is an element of *discontinuity,* then it would not be fruitful to look for the preconditions for its possibility in nature. On the other hand, if temporal co-presence is an element of *continuity,* then it might be fruitful to look for preconditions for its possibility in nature today.

Let us explore the options for such preconditions in more detail. Our first option is to suppose that time in nature actually is Hausdorff. Then our routine assumption that it is so, as embodied in the theories of physics and cosmology, is correct, and this fact in turn helps explain why these sciences are so successful. In this case, the theological conclusion would be that time will be radically transformed by God into a non-Hausdorff form of temporality in the New Creation. The Hausdorff character of time as we now know it would then represent an element of eschatological discontinuity, and we would not expect to find suggestions here for new research in physics or cosmology.

But now for our second option: suppose instead that time in nature is in fact non-Hausdorff even though it is treated as Hausdorff in most of physics and cosmology. In this case, the non-Hausdorff character of time in nature represents an element of continuity between creation and the New Creation. This alternative, unlike the previous one, could lead to new research in science aimed at the discovery of the actual, non-Hausdorff character of time in nature. If this research is successful it might shed light on a further question: Why is the non-Hausdorff character of time so frequently suppressed in the theories of physics and cosmology, which in most cases have successfully represented time as Hausdorff? (This question will be picked up again in a different form in section E below.) Let us now explore ongoing research reflecting this second option.

We start with Mark Sharlow's 1998 paper "A New Non-Hausdorff Spacetime Model for Resolution of the Time-Travel Paradoxes." Here Sharlow discusses the problem of time travel paradoxes that can be found in chronology-violating spacetimes. If these spacetimes are given a non-Hausdorff topology, the paradoxes might be resolved, but this kind of topology, in turn, can lead to observers with multiple futures or pasts. Sharlow solves this problem by constructing a *non-separable,* non-Hausdorff spacetime. He also provides extensive references to non-Hausdorff spaces in mathematical physics.[57]

A decade later Sharlow published "The Quantum Mechanical Path Integral: Toward a Realistic Interpretation." This is a fascinating example of additional possibilities for non-Hausdorff manifolds in contemporary physics, examples that go against the usual Hausdorff formulations. First, however, a brief historical comment on the "path integral" approach to QM might be helpful. The path integral approach Sharlow appropriates was first developed by Richard Feynman and colleagues in the 1960s. Here the transition amplitude for a particle changing from quantum state A to B is represented by a mathematical sum over all possible paths between these states. So, in the famous example of the "double slit" experiment, we sum over the paths for the particle going through both the upper and the lower slit before reaching the detector. In his 2008 paper, Sharlow offers a realist interpretation of the path integral approach "according to which the particle actually follows the paths that contribute to the integral."[58] This, in turn, leads Sharlow to propose that spacetime has multiple branches and is, therefore, non-Hausdorff in structure.

Other relevant studies on non-Hausdorff spaces include Bertram Yood's work on non-Hausdorff spaces with unusual algebras and P. Hajicek's research on causal anomalies in non-Hausdorff spacetimes.[59] Whether any of this research will eventually lead to a *temporal* non-Hausdorff interpretation in current physics and cosmology is an open question and one which is worthy of future study.

D. ESCHATOLOGY: SRP 3, THE SEARCH FOR THE RETROACTIVE CAUSALITY OF THE IMMEDIATE FUTURE AND THE TOPOLOGY OF ESCHATOLOGICAL CAUSALITY IN PHYSICS AND COSMOLOGY

In chapter 1, I suggested that Pannenberg combines two new and distinct views of time: (1) the causal priority of the immediate future as a determining factor in what happens in the present (along, of course, with the causal efficacy of the past) and (2) the proleptic character of the eschaton appearing in historical time (i.e., the first Easter) as a transhistorical event "reaching back" from the eschaton of the New Creation to the event of Easter in history. In chapter 2, I offered the rough outlines of a topological model for prolepsis as a way to think about the transtemporal connection between the New Creation and historical events beginning with Easter.

My goal in the current section is to explore ways in which some areas in physics and cosmology might be viewed as reflective of the preconditions for the possibility of Pannenberg's two views of time. At the end of this chapter (section E) I will also seek to account for why time as we routinely know it in everyday life and in science lacks Pannenberg's sense of the causal efficacy of the future and, more crucially, the topological presence of the eschaton. I will do so by offering a theological rationalization for the "hiddenness" of eschatological time in present time.

1. Causal Efficacy of the Immediate Future: The Search for Backward Causality

In the overview to special relativity (SR) in chapter 4, we saw that Einstein's first postulate, the principle of relativity—namely that the laws of physics take the same form for observers in uniform relative motion—is rooted in the Galilean/Newtonian assumption of relativity underlying the laws of classical mechanics. With SR Einstein extended the principle of relativity to include Maxwell's electromagnetism and thus light, whose speed, c, is stipulated by the second postulate of SR to have the same value for all observers. To enact these postulates Einstein replaced the Galilean transformation with the Lorentz transformation, leading to the Minkowskian, four-dimensional spacetime interpretation of SR almost universally held by scientists today. The challenge to construct a "flowing time" interpretation of the spacetime interpretation of SR was the subject of chapter 5.

Now we turn to the search for "backward causality" (or the causality of the future on the present) in contemporary physics as one way of conceptualizing an element of continuity between the present creation and the New Creation following Pannenberg's idea of the "causality of the future."[60] As we shall see, electromagnetism can be given an unusual time symmetric formulation. Time is usually seen to flow irreversibly forward from the past to the future, reflecting the efficient causality of the past on the present as assumed in ordinary experience, even if the equations of motion are symmetric in time and allow, mathematically, for both forward (from the past) and backward (from the future) motion in time. In the unusual kind of time symmetric account we will explore here, however, the present emerges out of an underlying combination of forward and backward causality in which physical processes actually flow in both directions of time.

We begin with the standard interpretation of Maxwell's equations that summarize the complete phenomena of classical electromagnetism. According to this interpretation an accelerated charged particle emits light waves that move outward in space and forward in time. When absorbed by a second particle these waves cause it to accelerate, thus completing the picture of the efficient causality of the first particle on the second. But as just mentioned, classical electromagnetism can also be formulated in terms of a time symmetric account in which both forward-moving waves, that is, waves moving forward in time, and backward-moving waves, that is, waves moving backwards in time, are real processes exchanged between particles.[61] In this picture, the first particle both affects *and* is affected by the second particle. It is this "backward causality" of the second on the first particle which, I am suggesting, offers a very rough idea of a possible precondition in nature for Pannenberg's idea of the "causality of the future."[62]

First, however, a brief recounting of the history of the unification of electricity and magnetism in Maxwell's equations seems appropriate in order to understand how this possibility arises in physics. Beginning with their discovery in ancient Greece, electricity and magnetism were understood as entirely separate phenomena. In the seventh century BCE, Thales of Miletus discovered that amber, when rubbed, attracts straw electrostatically. Six centuries later Pliny the Elder found that magnetite, more commonly called lodestone (Fe_3O_4), attracts iron magnetically. As recently as the seventeenth century, scientists such as William Gilbert still described electricity and magnetism separately. Even a century later, Charles-Augustin de Coulomb could discuss electrostatics without reference to magnetism, showing that the electrostatic force between stationary charges varies as the inverse square of the distance between them.

Benjamin Franklin's experiments in the mid-eighteenth century, however, pointed to a relationship between electricity and magnetism, and in 1820 Hans Christian Oersted discovered a pivotal connection between them: electric current in a wire deflects a magnetic compass needle. In that same year, André-Marie Ampère showed that parallel wires containing electric currents attract or repel each other by the force of magnetism.

Eleven years later Michael Faraday demonstrated the principle of induction: when a magnet is moved past an electric coil it produces an electric current in the coil. Finally in 1873 James Clerk Maxwell published a set of four equations that mathematically unified all that was known about electricity and magnetism based on his concept of the electromagnetic field. It was a concept derived, in part, from Faraday. Maxwell also proposed that light is a wave of the electromagnetic field.

Maxwell's equations can be presented in a succinct style by using standard vector notation:[63]

$$\nabla \cdot \mathbf{E} = 4\pi\rho \tag{1a}$$
$$\nabla \times \mathbf{E} + (1/c)\, \delta\mathbf{B}/\delta t = 0 \tag{1b}$$
$$\nabla \cdot \mathbf{B} = 0 \tag{1c}$$
$$\nabla \times \mathbf{B} - (1/c)\, \delta\mathbf{E}/\delta t = (4\pi/c)\mathbf{J} \tag{1d}$$

Here $\mathbf{E}$ is the electric field, $\mathbf{B}$ the magnetic field, ρ is the charge density (i.e., the electric charge per unit volume), $\mathbf{J}$ the current density (i.e., the current per unit cross-sectional area), and c is the speed of light in a vacuum. Equation 1a is Coulomb's law of electrostatics. Equation 1b is Faraday's law of induction. Equation 1c rules out magnetic monopoles in nature—the magnetic analog to electric charge.[64] Equation 1d expresses the connection between the magnetic field and changes in time in the electric field and the current density.[65]

One can show that Maxwell's equations lead to the classical wave equation for $\mathbf{E}$ in the vacuum (i.e., where $\rho = 0$ and $\mathbf{J} = 0$):[66]

$$\delta^2\mathbf{E}/\delta t^2 = c^2\nabla^2\mathbf{E} \tag{2}$$

Let us apply the wave equation to a simple problem in physics. We know that when a charged particle is accelerated (i.e., nudged or wiggled) it emits light waves. Solutions to equation 2 can be used to describe these waves. Some solutions represent the way we believe light waves actually behave: they travel outward in time from the source (the accelerated charge) along the light cone. Think, for example, of light emitted from the sun. The light travels outwards in space and forward in time until it reaches the Earth some eight minutes later. Technically these solutions are called "retarded waves," but to avoid confusion I will call them "forward moving waves." But other mathematical solutions to equation 2 are possible: so-called advanced waves, which move outward from the source but backward in time, waves that move from the present into the past. Think of light waves that arrive at Earth eight minutes *before* they are emitted from the sun! As indicated above I will call them backward moving waves.

The *mathematical* possibility of both kinds of solutions to equation 2 is due to the fact that the wave equation is "time symmetric" or "time reversible": its form is unchanged by the substitution of "$-t$" for "t". This is a general result in classical physics: the equations of classical mechanics and the equations that describe the two classically known fundamental forces—gravity and electromagnetism—are time reversible. Compare this with classical thermodynamics. Its equations are not time symmetric; their solutions only depict processes flowing forwards in time.[67] This means that classical thermodynamics provides a possible basis for an "arrow of time" in nature: to put it succinctly, hot coffee, left alone, always cools and never spontaneously warms.

Now an interesting question arises: Is the time symmetry of electromagnetism merely a formality, an artifact of the equations, or could there really be "backward moving waves" in nature? Put another way, does the time symmetry that includes backward moving waves allow us to interpret Maxwell's portrayal of nature as involving both the usual efficient causes acting from the past and what we can call retroactive causes acting from the future? And if we take seriously what we can call this double causality, how will we "save the phenomena"—how will we explain why our everyday experience is always of light waves that move outward from their source into the future (e.g., think "light radiates out from the sun")? Why is there a unique "arrow of time" in nature in which only efficient causes seem to act from the past on the present?

These and other key questions were addressed in 1945 when John Archibald Wheeler and his then graduate student Richard Feynman published a fully time symmetric formulation of electromagnetism that explicitly included backward moving as well as forward moving electromagnetic waves.[68] The key to their formulation was to take into account what they called the "response" of the rest of the universe to waves emitted forward *and* backward in time.

Consider an electron placed at the origin, O (fig. 6.1):[69]

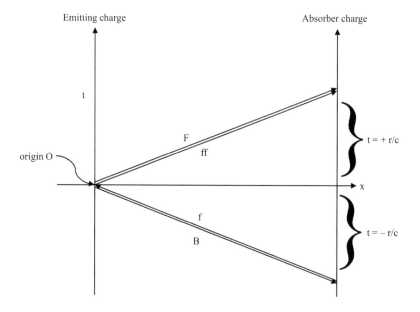

Figure 6.1. Forward Moving and Backward Moving Waves Emitted by the Charge at the Origin and Reflected by the Distant Absorber Cancel to Give Only Forward Moving Waves Expanding Outward from the Charge

If we nudge the electron at time t = 0, a forward moving wave "F" and a backward moving wave "B" will radiate outward at the speed of light.[70] Eventually another electron at a distance r will absorb the backward moving wave at a time t = −r/c. This electron then creates a reaction wave "f" that propagates forward in time and arrives at the origin at a time t = r/c − r/c = 0, that is, at *exactly* the moment when the charge at the origin is first nudged.[71] The forward moving wave "f" then passes by the electron at the origin and expands outward, where it is labeled "ff."[72] The net result is that the combination of the forward moving wave "F" and the forward moving wave "ff" yields the complete empirical wave moving forward from the charge at the origin as found in ordinary experience. Meanwhile, the backward moving wave "B" from the source and the forward moving wave "f" from the absorber *cancel each other out,* leaving no empirical evidence of light traveling backward in time—except for one fact. We know that when a charge emits light it experiences what is called radiative damping: a deceleration that slows down its motion. Moreover, the kinetic energy lost by the charge as it slows down equals the energy carried off by the light it emits in accordance with the conservation of energy.

But what produces radiative damping and why is its magnitude just right to satisfy the conservation of energy? In classical electromagnetism it was called a "self-force," but this term did not really explain either why it occurs or its magnitude. Wheeler and Feynman, though, showed that the self-force is actually due to the response "f" of the absorber on the charge at the origin and that this response is just the right amount to guarantee the conservation of energy. They then generalized this result and applied it to a universe filled with charges that constitute the future and past absorber. They called the combined reaction of these absorbers the "response of the universe."

Fred Hoyle and J. V. Narlikar took up the Wheeler-Feynman interpretation of electromagnetism, extended it to the physics of gravity in a relativistic spacetime, and developed it into a full cosmology, the so-called steady state cosmology.[73] This cosmology was aimed specifically at addressing the problem of t = 0 in Einstein's big bang model. Developed independently by Hermann Bondi, Thomas Gold, and Fred Hoyle, the steady state cosmology depicts the universe as expanding exponentially in size from an infinite past into an infinite future. In order that the distribution and density of galaxies remain constant in time, Hoyle and colleagues stipulated that matter was created continuously and spontaneously in space as it expands.[74] The steady state model was supported by many cosmologists for roughly two decades. In 1964, however, the discovery of the cosmic microwave background radiation by Arno Penzias and Robert Wilson marked the end of this model for the majority of cosmologists. From then on, Einstein's big bang model rapidly gained predominance even though its initial singularity, t = 0, has continued to be seen as problematic by many scientists and philosophers of science.[75]

Despite this setback, Hoyle and Narlikar continued to pursue an alternate relativistic theory of gravity whose conceptual roots lie in both the steady state model and, in turn, in the Wheeler-Feynman reformulation of electromagnetism.[76] One of the key features of their work is the distinctive way it incorporates Mach's Principle, in which the existence of the universe affects, even produces, what we think of as intrinsic properties of matter such as mass.[77] They then proposed that the redshift of distant galaxies can best be understood not as evidence of an *expanding* universe as measured by "rulers" of constant length, as is assumed in big bang cosmology, but instead as evidence of a *static* universe as measured by "rulers" whose length decreases, in time.[78] The decrease in length, in turn, is due to a proposed increase in mass M in time. This, in turn, has led Hoyle and Narlikar to a new cosmology that depicts an infinitely old and static universe without an initial absolute singularity, $t = 0$, as found in big bang cosmology, but one which can still account for all the redshift data from distant galaxies.[79]

Meanwhile, in more recent technical work, Hoyle and Narlikar show that in both open and closed big bang cosmologies, the Wheeler-Feynman proposal for combining forward moving and backward moving electromagnetic waves does not yield the correct empirical result—strictly forward moving waves—because the future and past distributions of matter in these cosmologies are not symmetric.[80] They also suggest a way to view the singularity $t = 0$ in big bang cosmology merely as a mathematical artifact. In their new cosmology, the past history of the universe is much vaster than big bang scenarios allow; indeed it could be infinitely old, a claim that clearly reflects the roots of their recent approach in the original steady state model.[81]

The work of Hoyle and Narlikar, in turn, has been studied widely over the past half century. Early papers include those by Stephen Hawking and Paul Davies.[82] More recently, G. F. R. Ellis has compared their work with other alternatives to GR, and Ignazio Ciufolini and John Archibald Wheeler have assessed it in detail in relation to the way Mach's Principle is incorporated into GR.[83] In addition, Helge Kragh gives a detailed historical overview of the many directions these ideas have taken in his masterful work *Cosmology and Controversy*.[84]

While I have been focusing on time symmetric approaches to electromagnetism and gravity, it is worth noting that a similar approach is being pursued regarding quantum mechanics. In section A of this chapter I referenced Roderick Sutherland's work on a time symmetric formulation of a Bohmian approach to QM. A time symmetric approach to Bohmian QM was also published in 2008 by Davide Fiscaletti and Amrit Srecko Sorli.[85] Additional research on a time symmetric formulation of QM is under way by a team of scientists and philosophers including Yakir Aharonov, Jeff Tollaksen, as well as David Albert, Paul Davies, Brian

Greene, and Maulik Parikh.[86] Its significance is reflected in part by the fact that Aharonov won the immensely prestigious 2010 National Medal of Science.

In sum, then, theistic scientists might well chose to explore these time symmetric approaches to electromagnetism, quantum mechanics, gravity, and cosmology because they offer something like the preconditions in nature that are suggestive of and might flow from Pannenberg's concept of the "causality of the future." To put this more tentatively, the fact that physics and cosmology can be construed in this way offers us a modest "proof of concept" that theology can indeed suggest fruitful directions for scientific research.

There is something deeply satisfying in all this. As is well known, one of the reasons that originally led Hoyle to construct and defend the steady state cosmology as an alternative to big bang cosmology was his ardent commitment to atheism and thus his strident rejection of t = 0 in big bang cosmology, a rejection exacerbated by the fact that many prominent Christians (including Pope Pius XI) took t = 0 as supportive of theism.[87] Instead, for Hoyle, a universe with no absolute beginning favored atheism. Surprisingly now, over a half a century later, time symmetric approaches to electromagnetism, quantum mechanics, gravity, and cosmology might be consonant with the kind of eschatology that a theist such as Pannenberg supports. In addition, time symmetric formulations of causality could be pursued within the framework of GR. If so, these formulations might preserve the feature of t = 0 that is seen by many as "friendly" to theism (*pace* the scientific challenge to it by inflationary and quantum cosmology) and as adding to it the kind of complex forward and backward causality that Pannenberg might well endorse.

And that, no doubt, would be satisfying indeed!

2. Prolepsis: The Search for a Topology Appropriate to an Eschatological Causality

If by following Pannenberg we believe that the eschaton, as "now" and "future," has already appeared proleptically in the events of Easter while yet remaining still to come, we might expect that the universe already has a more complex topology than that of ordinary spacetime in which simple worldlines trace out the history of particles. We might even expect to find hints of this more complex topology prefigured within the

theories of physics and cosmology. In chapter 2, I suggested that the prolepsis of Easter might lead to a continual series of proleptic connections
between the eschaton of the New Creation and the ongoing events in the
world today, as indicated in highly rudimentary and schematic terms by
figure 6.2a (Easter) and figure 6.2b (multiple prolepses) reprinted here:

CREATION   NEW   CREATION

prolepsis: resurrection of Jesus as the manifestation of the eschaton

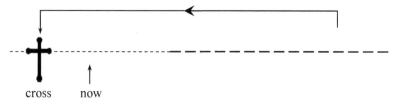

Figure 6.2a. Resurrection as Prolepsis

CREATION   NEW   CREATION

multiple prolepses: the manifestation of the eschaton at all
events in the future of this creation

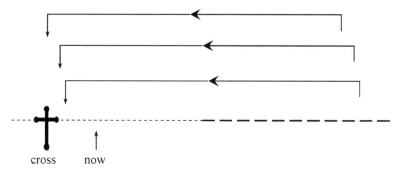

Figure 6.2b. Multiple Prolepses and the Continuity of Personal Identity

What implications might this have for our scientific view of nature?

One suggestion is that every spacetime event might be subject to a
"splitting": a topological feature not unlike that which we explored in
chapter 2 as a way to illustrate co-presence. One idea there was to rep-

resent co-presence by way of a non-Hausdorff manifold, such as the "complete feather," reprinted here as figure 6.3:

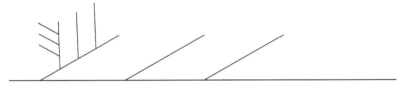

Figure 6.3. The "Complete Feather" under Construction

Note: When complete, the "spine" of the feather—the real line—has an infinity of branches, each of which has an infinity of branches and so on endlessly.

According to this idea, as we move into the future along a worldline (the "spine" of the feather), each event is an intersection of this single, regular path leading into the ordinary future and a proleptic path (a branch of the feather reaching back from the eschatological future) connecting that event with the eschaton located within the New Creation. Of course the feather model does not capture the element of the transformation of the ordinary future into the eschatological future, in which the "backward reaching" feathers are the transformed effects of the future New Creation. Nevertheless it captures something of the claim that each present moment now is both a moment leading to the ordinary future and a moment impacted by the arrival of the truly eschatological future.

Put more carefully, the branches of the feather lead back in multiple ways from the future eschaton of the transformed New Creation into each present event in nature. We begin with the Easter event as the normative intersection of ordinary history moving forward in time and the eschaton moving back from the New Creation to the Easter event. I propose we then extend this model of prolepsis to include all moments of time proceeding and following Easter as proleptic moments, even though their "power and authority" as proleptic connections between ordinary history and the Easter event is based strictly and entirely on Easter as the normative proleptic ("final") event. This model could capture something of the "now" and "future" double meaning of eschatology that Pannenberg affirms. In essence, something like a non-Hausdorff topology for physical time might be a precondition for the claim that the world as we know it is in fact being continually taken up into the presence of the Resurrected

Lord from the eschatological future even as the eschatological future is transformed out of the ordinary future as opposed to the mere continuation of ordinary, physical time.

This idea also goes a long way in responding to a crucial challenge raised by natural theodicy (see the appendix to the introduction, sections A.3 and E.1). Recall that following our guidelines a criterion for an acceptable eschatology is that it address the specific challenges raised by natural theodicy. One of these challenges is the ubiquity of death in nature long before the arrival of homo sapiens. Not only is death not the consequence of the Fall, but instead it is constitutive of the processes underlying the evolution of species. This fact seems to pose an insurmountable problem for faith in the seemingly unmitigated degree of suffering, disease, death, and extinction in nature. My response in general is to turn the problems raised by theodicy into criteria of theory choice in considering various eschatologies. Thus an eschatology based in part on the idea of a non-Hausdorff spacetime responds, at least in a preliminary way, to this challenge of natural theodicy by suggesting that the moment of death for every creature in nature is a proleptic moment of encounter with the Risen, saving Lord. Thus at the moment of death each creature passes bodily into the ordinary future but its totality as that creature, now transformed through the power of Christ, passes into the eschatological future, the New Creation, to live forever in joy. I would argue, then, that such an eschatology in which each creature has an eschatological destiny ("every sparrow that falls," Matt. 10:29, Lk. 12:6) goes a long way in responding to this challenge from natural theodicy.

I would like to pause for a moment to compare the way I used non-Hausdorff manifolds in chapter 2 and with the way I am using them here in chapter 6. In chapter 2 I used non-Hausdorff manifolds to represent what I called Pannenberg's idea of the co-presence of all events in eternity and to indicate the connection between individual death and the eschaton through multiple prolepses. In the present chapter I take non-Hausdorff manifolds to represent Pannenberg's eschatological idea of prolepsis as the "arrival of the future," and to extend prolepsis back into nature to include multiple prolepses taking place everywhere throughout the universe because of the normative proleptic event of Easter. I am also suggesting that the non-Hausdorff character of spacetime as presented above in section C.2 might lead to further research directions in physics and cosmology, such as we find in the research of Sharlow, Yood, and Hajicek.

But all this, in turn, raises a pressing question: Why is there so little evidence that time is anything other than our everyday view of time, the "spine" of the non-Hausdorff feather? Why is eternal, eschatological time, if it is truly real in nature, neither self-evident to us in ordinary life nor an obvious factor in the fundamental theories of physics and cosmology?

E. TIME VERSUS ETERNITY: A THEOLOGICAL JUSTIFICATION FOR THE PHYSICAL HIDDENNESS OF ETERNAL TIME WITHIN CREATURELY TIME

In this chapter I have sought to probe into what might be hints and glimmers of duration, co-presence, the causality of the future, and prolepsis in the ways we currently describe the universe via physics and cosmology. However, an underlying question remains: Why are these complex temporal structures—if true features of nature and not just subjective anticipations of the temporality of the eschatological eternity of the New Creation—not in some much more straightforward way a part of the treatment of time in contemporary physics? Indeed, why are there at most, and then only dimly, hints of what might be their preconditions in today's universe?

I offer two types of theological responses to this crucial question. The first is based on Luther's theology. According to Luther one can find two quite distinct theologies of revelation in the New Testament texts, and Luther was adamant in supporting one over the other. The first is the "theology of glory" (*Theologia Gloriae*) and the second is the "theology of the cross" (*Theologia Crucis*). On the one hand, the theology of glory places confidence in the power of things as seen by human senses and as known through reason to reveal God's true, but unseen, action. Luther cites Romans 1:20a as the classic text here: "Ever since the creation of the world [God's] invisible nature, namely, his eternal power and deity, has been clearly perceived in the things that have been made."[88] On the other hand, the theology of the cross argues that the cross is the only place where we find revealed who God truly is and how God truly saves.[89] According to Ted Peters, such a theology "begins with the disappointment of Jesus' followers . . . [when their hopes were] smashed on the rocks of Golgotha."[90] It is rooted in Luther's recognition of the theological absurd-ity of a crucified messiah: "It is most difficult . . . to recognize as King one who has died such a desperate and shameful death. . . . Plainly this will be folly to the Gentiles and a stumbling block to the Jews (1 Cor. 1:23) unless you raise your thoughts above all this."[91]

Taking up Luther's theology of the cross I suggest that the true char-acter of time—its co-present, proleptic, retroactively causal, and es-chatological character—is hidden within the ordinary features of time found in our daily experience of time and enshrined in physics and cos-mology. To use terms shared by Jürgen Moltmann and Peters, I describe this hiddenness by saying that *adventus*, the eschatological power of the true future (i.e., the coming and real presence of Christ), is hidden under *futurum*, the non-eschatological pretense of the ordinary and predictable future (i.e., the appointments in our personal calendars).[92] But why should the eschatological future be so hidden as to be almost invisible? Again, I return to Luther: I believe that this hiddenness is rooted in, and explained as a result of, God's decision/eternal decree to save the world through the absurdity of the cross. If so, the full temporal character of the world, as a precondition for the possibility of God's eschatological action in sav-ing the world and transforming it into the endless New Creation, is hid-den under the ordinariness and futility of temporal processes that lead into a future just like our present, one in which mortal life is in the end overwhelmed by death and the grave.

The second response to the question about the hiddenness of the world's real temporal complexity comes through a careful analysis of the theodicy offered by John Hick, one which helps, to a certain extent, with the problem of suffering in nature. In his well-known book *Evil and the God of Love,*[93] Hick argues that ours is a "soul-making" world, drawing on his reading of both Irenaeus and Frederick Schleiermacher. If part of God's purpose in creating our world is to enable the development of moral character and virtue through the tested decisions of free individu-als, then our world must be one in which both pain and pleasure, sorrow and joy, occur in everyone's life. Indeed, their "haphazard and unjust dis-tribution" is objectively beneficial because "the right must be done for its own sake rather than as a reward." One of the crucial preconditions for ours to be a soul-making world is therefore what Hick calls "epistemic distance." In essence, God must be hidden behind the veil of nature and history if God's presence is not to be so overwhelmingly evident that our moral choices are no longer freely taken but forced by the self-evident and overwhelming majesty of the divine presence. Thus, in a famous phrase drawn from Dietrich Bonhoeffer, Hick writes that the world must be "*etsi deus non daretur*" (as if there were no God).[94] And I would add to this that God's presence must not be so completely hidden that atheism is self-evident nor so convincingly present that moral development as the product of freely chosing the good for its own sake would be undermined. In short, the "as if" qualifier by Hick is a vital feature of the epistemic ambiguity of the world, since it leaves open the possibility of genuine moral development and an ongoing life of discipleship.[95]

In a previous analysis of Hick, I showed that epistemic distance is also a requirement for the possibility of studying nature through the natural sciences. The reason is that natural science is based on methodological naturalism: a scientific explanation of the processes of nature should rely on natural causes alone without the introduction of divine causation. Epistemic distance seems to be precisely what is required for natural science, based on methodological naturalism, to be possible: if the world is as if there were no God, it would be possible to obtain valid (though limited) knowledge of it without appeal to God. In short, then, there is an intimate and as yet unnoticed connection between the epistemic prerequisites for a world open to moral growth and for a world open to scientific research.[96] We can take this one step further. If ours is a world of epistemic distance, then just as historical and personal events can be interpreted both theistically and atheistically, so should science, the theistically neutral epistemic tool for discovery in nature, be interpretable both theistically and atheistically. Thus ours is a world in which "theology and science" is not only intellectually possible but actually required if we are to make a scholarly and convincing case for a theistic interpretation of natural science that is more robust than its atheistic interpretation. In sum, a world created by God for the possibility of moral growth is also a world in which science is possible and a world in which the field of "theology and science" is required to produce a convincing theistic interpretation of science against its competitors.

This connection has a direct implication for our question here about the hiddenness of the full temporal complexity of the world. On the one hand, our world must be such that the essential structure of eternal, eschatological time in its relation to creaturely time must be hidden within the structures of creaturely time. A world adequate for moral growth is a world in which time will not seem, at least in its explicit form in everyday experience and natural science, self-evidently to have the complex structure reflecting its true relation to eternity both in the present and in the eternal eschaton, even if it does in fact have such a structure, or at least the preconditions for its possibility for being transformed into this temporality in the New Creation. On the other hand, as I have already said, if it is to be one in which theistic moral growth can happen, our world cannot be one in which atheism is a completely self-evident and foregone conclusion. Nature must, instead, be ambiguous about the "God question." If we extend this to the question about time in science, it would seem that science cannot be totally incapable of "detecting" something of the more complex structure of time than we find in ordinary physics, even if we only find hints of this structure in its theories. In essence the full complexity of time that theology points to will be mostly hidden under the contours of ordinary time and yet some vestiges of it will be detectable, at least in glimmers, in, under, and through those contours. Thus, science should be open to the possibility of looking fruitfully for a more complex temporal structure in nature, the task begun in this chapter.

NOTES

Introduction

1. For this reason many theological treatments of "time and eternity," being limited to the way the world is now and not as it will become eschatologically, are inadequate. This problem arises in force when the context is shifted from systematic and philosophical theology and restricted to philosophy per se.

2. For an excellent study of the diversity of positions on the doctrine of the Trinity in twentieth-century theology and the articulation of his own unique perspective, see Ted Peters, *God as Trinity: Relationality and Temporality in the Divine Life* (Louisville, KY: Westminster/John Knox Press, 1993).

3. Julian of Norwich, *Showings,* trans. Edmund Colledge, O.S.A., and James Walsh, S.J. (New York: Paulist Press, 1978), 225.

4. The other fundamental theory is quantum mechanics (QM).

5. Albert Einstein, in a letter to the wife of Michele Besso, quoted in Ilya Prigogine, *From Being to Becoming: Time and Complexity in the Physical Sciences* (San Francisco: W. H. Freeman, 1980), 203.

6. The title of Moltmann's Gifford Lectures, *God in Creation,* which emphasizes the coming of God into our world, was one of several sources inspiring the title of this book, which emphasizes the temporality of our world as sourced within the divine eternity. Jürgen Moltmann, *God in Creation: A New Theology of Creation and the Spirit of God,* Gifford Lectures 1984–85 (San Francisco: Harper and Row, 1985).

7. Peters offers an excellent definition of prolepsis: it is "the pre-actualization of the future consummation of all things in Jesus Christ . . . who died on Good Friday and rose from the dead on Easter Sunday . . . [Jesus] anticipates in his person the new life that we humans and all creation are destined to share." See Ted Peters, *God—the World's Future: Systematic Theology for a Postmodern Era* (Minneapolis: Fortress Press, 1992), xi.

8. That such an interaction is possible is based on the fruits of some six decades of research on the kinds of constructive relations between theology and science. It presupposes and builds upon the pioneering scholarship of Ian G. Barbour, particularly his proposal of a detailed analogy between scientific and theological method. It also includes the immense contribution of Arthur Peacocke in delineating an epistemic hierarchy between academic disciplines starting with

physics and ending with cultural studies, one in which epistemic emergence includes both constraints and non-reducibility. It is indebted to the arguments by Nancey Murphy that further address epistemic holism and support both causal non-reducibility and post-foundational epistemology. The result is CMI. For details on the development of CMI, see Robert John Russell, *Cosmology from Alpha to Omega: The Creative Mutual Interaction of Theology and Science* (Philadelphia: Fortress Press, 2008), introd.

9. Cf. Russell, *Cosmology from Alpha to Omega,* chap. 10.

10. Wolfhart Pannenberg, "Theological Questions to Scientists," in *The Sciences and Theology in the Twentieth Century,* ed. A. R. Peacocke (Notre Dame, IN: University of Notre Dame Press, 1981), 3–16, cf. 12.

Appendix to the Introduction

Portions of section B are taken from Robert John Russell, "Bodily Resurrection, Eschatology, and Scientific Cosmology: The Mutual Interaction of Christian Theology and Science," in *Resurrection: Theological and Scientific Assessments,* ed. Ted Peters, Robert John Russell, and Michael Welker (Grand Rapids, MI: Eerdmans, 2002), 3–30. A previous version of section C was published as Robert John Russell, "Scientific Insights into the Problem of Personal Identity in the Context of a Christian Theology of Resurrection and Eschatology," in *Personal Identity and Resurrection,* ed. Georg Gasser (Ashgate, 2010), 241–58. Portions of sections D.4 and D.5 were previously published in Russell, *Cosmology from Alpha to Omega: The Creative Mutual Interaction of Theology and Science* (Philadelphia: Fortress Press, 2008), chap. 10.

1. In this sense Pannenberg's method seems to me to be unfortunately similar to that of Intelligent Design where secular science is challenged for not including agency in its explanatory framework.

2. I eventually developed a complex typology of forms of contingency based primarily on an analysis of Pannenberg's writings. These included the ontological contingency of the universe as a whole and of each part within it, referring to their sheer existence, and "existential contingency" of the universe as a whole and of each of its parts, referring to their mode of existence. They also included "first instantiation contingency" as one of several forms of what I called "nomological contingency." See Russell, *Cosmology from Alpha to Omega,* introd., pp. 14–16, and chap. 1.

3. Ernan McMullin, "How Should Cosmology Relate to Theology?" in *The Sciences and Theology in the Twentieth Century,* ed. A. R. Peacocke (Notre Dame, IN: University of Notre Dame Press, 1981), 17–57.

4. See Russell, *Cosmology from Alpha to Omega,* 11–14.

5. See Robert John Russell, "Five Key Topics on the Frontier of Theology and Science Today," *Dialog: A Journal of Theology* 46, no. 3 (Fall 2007): 199–207.

6. For a recent scholarly survey and series of proposals on theistic evolution, see Ted Peters and Martinez Hewlett, *Evolution from Creation to New Creation: The Controversy in Laboratory, Church, and Society* (Nashville: Abingdon Press, 2003).

7. Russell, *Cosmology from Alpha to Omega,* chaps. 4–6.

8. See George F. R. Ellis, "Ordinary and Extraordinary Divine Action: The Nexus of Interaction," in *Chaos and Complexity: Scientific Perspectives on Divine Action,* ed. Robert J. Russell, Nancey C. Murphy, and Arthur R. Peacocke (Vatican City State: Vatican Observatory; Berkeley, CA: Center for Theology and the Natural Sciences, 1995), 359–96; George F. R. Ellis, "Quantum Theory and the Macroscopic World," in *Quantum Physics: Scientific Perspectives on Divine Action,* ed. Robert John Russell, Philip Clayton, Kirk Wegter-McNelly, and John Polkinghorne (Vatican City State: Vatican Observatory; Berkeley, CA: Center for Theology and the Natural Sciences, 2001), 259–91; Nancey Murphy, "Divine Action in the Natural Order: Buridan's Ass and Schrödinger's Cat," in *Chaos and Complexity,* 325–58; Robert John Russell, "Divine Action and Quantum Mechanics: A Fresh Assessment," in *Quantum Mechanics: Scientific Perspectives on Divine Action,* ed. Robert John Russell, Philip Clayton, Kirk Wegter-McNelly, and John Polkinghorne (Vatican City State: Vatican Observatory; Berkeley, CA: Center for Theology and the Natural Sciences, 2001), 293–328; Thomas F. Tracy, "Particular Providence and the God of the Gaps," in *Chaos and Complexity,* 289–324; Thomas F. Tracy, "Creation, Providence, and Quantum Chance," in *Quantum Mechanics,* 235–58. See also Ian G. Barbour, "Five Models of God and Evolution," in *Evolutionary and Molecular Biology: Scientific Perspectives on Divine Action,* ed. Robert John Russell, William R. Stoeger, and Francisco J. Ayala (Vatican City State: Vatican Observatory; Berkeley, CA: Center for Theology and the Natural Sciences, 1998), 419–42; Philip Clayton, "Tracing the Lines: Constraint and Freedom in the Movement from Quantum Physics to Theology," in *Quantum Mechanics,* 211–34.

9. This is a strictly theological answer using the philosophical option for NIODA. It presupposes and accepts the standard theories in evolutionary biology and interprets them in terms of a theology of divine action. It is completely distinct from Intelligent Design which challenges these standard theories, trying to find evidence for their failures and inadequacies, and then arguing that we should change them by changing the methodology underlying the natural sciences (i.e., methodological naturalism) and by introducing an "intelligent agent" as part of the *scientific* explanation of evolution.

10. Natural evil can be defined as both biological evil (e.g., suffering, disease, death, and extinction) and physical evil (e.g., hurricanes, tsunamis,

earthquakes, meteor impact). For an extensive discussion see Christopher Southgate, *The Groaning of Creation: God, Evolution and the Problem of Evil* (Westminster/John Knox Press, 2008).

11. Michael Ruse, *Can a Darwinian Be a Christian? The Relationship between Science and Religion* (Cambridge: Cambridge University Press, 2001).

12. Russell, *Cosmology from Alpha to Omega,* chaps. 7, 8.

13. Reinhold Niebuhr, *The Nature and Destiny of Man: A Christian Interpretation,* vol. 1, *Human Nature* (New York: Charles Scribner's Sons, 1941, repr. 1964).

14. For a discussion, see *The Evolution of Evil,* ed. Gaymon Bennett, Martinez J. Hewlett, Ted Peters, and Robert John Russell (Vandenhoeck and Ruprecht, 2008).

15. Robert John Russell, "Physics, Cosmology, and the Challenge to Consequentialist Natural Theodicy," in *Physics and Cosmology: Scientific Perspectives on Natural Evil,* ed. Nancey Murphy, Robert John Russell, and William R. Stoeger, S.J. (Vatican City State: Vatican Observatory; Berkeley, CA: Center for Theology and the Natural Sciences, 2007), 109–30.

16. Thomas F. Tracy, "Evolution, Divine Action, and the Problem of Evil," in *Evolutionary and Molecular Biology: Scientific Perspectives on Divine Action,* ed. Robert John Russell, William R. Stoeger, and Francisco J. Ayala (Vatican City State: Vatican Observatory; Berkeley, CA: Center for Theology and the Natural Sciences, 1998), 511–30.

17. Karl Barth, *Church Dogmatics,* 4 vols. in 12, ed. G. W. Bromiley and T. F. Torrance, trans. G. T. Thomson et al. (T and T Clark: Edinburgh, 1936–77), 3.1, 388 ff.

18. John Hick, *Evil and the God of Love,* rev. ed. (San Francisco: Harper and Row, 1966).

19. Denis Edwards, "Every Sparrow That Falls to the Ground: The Cost of Evolution and the Christ-Event," *Ecotheology* 11, no. 1 (March 2006): 103–23.

20. Marilyn McCord Adams, *Horrendous Evil and the Goodness of God* (Ithaca, NY: Cornell University Press, 2000).

21. John Polkinghorne, *The Faith of a Physicist: Reflections of a Bottom-up Thinker* (Princeton: Princeton University Press, 1994), 83–85.

22. In his response to critics, Hick too tells us that the only adequate response to evil is eschatology. See John Hick, *Evil and the God of Love,* esp. 362–64.

23. Ted Peters, *Anticipating Omega: Science, Faith, and Our Ultimate Future* (Gottingen: Vandenhoeck and Ruprecht, 2006).

24. John 11:38–44. The New Testament also documents the resuscitation of Jairus's daughter in Mark 5:21–43 (see also Matt. 9:18–26 and Luke 8:40–56) and the widow's son at Nain (Luke 7:11–15).

25. N. T. Wright, *The Resurrection of the Son of God* (Minneapolis: Fortress Press, 2003), 8.

26. Rudolf Bultmann, *Jesus Christ and Mythology* (New York: Charles Scribner's Sons, 1958), 14–15.

27. Gerald O'Collins, S.J., "The Resurrection: The State of the Questions," in *The Resurrection: An Interdisciplinary Symposium on the Resurrection of Jesus* (Oxford: Oxford University Press, 1997), 6. I am using O'Collins' "what/why" terminology here. Note that O'Collins is also drawing on John Hick's terminology in *The Metaphor of God Incarnate* (London: SCM Press, 1993); see O'Collins, "The Resurrection," 6n3.

28. O'Collins, "The Resurrection," 6; Hick, *Metaphor of God Incarnate,* 24.

29. Norman Perrin, *The Resurrection Narratives: A New Approach* (Philadelphia: Fortress Press, 1977), 82–83.

30. Willi Marxsen, *The Resurrection of Jesus of Nazareth,* trans. Margaret Kohl (Philadelphia: Fortress Press, 1970), 77, 156.

31. Sallie McFague, *Models of God: Theology for an Ecological, Nuclear Age* (Philadelphia: Fortress Press, 1987), 59; McFague is drawing here on Norman Perrin's work; see *Models of God,* 199n1.

32. Hans Küng, *On Being a Christian* (Garden City, NY: Doubleday, 1984), 364–66; see also Hans Küng, *Credo* (London: SCM Press, 1993), 104–5.

33. Stephen Davis, *Risen Indeed: Making Sense of the Resurrection* (London: SPCK, 1993), 40.

34. This certainly holds for Barth circa the *Church Dogmatics* where the Resurrection is an historical event separate from the Crucifixion, and the ET is essential to the meaning of the Resurrection. The early Barth circa *Romans* tends to treat the Resurrection entirely as revelation, as nonhistorical and not separate from the Crucifixion, and Barth tends, there, to view the ET as irrelevant. For O'Collins's comparison of Barth and Pannenberg on these and other issues, see Gerald O'Collins, S.J., *Jesus Risen: An Historical, Fundamental and Systematic Examination of Christ's Resurrection* (New York: Paulist Press, 1987).

35. For a list of thirty-seven such scholars, see O'Collins, "The Resurrection," 13–17. O'Collins reprints a list of thirty scholars from his *Jesus Risen* (p. 123) and adds seven more names to it (p. 14).

36. Raymond E. Brown, *The Virginal Conception and Bodily Resurrection of Jesus* (New York: Paulist Press, 1973), 72.

37. Davis, *Risen Indeed,* 40.

38. Rudolf Bultmann, *History of the Synoptic Tradition* (1931; Oxford: Blackwell, 1963), 287–91. The roots of Bultmann's work on demythologizing the Bible are found in his earlier work on the New Testament. See Rudolf Bultmann, *Theology of the New Testament: Complete in One Volume,* trans. Kendrick Grobel (New York: Charles Scribner's Sons, 1951, repr. 1955); and Bultmann, *Jesus Christ and Mythology.* In *History,* Bultmann applies these earlier methods of form criticism to the Synoptic Gospels, producing a detailed analysis and classification of this material.

39. One of the most thorough examinations of the biblical material regarding the resurrection of Jesus and the eschatological New Creation can be found in David Wilkinson, *Christian Eschatology and the Physical Universe* (New York: T and T Clark, 2010), chap. 5, pp. 89–114.

40. Gerald O'Collins, S.J., *The Resurrection of Jesus Christ* (Valley Forge, PA: Judson Press, 1973), 95.

41. Arthur Peacocke, *Theology for a Scientific Age: Being and Becoming— Natural, Divine and Human,* enlarged ed. (Minneapolis: Fortress Press, 1993), 279–88.

42. Pheme Perkins, *Resurrection: New Testament Witness and Contemporary Reflection* (London: Geoffrey Chapman, 1984); Küng, *On Being a Christian,* 364–66. This is precisely the reference O'Collins uses in his argument against Küng. See O'Collins "The Resurrection," 13.

43. Peacocke, *Theology for a Scientific Age,* 332.

44. Ibid., 344–45. The position developed by Thorwald Lorenzen represents an unusual hybrid in supporting the bodily resurrection without reliance on the historicity of the ET, a hybrid somewhat like that of Peacocke. See Lorenzen, *Resurrection and Discipleship: Interpretive Models, Biblical Reflections, Theological Consequences* (Maryknoll: Orbis Books, 1995), esp. chap. 8. On balance I am skeptical about whether this position is internally coherent.

45. See John Polkinghorne, *The Way the World Is* (Grand Rapids, MI: Eerdmans, 1983), chap. 8, esp. p. 91; Polkinghorne, *Faith of a Physicist,* chaps. 6, 9, esp. 114–15 and 163–70; John C. Polkinghorne, *Serious Talk: Science and Religion in Dialogue* (Valley Forge, PA: Trinity Press International, 1995), chap. 7.

46. Brown, *Virginal Conception,* 108–11. Brown offers this hypothesis in part because he rejects the attempt to harmonize the various appearance traditions. Instead, following the argument of A. Descamps, Brown supports the "single appearance" interpretation, and his hypothesis suggests an historical rendering of how a single appearance to the Twelve can lead to multiple appearances in Jerusalem and in Galilee. Pannenberg cites von Campenhausen as rejecting the "flight" hypothesis as a "legend of the critics," but Pannenberg, in turn, argues against von Campenhausen's psychological interpretation of the flight. See JGM, 104–5. For a recent extensive and careful defense of the "flight" hypothesis, including his criticism of Hans Conzelmann's attempt to refute it, see Lorenzen, *Resurrection and Discipleship,* 119–22.

47. JGM, pt. 1, sec. 4, "Jesus' Resurrection as a Historical Problem," pp. 88–106.

48. JGM, 74.

49. JGM, 98.

50. Brown, *Virginal Conception,* 128–29.

51. Janet Martin Soskice, "Resurrection and the New Jerusalem," in *The Resurrection: An Interdisciplinary Symposium on the Resurrection of Jesus* (Oxford: Oxford University Press, 1997), 48–49.

52. Soskice, "Resurrection," 57.

53. O'Collins, *Jesus Risen,* 154–57.

54. Ibid., 178–79.

55. O'Collins, "The Resurrection," 23–25.

56. N. T. Wright, *Surprised by Hope: Rethinking Heaven, the Resurrection, and the Mission of the Church* (New York: HarperOne, 2008), 151–62. For extensive technical details, see Wright, *Resurrection of the Son of God.*

57. Additional NT scholars who deal with the bodily resurrection and its entailments for eschatology and cosmology include Davis, *Risen Indeed*; Kenan B. Osborne, *The Resurrection of Jesus: New Considerations for Its Theological Interpretation* (New York: Paulist Press, 1997), 14–21; Sandra Schneiders, "The Resurrection of Jesus and Christian Spirituality," in *Christian Resources of Hope,* ed. Maureen Junker-Kenny (Dublin: Columba, 1995), 81, 109.

58. The following are four representative examples of the interrelated issues arising from eschatological discussion: 1) In what ways are the resurrection of Jesus and the general resurrection at the end of this world similar (i.e., both are bodily) but yet different (i.e., his accomplishes ours; his happened "in three days")? This leads to fundamental issues in Christology and soteriology, particularly atonement theory. 2) What is the basis for the continuity of identity between us at our death and us in the general resurrection? This leads to fundamental issues in theological anthropology. 3) In what way do all of earth's creatures participate in the new heaven and earth? This will lead to the doctrine of creation, the problem of "natural evil," and theodicy, and their relation to eschatology. 4) As an act of the eternal God, what is the relation between these events at "the end of time" and ordinary, daily time? This takes us to the problem of time and eternity in the doctrine of God. I will not pursue the first three issues in detail in this book, though I will return to them briefly below as offering resources for my approach to the problem of eschatology and cosmology, particularly the problem of continuity and discontinuity.

59. Being a transformation—or what Polkinghorne calls a creation *ex vetere*—and not a second creation, it is embedded in the underlying doctrine of creation *ex nihilo.*

60. Wolfhart Pannenberg, "Theological Questions to Scientists," in *The Sciences and Theology in the Twentieth Century,* ed. A. R. Peacocke (Notre Dame, IN: University of Notre Dame Press, 1981), 3–16; repr. in Wolfhart Pannenberg, *Toward a Theology of Nature: Essays on Science and Faith,* ed. Ted Peters (Louisville, KY: Westminster/John Knox Press, 1993), 15–28.

61. Some discussion of cosmology can be found in Hans Schwarz, *Eschatology* (Grand Rapids, MI: Eerdmans, 2000), esp. chap. 7 pt. 2a, pp. 387–90; Ulrich H. J. Körtner, *The End of the World: A Theological Interpretation,* trans. Douglas W. Stott (Louisville: Westminster John Knox Press, 1995), chap. 4, sec. 6, pp. 138–48; Richard Bauckham and Trevor Hart, *Hope Against Hope: Christian Eschatology at the Turn of the Millennium* (Grand Rapids, MI: Eerdmans,

1999), 127–32; and George L. Murphy, "Hints from Science for Eschatology—and Vice Versa," in *The Last Things: Biblical and Theological Perspectives on Eschatology,* ed. Carl E. Braaten and Robert W. Jenson (Grand Rapids, MI: Eerdmans, 2002), 146–68.

62. For a helpful online survey, see "Christian Eschatology," *Wikipedia,* last modified August 16, 2011, http://en.wikipedia.org/wiki/Christian_eschatology. See also Robert John Russell, "Cosmology and Eschatology," in *Oxford Handbook of Eschatology,* ed. Jerry Walls (Oxford: Oxford University Press, 2006), 563–80.

63. Albert Schweitzer, *The Quest of the Historical Jesus,* trans. James M. Robinson (New York: Macmillan, 1968). See also Albert Schweitzer, *The Mystery of the Kingdom of God: The Secret of Jesus' Messiahship and Passion,* trans. Walter Lowrie (New York: Macmillan, 1950).

64. Karl Barth, *The Epistle to the Romans,* trans. Edwyn C. Hoskyns, 6th ed. (London: Oxford University Press, 1933), 314.

65. Paul Tillich, *Systematic Theology: Three Volumes in One* (Chicago: University of Chicago Press, 1967), 394–96; italics mine.

66. Walter M. Abbott, S.J., ed., *The Documents of Vatican II* (Louisville, KY: America Press, 1966), no. 48, pp. 78–79.

67. See Jürgen Moltmann, *The Coming of God: Christian Eschatology,* trans. Margaret Kohl (Minneapolis: Fortress Press, 1996); see also Moltmann, *Theology of Hope: On the Ground and the Implications of a Christian Eschatology,* trans. James W. Leitch (New York: Harper and Row, 1967), esp. introd., pt. 1.

68. Karl Barth, *Church Dogmatics,* 3.2, 624. Wolfhart Pannenberg argues for both the entry of each person into eternity immediately at death and at the end of the age. ST, 606–7. See also Moltmann, *Coming of God,* 71, 84–85.

69. Bauckham and Hart, *Hope against Hope,* 132.

70. See ST3, chap. 5, esp. pt. 3. See also TKG.

71. For nontechnical introductions, see James Trefil and Robert M. Hazen, *The Sciences: An Integrated Approach,* 2nd ed. (New York: John Wiley and Sons, 2000), chap. 15; Donald Goldsmith, *Einstein's Greatest Blunder? The Cosmological Constant and Other Fudge Factors in the Physics of the Universe* (Cambridge, MA: Harvard University Press, 1995); and George F. Ellis and William R. Stoeger, S.J., "Introduction to General Relativity and Cosmology," in *Quantum Cosmology and the Laws of Nature: Scientific Perspectives on Divine Action,* ed. Robert J. Russell, Nancey C. Murphy, and Chris J. Isham (Vatican City State: Vatican Observatory; Berkeley, CA: Center for Theology and the Natural Sciences, 1993), 33–48. For a more technical introduction, see Charles W. Misner, Kip S. Thorne, and John Archibald Wheeler, *Gravitation* (San Francisco: W. H. Freeman, 1973), pt. 6.

72. Minkowski's geometrical interpretation quickly became the standard interpretation, although it is still disputed because it suggests a "block uni-

verse" view in which all time is equally real and our experience of "flowing time" is an illusion. See chapter 5 where I offer instead a "flowing time" interpretation of special relativity.

73. Misner, Thorne, and Wheeler, *Gravitation,* 5.

74. For details of the very complex debate over the interpretation of the redshift, etc., see Helge Kragh, *Cosmology and Controversy: The Historical Development of Two Theories of the Universe* (Princeton, NJ: Princeton University Press, 1996).

75. Mathematical theorems by Roger Penrose, Stephen Hawking, and others in the 1960s proved that standard big bang models include an initial singularity, $t = 0$, provided that basic energy conditions hold for the mass-energy of the universe. See Misner, Thorne, and Wheeler, *Gravitation,* sec. 34.6, pp. 964–70.

76. See Misner, Thorne, and Wheeler, *Gravitation,* box 27.5, pp. 746–47, for additional cosmologies with nonzero values of the cosmological constant.

77. Although a competing cosmology, the "steady state" model, was developed in the 1940s by Fred Hoyle and colleagues, it was effectively abandoned in the 1960s in light of the big bang's explanation of the relative abundances of hydrogen and helium, the cosmic microwave background radiation, and other facts. For an excellent and detailed account, see Kragh, *Cosmology and Controversy.* Hoyle's approach as a general direction is still followed today by some cosmologists. I discuss Hoyle's work in some detail in chapter 6, section D.1.

78. These also include the horizon problem (Why does light emitted in the early universe and reaching us now from all directions look the same?) and the matter/antimatter ratio (Why is there slightly more matter than antimatter in the universe?).

79. See Edward W. Kolb and Michael S. Turner, *The Early Universe* (Reading, PA: Addison-Wesley, 1994); and John D. Barrow, *Impossibility: The Limits of Science and the Science of Limits* (Oxford: Oxford University Press, 1998), chap. 6, esp. p. 181. Basically, it is not clear whether the energy conditions hold in various inflationary cosmologies, and thus whether the singularity theorems mentioned above apply. See George F. R. Ellis, "Issues in the Philosophy of Cosmology," *Philosophy of Physics,* ed. Jeremy Butterfield (Elsevier, 2007), 1183–1286; preprint, submitted March 29, 2006, http://arxiv.org/abs/astro-ph/0602280, sec. 2.5.1; and George F. R. Ellis and Roy Maartens, "The Emergent Universe: Inflationary Cosmology with No Singularity," *Classical and Quantum Gravity* 21 (2004): 223–32; preprint, submitted October 25, 2003, http://arxiv.org/abs/gr-qc/0211082. However a revised version of the Penrose theorems opens up the possibility that the existence of $t = 0$ can still be attributed to inflationary cosmologies. See Arvind Borde and Alexander Vilenkin, "Eternal Inflation and the Initial Singularity," *Physical Review Letters* 72, no. 21 (1994): 3305–3308; and Arvind Borde, Allan H. Guth, and Alexander Vilenkin, "Inflationary Space-times Are Incomplete in Past Directions," *Physical Review Letters*

90, no. 15 (2003): 151301. I am grateful to Dr. Kirk Wegter-McNelly for calling my attention to this work.

80. Andrei D. Linde, "Particle Physics and Inflationary Cosmology," *Physics Today* 40, no. 9 (1987): 61–68. See figure 2.33 in chapter 2 of this volume.

81. J. B. Hartle and S. W. Hawking, "Wave Function of the Universe," *Physical Review D* 28 (1983): 2960–75. For an excellent presentation of the model and analysis of its philosophical and theological implications, see Chris J. Isham, "Creation of the Universe as a Quantum Process," in *Physics, Philosophy, and Theology: A Common Quest for Understanding,* ed. Robert J. Russell, William R. Stoeger, S.J. and George V. Coyne, S.J. (Vatican City State: Vatican Observatory, 1988), 375–408.

82. Alex Vilenkin, "Quantum Cosmology and the Initial State of the Universe," *Physical Review D37* (1988): 888f. For another excellent presentation of the model and analysis of its implications see Chris J. Isham, "Quantum Theories of the Creation of the Universe," in *Quantum Cosmology and the Laws of Nature: Scientific Perspectives on Divine Action,* ed. Robert J. Russell, Nancey C. Murphy, and Chris J. Isham (Vatican City State: Vatican Observatory; Berkeley, CA: Center for Theology and the Natural Sciences, 1993), 49–90.

83. For nontechnical introductions, see Goldsmith, *Einstein's Greatest Blunder?*; Willem B. Drees, *Beyond the Big Bang: Quantum Cosmologies and God* (La Salle, IL: Open Court, 1990), appendices 3, 4. For a more technical introduction, see Isham, "Creation of the Universe."

84. Gerson Goldhaber and Saul Perlmuter, "A Study of 42 Type Ia Supernovae and a Resulting Measurement of Omega(M) and Omega(Lambda)," *Physics Reports—Review Section of Physics Letters* 307, nos. 1–4 (December 1998): 325–31. For a brief online introduction, see "Accelerating Universe," last modified August 15, 2011, *Wikipedia,* http://en.wikipedia.org/wiki/Accelerating_universe.

85. One problem is that, by increasing the total effective energy density of the universe, a positive value for Λ could actually close the universe. Curiously, though, even in this case the visible portion of the universe can be expected to expand forever. It is still possible, however, that the universe will recollapse if the cosmological "constant" is actually a variable which dies away in the far future and if the curvature of the universe is closed. George F. R. Ellis, private communication with the author, June 2009.

86. For the limitations of this book I will not consider "nonbiological" scenarios such as artificial life, computer-based immortality, and so on, except for a brief discussion of the views of Freeman Dyson and of Frank Tipler and John Barrow below.

87. In section C of this appendix, I focused on the theological implications of t = 0 in the early universe. Here the differences in these implications clearly depended on the varying scientific status of t = 0 in standard versus inflationary

big bang cosmology, etc. However the differences in the scientific status of t = 0 in the early universe are relatively unimportant when it comes to scientific predictions for the far future of the universe and life in it. Thus, while a detailed understanding of the early universe requires clear answers to as yet unsettled theoretical questions in fundamental physics, we may discuss the far future of the universe with more confidence. Here we can work strictly with general relativity and disregard the effects of quantum mechanics on the universe as a whole—at least until the universe, if it is closed, recollapses back to Planck scales. Hence the far future predictions based on science are much more reliable than the status of t = 0 and, in turn, much less avoidable by theological eschatology.

88. John D. Barrow and Frank J. Tipler, *The Anthropic Cosmological Principle* (Oxford: Clarendon Press, 1986), 653–54; Barrow and Tipler assume that $\Lambda = 0$. See also William R. Stoeger, S.J., "Scientific Accounts of Ultimate Catastrophes in Our Life-Bearing Universe," in *The End of the World and the Ends of God: Science and Theology on Eschatology,* ed. John Polkinghorne and Michael Welker (Harrisburg: Trinity Press International, 2000), 19–27. However, a nonzero cosmological constant Λ can make even a closed universe expand forever. See Misner, Thorne, and Wheeler, *Gravitation,* box 27.5, p. 747, the fifth example, the "hesitation" universe for $\Lambda > $ CRIT.

89. Barrow and Tipler, *Anthropic Cosmological Principle,* 648.

90. Bertrand Russell, "A Free Man's Worship," *Mysticism and Logic and Other Essays* (1903; London: Allen and Unwin, 1963), 41.

91. Steven Weinberg, *The First Three Minutes: A Modern View of the Origin of the Universe* (New York: Basic Books, 1977), 154–55.

92. John Macquarrie, *Principles of Christian Theology,* 2nd ed. (1966; New York: Charles Scribner's Sons, 1977), chap. 15, esp. 351–62.

93. Ted Peters, *God as Trinity: Relationality and Temporality in the Divine Life* (Louisville, KY: Westminster/John Knox Press, 1993), 175–76. See also George L. Murphy, "Cosmology and Christology," *Science and Christian Belief* 6, no. 2 (October 1994).

94. Arthur Peacocke, *Creation and the World of Science* (Oxford: Clarendon Press, 1979), 329. To his credit, Peacocke also wrote that "our grounds for hope [cannot be] generated from within the purely scientific prospect itself" (ibid).

95. Pannenberg, "Theological Questions to Scientists," 12, 14–15; repr. Pannenberg, *Toward a Theology of Nature,* 24, 26–27. My response is that while we should indeed avoid an "easy solution," we should nevertheless attempt a "hard" one, and I hope to have outlined one such as this in the volume.

96. Freeman Dyson, "Time without End: Physics and Biology in an Open Universe," *Reviews of Modern Physics* 51 (1979): 447, 453, 459–60. For a nontechnical discussion, see Freeman Dyson, *Infinite in All Directions* (New York: Harper and Row, 1988).

97. Barrow and Tipler, *Anthropic Cosmological Principle,* chaps. 2–4, for an excellent introduction to the history of teleological interpretations of cosmology. See Mark W. Worthing, *God, Creation, and Contemporary Physics* (Minneapolis: Fortress Press, 1996), chap. 6, for a summary of this history.

98. Freeman Dyson, *Disturbing the Universe* (New York: Harper and Row, 1979), chap. 23; Dyson, *Infinite in All Directions,* chap. 6. Even in Dyson's opening remark one can already hear the reductionist assumptions that while making his approach possible technically are highly problematic theologically: "I hope to hasten the arrival of the day when eschatology . . . will be a respectable scientific discipline and not merely a branch of theology."

99. Frank J. Tipler, *The Physics of Immortality: Modern Cosmology, God, and the Resurrection of the Dead* (New York: Doubleday, 1994). For an earlier essay, see Frank J. Tipler, "The Omega Point Theory: A Model of an Evolving God," in *Physics, Philosophy, and Theology,* 313–32.

100. Drees, *Beyond the Big Bang,* chap. 4. See also Fred W. Hallberg, "Barrow and Tipler's Anthropic Cosmological Principle," *Zygon: Journal of Religion and Science* 23, no. 2 (June 1988): 139–57.

101. John C. Polkinghorne, *Science and Providence: God's Interaction with the World* (Boston: Shambhala, 1989), 96; Ian G. Barbour, *Religion in an Age of Science,* Gifford Lectures 1989–1990 (San Francisco: Harper and Row, 1990), 151–52; Peacocke, *Theology for a Scientific Age,* 345; Philip Clayton, *God and Contemporary Science* (Grand Rapids, MI: Eerdmans, 1997), 132–36; and Worthing, *God, Creation, and Contemporary Physics,* chap. 5. See also the two-part book symposium on Tipler's *Physics of Immortality* in the June and September 1995 issues of *Zygon: Journal of Religion and Science.*

102. On the interaction between Pannenberg and Tipler, see Frank J. Tipler, "The Omega Point as Eschaton: Answers to Pannenberg's Questions for Scientists," *Zygon: Journal of Religion and Science* 24, no. 2 (June 1989): 217–53; and Wolfhart Pannenberg, "Theological Appropriation of Scientific Understandings: Response to Hefner, Wicken, Eaves, and Tipler," *Zygon: Journal of Religion and Science* 24, no. 2 (June 1989), 255–71. For responses by Drees, see Willem B. Drees, "Contingency, Time, and the Theological Ambiguity of Science," in *Beginning with the End: God, Science, and Wolfhart Pannenberg,* ed. Carol Rausch Albright and Joel Haugen (Chicago: Open Court, 1997), 217–47. For responses by Russell, see Robert John Russell, "Cosmology and Eschatology: The Implications of Tipler's 'Omega Point' Theory to Pannenberg's Theological Program," in *Beginning with the End,* ed. Carol Rausch Albright and Joel Haugen, 195–216; see also Robert John Russell, "Cosmology, Creation, and Contingency," in *Cosmos as Creation: Theology and Science in Consonance,* ed. Ted Peters (Nashville: Abingdon Press, 1989), 201–4; and Robert John Russell, "Cosmology from Alpha to Omega," *Zygon: Journal of Religion & Science* 29, no. 4 (December 1994): 570–72. For additional responses, see the March 1999 issue of *Zygon.* For criticism of Tipler, see William R. Stoeger, S.J., and

George Ellis, "A Response to Tipler's Omega-Point Theory," *Science and Christian Belief* 7, no. 2 (1995): 163–72; see also Hyung Sup Choi, "A Physicist Comments on Tipler's 'The Physics of Immortality,'" *CTNS Bulletin* 15, no. 2 (spring 1995): 21–22.

103. Peacocke, *Theology for a Scientific Age,* 344–45; Peacocke, *Creation and the World of Science,* 353.

104. John Haught can be read as taking this sort of approach occasionally. According to Haught, scientific scenarios of the cosmic future are based on "emaciated mathematical abstractions that ignored the contingent openness of nature's *de facto* historicity." Such scenarios are temporary and we need not be too concerned with them. See John F. Haught, *Science and Religion: From Conflict to Conversion* (New York: Paulist Press, 1995), 174–78. See also John F. Haught, *The Promise of Nature: Ecology and Cosmic Purpose* (New York: Paulist Press, 1993), 124–25, 130. I return to Haught's views in more detail below.

105. For a basic introduction, see John B. Cobb, Jr. and David Ray Griffin, *Process Theology: An Introductory Exposition* (Philadelphia: Westminster Press, 1976).

106. Barbour, *Religion,* 241, for references to this by other process thinkers, 288n37; Marjorie Hewitt Suchocki, *God, Christ, Church: A Practical Guide to Process Theology* (New York: Crossroad, 1982), esp. chaps. 11, 17.

107. Cobb and Griffin, *Process Theology,* 123. For a recent assessment, see David Wheeler, "Toward a Process-Relational Christian Eschatology," *Process Studies* 22, no. 4 (Winter 1993): 227–37.

108. Suchocki, *God, Christ, Church,* esp. 114–15, 184–85. For further discussion of subjective immortality see Marjorie Suchocki, *The End of Evil: Process Eschatology in Historical Context* (Albany: State University of New York Press, 1988).

109. The far future of the universe might be quite relevant to process theology's doctrine of God even if not to its eschatology. In its rejection of *creatio ex nihilo,* process thought conceives of God as necessarily related to the world; God cannot be God without a world. But in the closed big bang model, the universe returns to a singular event similar to t = 0, which marked its birth. In this model, at least one could say that there would not be a universe "after" ours, just as there was none "before" ours. Such a situation would seem to undermine a process conception of God. Still, an important caveat here is needed. According to process thought, God does not need to be related uniquely to our universe; as long as there is *a* universe God is God, and cosmology cannot rule out "many worlds" as we have already seen. In fact some process theologians have taken these possibilities quite seriously. Lewis Ford writes,

> I envision an endless series of expansions and contractions of the universe, in which all the outcomes and achievements of each cosmic epoch are crushed to bits in a final cataclysmic contraction, to provide a mass/energy

capable of assuming a novel physical organisation in the next expansion. . . .
Only in some such fashion will it be possible for God to pursue his [sic]
aim at the actualization of all pure possibilities, each in its due season.

Lewis S. Ford, *Hope and the Future of Man,* ed. E. H. Cousins (Philadelphia:
Fortress Press, 1972), quoted in Peacocke, *Creation and the World of Science,*
349. I view this comment as quite remarkable since it basically takes the form
of an empirical prediction: the universe *must* continue endlessly beyond the
"big crunch." This prediction is empirical (it refers to the actual universe) but it
is based on metaphysics—a rare occurrence in philosophical theology reflecting what I call paths 7 and 8 in CMI (see sec. D.5 below).

110. To compound the problem, all three of these views are found in the
same book, as the following references indicate.

111. John F. Haught, *God after Darwin: A Theology of Evolution* (Boulder,
CO: Westview Press, 2000), 43. See also Haught, *Promise of Nature,* 128–35,
where it is clear he is referring to objective immortality. Whether Haught extends
this to include subjective immortality in the remaining portion of chapter 5, or
whether he is merely rejecting a body/soul dichotomy, is, however, not at all
clear, and that makes his position even less helpful.

112. Haught, *God after Darwin,* 115.

113. Ibid.; the second perspective is intermingled with the first on p. 115.

114. Ibid., 160–64.

115. Moltmann, *Coming of God,* 259–60.

116. Ibid., 261.

117. Edwards, "Every Sparrow."

118. Peters, *God as Trinity,* 168–70.

119. Ted Peters, *God—the World's Future: Systematic Theology for a
Postmodern Era* (Minneapolis: Fortress Press, 1992), 134–39, esp. 134; and
Peters, *God as Trinity,* 173.

120. Peters, *God—the World's Future,* 308–9.

121. ST2, chap. 7, pt. 3, sec. 2, esp. pp. 158–61.

122. One thing seems clear at the outset: if we are to use Pannenberg's
suggestions about the theological value of the underlying ideas of Barrow and
Tipler, we must recognize the fact that the open universe which science now
strongly favors does not have what could be called an Omega Point, and without this crucial piece in the scientific picture it is hard to see how the theologically valuable ideas that might be found in Barrow and Tipler can be reclaimed.

123. ST3, 589. In volume 2, Pannenberg reports that scientists favor the
open (i.e., flat) model but that new evidence might support the closed model
(ST2, 158). In volume 3, however, Pannenberg claims that scientific opinion
"no longer upholds" the open model but "teaches" the closed model. In fact,
however, the open model has been generally supported since the late 1970s.

124. John C. Polkinghorne, *Faith of a Physicist,* esp. chap. 9.

125. Ibid., 167.

126. Ibid., 167–70.

127. John Polkinghorne, "Eschatology: Some Questions and Some Insights from Science," in *The End of the World and the Ends of God: Science and Theology on Eschatology,* ed. John Polkinghorne and Michael Welker (Harrisburg: Trinity Press International, 2000), 29–30. For a more detailed discussion of eschatology, see John Polkinghorne, *The God of Hope and the End of the World* (New Haven, CT: Yale University Press, 2002).

128. See Russell, "Bodily Resurrection"; and Robert John Russell, "Eschatology and Physical Cosmology: A Preliminary Reflection" in *The Far Future: Eschatology from a Cosmic Perspective,* ed. George F. R. Ellis (Philadelphia: Templeton Foundation Press, 2002), 266–315.

129. Ian G. Barbour, *Issues in Science and Religion* (1966; New York: Harper and Row, 1971); Barbour, *Religion*; Francisco J. Ayala, Introduction, *Studies in the Philosophy of Biology: Reduction and Related Problems,* ed. Francisco J. Ayala and Theodosius Dobzhansky (Berkeley: University of California Press, 1974); Arthur Peacocke, "Reductionism: A Review of the Epistemological Issues and Their Relevance to Biology and the Problem of Consciousness," *Zygon: Journal of Religion and Science* 11, no. 4 (December 1976): 307–34; Peacocke, *Theology for a Scientific Age,* see esp. fig. 3, p. 217.

130. A growing number of scholars are developing the case for emergence in a variety of ways. These include Philip Clayton, Terrence Deacon, Niels Gregersen, and Nancey Murphy. A thorough survey would take us beyond the scope of this introduction.

131. Barbour, *Issues in Science and Religion*; Ian G. Barbour, *Myths, Models, and Paradigms: A Comparative Study in Science and Religion* (New York: Harper and Row, 1974); and Barbour, *Religion,* esp. fig. 1, p. 32 and fig. 2, p. 36.

132. For a detailed exposition and analysis of Barbour's contributions to theology and science, see Robert John Russell, "Ian Barbour's Methodological Breakthrough: Creating the 'Bridge' between Science and Theology," in *Fifty Years in Science and Religion: Ian G. Barbour and His Legacy,* ed. Robert John Russell (Aldershot: Ashgate, 2004), 45–59, and other chapters in this volume.

133. Russell, *Cosmology from Alpha to Omega,* 20–24. For simplicity, I limit the conversation to physics and its effects on theology. Clearly, many other fields can and should be added to produce a fuller account.

134. Integration as Barbour's fourth way includes natural theology and the theology of nature along with systematic synthesis. See Barbour, *Religion,* chap. 1.

135. For a careful assessment of CMI, see Philip Clayton, "'Creative Mutual Interaction' as Manifesto, Research Program, and Regulative Ideal"; Nancey Murphy, "Creative Mutual Interaction: Robert John Russell's Contribution to

Theology and Science Methodology"; William R. Stoeger, S.J., "Relating the Natural Sciences to Theology: Levels of Creative Mutual Interaction"; and J. Wentzel Van Huyssteen, " 'Creative Mutual Interaction' as an Epistemic Tool for Interdisciplinary Dialogue," in *God's Action in Nature's World: Essays in Honor of Robert John Russell,* ed. Ted Peters and Nathan Hallanger (Aldershot, England: Ashgate, 2006). Nancy Wiens used CMI in her recent dissertation on Christian spiritual discernment in light of the "theology and science" conversation. She adds a ninth path to represent spirituality's potential influence on science. Instead of beginning with theories or doctrines as my paths 6, 7, and 8 do, Nancy's ninth path starts in the context of discovery and moves downward to the philosophical assumptions that shape the observation of scientific data. See Nancy Wiens, "Discernment and Nature: Exploring Their Relationship through Christian Spirituality and the Natural Sciences," PhD diss., Graduate Theological Union, 2007.

136. "I see that you believe these things are true because I say them. Yet, you do not see how. Thus, though believed, their truth is hidden from you." Dante Alighieri, *Paradiso,* 20.88–90; *The Divine Comedy,* trans. John Ciardi (New York: W. W. Norton, 1970).

137. It is worth recalling the NT text here, see 1 Corinthians 15:12–19 quoted above in section B.2.

138. It should be clearly noted that this directly contradicts the standard Enlightenment assumption that the laws of nature are static, known, prescriptive, and deterministic. Each of these claims are challenged by the agenda of this present volume.

139. Russell, *Cosmology from Alpha to Omega,* introd., chaps. 1, 10. See also Robert John Russell, "The Bodily Resurrection of Jesus as a First Instantiation of a New Law of the New Creation: Wright's Visionary New Paradigm in Dialogue with Physics and Cosmology," in *From Resurrection to Return: Perspectives from Theology and Science on Christian Eschatology,* ed. James Haire, Christine Ledger, and Stephen Pickard (Adelaide: ATF Press, 2007), 54–94.

140. See Russell, *Cosmology from Alpha to Omega,* chap. 8.

141. For a very compelling case, see Peters, *Anticipating Omega,* chaps. 1, 2.

142. Ibid., chap. 8.

143. These would *not* include revisions in the methodological naturalism that underlies the natural sciences. I definitely do *not* mean revisions such as these which would be supported by advocates of Intelligent Design.

144. For a particularly lucid, detailed, and compelling defense of these and related issues from the perspective of a Roman Catholic biblical scholar, see O'Collins, *Jesus Risen*; and O'Collins, "The Resurrection." See also Gerald O'Collins and Daniel Kendall, "Did Joseph of Arimathea Exist?" *Biblica* 75 (1994): 235–41.

Chapter 1. The Trinitarian Conception of Eternity
and Omnipresence in the Theology of Wolfhart Pannenberg

1. For pragmatic reasons, I rely primarily on standard English translations of Pannenberg's original German texts.

2. Pannenberg, "Eternity, Time and Space," in *The Historicity of Nature*; cited as ETS in the notes. Of the journal articles I reference, most notably, Pannenberg, "Eternity, Time and the Trinitarian God," cited as ETTG in the notes. For a slightly revised version of ETTG, and more conveniently available, see "Eternity, Time and the Trinitarian God," *CTI Reflections* 3 (1999): 48–61.

3. Pannenberg draws these primarily from the Psalms and Isaiah.

4. ST1, 401–2.

5. ST1, 402–3.

6. ST1, 403.

7. See M, 75.

8. Again, my intention in this volume is not a critical assessment of Pannenberg's interpretation of these classical sources. Instead I will assume that they are to some degree correct and adopt them here.

9. ST1, 403. As Pannenberg notes, Plotinus followed some elements in the thinking of both Plato and Aristotle. With Aristotle, Plotinus distinguished time from motion "because that which is not moved or remains stationary still remains within time." Following Plato, Plotinus thought of time as "the image of eternity," but, unlike Plato, the image was of "the totality of the eternal itself" and this image is mediated to us by the world soul. Ibid.; see also M, 76.

10. M, 76; ST1, 403.

11. ST2, 92–93. The English translation reads "if nothing sinks *for it* into the past" (emphasis mine). I have instead written "from it."

12. ST2, 95; see also 95n238.

13. ST1, 404; M, 76–77.

14. Pannenberg cites a particularly eloquent passage from Plotinus: "Instead of the completed infinite and whole, [there is only] the moment after moment into the infinite; instead of the unitary whole, [only] the partial and always merely future whole." M, 77n21.

15. M, 77.

16. ST1, 408.

17. M, 77. We find this already in the 1969 publication, *Theology and the Kingdom of God,* where, in chapter 1, Pannenberg writes that "we see the present as an effect of the future, in contrast to the conventional assumption that past and present are the cause of the future." TKG, 54, see also 59, 63, 67, and chap. 4, 139–41.

18. ST2, 95.

19. ST1, 404. Here and elsewhere Pannenberg notes that Augustine adopted a timeless view of eternity in order to protect the divine immutability. He argues that Augustine's reliance on the traditional exegesis of Exodus 3:14 ("I am who I am") reinforced his understanding of God's eternity as timeless. But today, in light of contemporary exegesis, Pannenberg stresses that we recognize that the Exodus passage carries a future-oriented interpretation, "I shall be who I shall be," and with this a future is attributed to God in relation to the creation. ETTG, 10.

20. This is Pannenberg's translation of Boethius's definition of eternity, found in ST1, 404. According to Boethius, eternity is "interminabilis vitae tota simul et perfecta possessio" (*Consolation of Philosophy* 5.6). Note that *interminabilis* can be translated as "unlimited", "unending," "everlasting," etc., each carrying a slightly different meaning. In a note, however, Pannenberg re-translates Boethius's definition as "the unending, total, and perfect possession of life" (ST1, 404n142).

21. ST1, 405.

22. ST1, 404; ST3, 595–96. Pannenberg is less enthusiastic about Paul Tillich's understanding of eternity (ST1, 407n153), and he outrightly rejects the view of philosopher Nelson Pike, who accepts the (Platonically based) antithesis between eternity and temporality (ST1, 405). He also writes that although a matrix of interlaced modes of time, such as explored by Jürgen Moltmann in *God in Creation,* might be helpful in understanding the nature of physical time or the historicity of our experience of time, it should not be confused with the idea of eternity as a time-bridging present. ST2, 92n238.

23. ST1, 405n146. Here Pannenberg cites Karl Barth, *Church Dogmatics,* 4 vols. in 12, ed. G. W. Bromiley and T. F. Torrance, trans. G. T. Thomson et al. (T and T Clark: Edinburgh, 1936–77), 2.1, 615.

24. ST1, 405. Pannenberg is here reiterating a principle first set down by Karl Rahner in *The Trinity* (New York: Herder and Herder, 1970), 46. In a series of two articles published in 1987, Ted Peters coined the term, "Rahner's Rule," for this principle, a term now widely used in Trinitarian literature. See Ted Peters, "Trinity Talk," *Dialog: A Journal of Theology* 26, no.1 (winter 1987): 44–48; and no. 2 (spring 1987): 133–38. See also Ted Peters, *God as Trinity: Relationality and Temporality in the Divine Life* (Louisville, KY: Westminster/ John Knox Press, 1993), 22, 96–103. Pannenberg credits Rahner explicitly with this insight in ETTG, 12.

25. ST1, 405; ST3, 595. Pannenberg, however, is critical of Barth's early crisis theology for making eternity the end of time at each moment instead of at the future end of the world, and thus for "detemporalizing of the eschatological expectation of primitive Christianity." ST3, 594.

26. Barth, *Church Dogmatics,* 3.2, 568: "There is no part of our time which is not as such also in [God's]. It is, so to speak, embedded in His eternity." See ST1, 406n149.

27. ST1, 405–6.

28. ST1, 406.

29. There is an important question of translation from the German to the English text here. My intention is to delineate Pannenberg's idea that each present has its own *distinct* past and future and Pannenberg's idea that the present *separates* or *divides* past from future. But Pannenberg's terminology for the latter—separation or division—can be somewhat ambiguous: Are they equivalent or not? Let us read the full text, part of which was cited above:

> Eternity is the undivided present of life in its totality . . . . We are not to think of this as a present separated from the past on the one hand or as the future on the other. Unlike our human experience of time, it is a present that comprehends all time that has no future outside itself. The present that has a future outside itself is limited. A present can be eternal only if it is not separate from the future and if nothing sinks for it into the past. . . . We are to think of time with its sequence of events—future, present, past—as proceeding from eternity and constantly comprehended by it. . . . To understand the eternal present as a present that comprehends time is not to exclude past and future but to include them. . . . In the eschatological consummation we do not expect a disappearance of the distinctions that occur in cosmic time, but the separation will cease when creation participates in the eternity of God.

ST2, 92, 92n238, 93, 95. So, for example, in the second sentence of the English translation above, the German word being translated as "separated" is *abgesonderten*. But in the fifth sentence of the English translation, the German word being translated as "not separate" is *ungeschieden,* which might better have been translated "not divided." In any case I will use "separate" and "divided" as more or less synonymous for what Pannenberg means by time as being divided into isolated moments in the past and future, and I will differentiate this from the concept that there are distinct pasts and futures associated with each present moment. Oliver Putz, private communication with the author, spring, 2010; and Johanne Stubbe Teglbjaerg, private communication with the author, summer, 2010.

30. M, 78.

31. In ETTG, 9, Pannenberg employs the felicitous passage, "the soul achieves a kind of enduring presence, while it moves through the flux of time, always remembering the past and anticipating the future. The soul does so by distending its floating presence over past and future."

32. M, 79–80.

33. ST3, 91.

34. ST3, 91; ST3, 598.

35. M, 80–81. Pannenberg seems to be agreeing here with Augustine that duration occurs not only in human experience but in nature, referring to the *Confessions* 11.28.37 in 80–81n28. However in the same note Pannenberg then

points to an apparent contradiction in Augustine by reporting that in the *Confessions* 11.23.30 Augustine argues that "time as such lacks duration." The translator of *Metaphysics and the Idea of God,* Philip Clayton, argues that, for Pannenberg, duration is indeed objective in nature. "In the duration of the extended present . . . we know something of the divine time. . . . We then move back down again to the world, recognizing that time as duration is actually the deeper and more powerful way to understand *all* temporal process, including physical . . . time" (Philip Clayton, private correspondence with the author, September, 2007). Hence, in this book, I assume that Pannenberg believes that duration as temporal extension occurs even at the "bottom rung" of nature: physics. I will also suggest that duration for Pannenberg is structured by what I will call co-presence.

36. ST1, 406; ST3, 596. In further analyzing Barth's concept of duration Pannenberg makes several crucial points. First Pannenberg claims that for Barth, "the time that God gives to the creatures [is] participation in his eternity. By giving us time God actually gives us eternity" (ST3, 596). Nevertheless, this participation in eternity is not due to the kind of duration which creaturely time involves. In a note, Pannenberg clarifies this point by distinguishing explicitly between the limited duration of creaturely time and the unlimited duration of eternity—a distinction which Barth did not make; see ST3, 596n222. Returning to the text, Pannenberg states that Barth "would not predicate duration of the creaturely time that we are 'given'" (ST3, 597). Instead, for Barth, the "now" of our present is a boundary between past and future, while the "now" of eternity has duration. It is our crossing through the now of creaturely time that we have a relation to the whole of time, that is, eternity.

37. ST3, 597; see also ST2, 95–96, 123–24. Pannenberg includes "the regulated order of nature" as also decisive for creatures. ST2, 71–72.

38. ST2, 96.

39. ST3, 597–98.

40. ST3, 599. Here Pannenberg offers a revealing critique of the attempts by Kant and Heidegger to replace the role of eternity as a basis for continuity in time with the subjectivity of human ego.

41. ST3, 600.

42. ST3, 601n244; the translation reads: "The finitude of creaturely being undoubtedly rules out unlimited existence but not the *present* of the whole of this limited existence in the form of duration as full participation in eternity" (emphasis mine). Oliver Putz's translation of the German version of ST concurs with "presence."

43. ST3, 601. Such an act of overcoming disintegration only comes when we stop wanting to be God and allow God's Spirit to work in us. See Pannenberg's critique of Tillich's use of the term "essentialization" drawn from Schelling, as well as a brief comment on Whitehead, in ST3, 602–3.

44. ST3, 643–44.

45. Pannenberg includes an additional layer of complexity in his analysis of the structure of creaturely time, namely the "brokenness of our experience of time for which all life is torn apart by the separateness of past, present, and future" is related to "the structural sinfulness of our life." In the eschatological consummation we will no longer experience time as a sequence of separated moments but as "the *totality* of our earthly existence," a totality which we now can only grasp as an "anticipation" (ST3, 561). I will not include a detailed analysis of this further complexity in Pannenberg's thought, as it would involve a lengthy discussion of Pannenberg's treatment of the problems of sin and death and his treatment of the doctrine of redemption, and these lie beyond the limitations of this volume as I indicated in the introduction.

46. ST3, 603.

47. Pannenberg cites 1 John 3:2: "It does not yet appear what we shall be." ST3, 603.

48. ST3, 604.

49. ST3, 605.

50. ST3, 606.

51. ST3, 607.

52. ST2, 95; the translation reads: "*the distinction of life's moments* in the sequence of time cannot be one of the conditions of finitude as such" (emphasis mine). However, I believe Pannenberg means "the separation" and not "the distinctions of life's moments" because the distinctions do, indeed, remain in the eschaton. On checking on the original German version of ST, Oliver Putz concurs.

53. ST1, 409.

54. ST1, 410.

55. Pannenberg contrasts this view with that of Spinoza, who "regarded extension as a divine attribute." ST1, 411; see also ST2, 85.

56. ST1, 411; ST2, 86.

57. ST1, 359ff.

58. ST2, 85.

59. ST1, 412.

60. I have introduced the term "relational space" for Pannenberg's description of Leibniz's view of space. For a detailed historical account of the difference between these two concepts of space, see Max Jammer, *Concepts of Space: The History of Theories of Space in Physics,* 3rd enlarged ed. (New York, NY: Dover, 1993). Pannenberg frequently cites Jammer, using the original 1953 edition.

61. ST2, 87; ETS, 164–67, 172–74.

62. ETS, 164–65.

63. ST1, 413.

64. ETS, 164–65.

65. ST1, 413. Here, Pannenberg uses the term "absolute" in two different senses: Newton's metaphysical view of absolute space as that which determines which motions are inertial, and Clarke's theological view of the space of God's omnipresence. In a note Pannenberg quotes Clarke as writing that "Infinite Space is One, absolutely and essentially indivisible; And to suppose it parted is a contradiction in terms, because there must be Space in the Partition itself" (ST1, 413n168). The same quotation from Clarke appears in ST2, 86n222. Pannenberg also reminds us that Kant makes a similar point in the *Critique of Pure Reason* (§ A 23), and that, for Kant and Descartes, space and time are ways of intuiting the Infinite. They precede the finite contents of experience and are without limit. ST2, 89; see also M, 83.

66. ST2, 86n221.

67. ETS, 165. As Pannenberg points out, Kant relied on the Clarke-Leibniz debate about Newton's concept of divine omnipresence, although he related space and time to human intuition. Pannenberg notes that in Kant's *Critique of Pure Reason,* our human intuition of space and time as infinite wholes serve as a presupposition for our conception of specific spaces and times. They also serve as a presupposition to geometry because geometry relies on measurement via spatial and temporal units which are themselves part of space and time (ETS, 164–65). According to Pannenberg, ten years before his *Critique* Kant shared Clarke's view. The infinity and indivisibility of space and time are the presuppositions for the unity of our spatial and temporal perceptions, and they are also the form of God's omnipresence and eternity. With the *Critique,* however, Kant reconceived the unity of infinite space and time as "based upon the unity of the human subject of experience." Kant's shift was apparently motivated by a concern to avoid pantheism, particularly if infinite and undivided space were identified with Newton's absolute space. Pannenberg claims that Kant could have avoided this problem by agreeing with Clarke on the *undivided* character of the infinite space of God's omnipresence, but then he could not equate it with Newton's absolute space. Kant resolved the problem by noting that Newton adopted a receptacle view of absolute space (ETS, 172). Kant then sided with Leibniz by interpreting space as a system of relations, and he relocated the system of relations from the mind of God to the human subject. Pannenberg's criticism here is that the human subject, as finite, cannot be the basis for "the *objective validity* of our conception of the *infinite unity* of space and time that is presupposed in all experience." ETS, 166.

68. Pannenberg states that it was Augustine's commitment to divine immutability which led him to reject "both the idea of infinite space outside our world and that of time before the creation of the world." Pannenberg disagrees with Thomas Torrance, who claimed that the early Patristic writers had a relational view of space and that it was eventually displaced by a container view due to Platonic influences on their thought. ST2, 86n224.

69. ST1, 413n168.

70. ST1, 414.

71. ST2, 90–91. It seems clear that Pannenberg is referring to *special* relativity here and to *general* relativity on the previous pages in ST2 (see section B.4 below).

72. ETS, 163.

73. ST2, 87–88. Pannenberg credits Torrance with a return to the theological roots of a relational view of space. See ST2, 88n228.

74. I take this to be what Pannenberg has in mind when writing that "Einstein's criticism of Newton's view of absolute space did not outdate these ideas of Newton, since Einstein . . . extended the function of the concept of space in Newton into a general field theory of space-time" (ST1, 413–14).

75. Pannenberg agrees with Moltmann's argument that these two concepts of space can be related through the concept of creation. However he rejects Moltmann's stipulation that "the space of creation precedes creation, and the spaces created in it, as a third thing between the divine omnipotence and the world of creatures along the lines of the Jewish doctrine of Zimzum." His rejection is based on his concern that it would compromise God's omnipresence to creatures, and this would be highly problematic because the divine omnipresence is constitutive for their existence (ST2, 89n29). In ST2, 90n232, Pannenberg offers a critical appraisal of Karl Heim's characterization of divine omnipresence in terms of "suprapolar space."

76. ETS, 166–67. Pannenberg's description of general relativity here is inaccurate when he writes that "physical objects are accounted for as effects of the gravitational field of space-time" (167). In general relativity the geometry of spacetime and the distribution of mass-energy are two independent concepts interrelated through Einstein's field equations. It is only later, with his search for a unified field theory, that Einstein attempted to remove matter as an independent concept and identify it with the curvature of spacetime.

77. Pannenberg argues that if Kant had had this distinction, he could have accepted Clarke's concept of an infinite and undivided space and time without fear of pantheism.

78. ETS, 167–68.

79. ETS, 170–71. Pannenberg's comment here about Augustine's commitment to divine immutability adds to his earlier discussion of Augustine's insistence on eternity as timeless in ST1, 404 (see section A.1 above).

80. See ETS, 170–71. In ETTG, 11, Pannenberg points to two limitations in the similarity between Plotinian and biblical conceptual frameworks: the former lacks an eschatology and it leads to a concept of God for whom the future is not constitutive.

81. ETS, 170.

82. ETS, 171–72.

83. ETS, 172; Pannenberg cites Thomas F. Torrance, *Space, Time and Incarnation* (London: Oxford University Press, 1969), 13 ff, cf. 7–8.

84. STE, 173. Here Pannenberg makes an important aside by pointing to the possibility of "alternative interpretations of relativity by Einstein and Neo-Lorentzians," citing the work of William Lane Craig. Nevertheless, he writes that "the insight into the interrelatedness of space and time with masses and energies will remain a lasting contribution to the understanding of the conditions of finite reality, even in the discourse of philosophers and theologians." See William Lane Craig, *The Tenseless Theory of Time: A Critical Examination* (Dordrecht, Holland: Kluwer, 2000), 105 ff.

85. ETS, 173.

86. ST1, 405. Pannenberg's claim here is crucial to his entire project. It consists of three closely related ideas: 1) Every event in time has its own individual past and future. 2) In order for every event to be taken up into the eternal presence of God with its unique past and future preserved requires that the unity of God is differentiated. 3) God as differentiated unity requires the doctrine of the Trinity. The question, of course, is whether Pannenberg succeeds in demonstrating the truth of this claim. I think the claim is true, but I do not believe Pannenberg convincingly demonstrates its truth. The purpose of this book in large measure is to take up this claim and make it more convincing by reconstructing it in light of mathematics and physics and then showing its fruitfulness in leading to insights in the philosophy of time and in physics. It goes without saying that there are many other ways to make it more convincing, some of which lie in the writings of theologians who are working with closely related doctrines of the Trinity.

87. ETTG, 1.

88. ST1, 415.

89. ST2, 87.

90. ETS, 170.

91. ST1, 275.

92. ST1, 278.

93. For an excellent summary and analysis of Pannenberg's writings on the Trinity, see Peters, *God as Trinity,* 135–42. Peters extends his summary of Pannenberg's arguments for the unity of the three persons beyond what I have cited above to include Pannenberg's claim that their unity is also grounded in the obedience of the Son to the Father, the assumption that "deity is dependent upon lordship," that "God is personal only through one or another of the three hypostases," etc. While these arguments are crucial to Pannenberg's overall theological agenda, they are of less relevance for the purposes of this present volume.

94. ST1, 279, 312.

95. ST1, chap. 6. The concept of infinity arises in several places in the *Systematics,* where Pannenberg typically describes it in words such as these: "The thought of the true Infinite . . . demands that we do not think of the infinite

and the finite as a mere antithesis but also think of the unity that transcends the antithesis" (ST1, 446). Interested readers should turn, for example, to Pannenberg's discussion of the "Proofs of God" (ST1, 91–92, 106–7). I will concentrate here on the way it contours the doctrine of God, specifically the discussion of the divine attributes, where it plays a much more decisive role than elsewhere in the *Systematics.*

96. ST1, 337–47.

97. ST1, 340.

98. ST1, 342.

99. ST1, 345.

100. ST1, 347.

101. ST1, 347–59.

102. Thomas Aquinas, *Summa Theologica,* trans. Fathers of the English Dominican Province (Westminster, MD: Christian Classics, 1981), I.2–11.

103. ST1, 349.

104. ST1, 114. As Pannenberg points out, Descartes understands the priority of the idea of the infinite as found originally in a primordial "nonthematic awareness in which God, world, and self are still not differentiated" and thus an awareness which does not involve an explicit concept of the infinite. It is only later, after experience and reflection, that we can identify this awareness of the infinite with an awareness of God (ibid.).

105. ST1, 352.

106. ST1, 397–422; 422–48.

107. ST1, 392. "When we say that God is kind, merciful, faithful, righteous, and patient, the word 'God' is the subject of the descriptions. . . . But what does it mean to say all these things of 'God'? The answer lies in terms that explain the word 'God' as such, e.g., terms like infinite, omnipresent, omniscient, eternal, and omnipotent."

108. ST1, 397.

109. ST1, 396; see also ST1, 432. Pannenberg notes that while the Bible does not refer directly to God's infinity, it is clearly implied in God's definitional attributes: God's holiness and infinity are directly related while eternity, omnipresence, and omnipotence are "concrete manifestations of his infinity" under the categories of time, space, and power.

110. ST1, 397n126. Pannenberg cites Georg W. Hegel, *The Science of Logic,* trans. J. N. Findlay, bk. 1, sec. 1, chap. 2c. See Hegel by Hypertext, http://www.marxists.org/reference/archive/hegel/works/hl/hlbeing.htm#HL1_136. He also cites Friedrich Schleiermacher, *The Christian Faith,* ed. H. R. Mackintosh and J. S. Stewart (Edinburgh: T. and T. Clark, 1968), 1.56.2.

111. ST1, 397–98.

112. ST1, 400.

113. ST1, 400.

114. ST1, 400. Interestingly, Pannenberg claims that Barth argued against the use of the concept of the infinite in discussing the divine attributes because Barth's conception of the infinite placed it in opposition to the finite, and thus it would not allow for the divine immanence which omnipresence entails. See ST1, 412.

115. ST1, 408.

116. ST1, 412.

117. ST1, 415. Pannenberg offers a similar analysis of the role of the true Infinite in structuring God's omnipotence in relation to creaturely power. See ST1, 415–22.

118. ST1, 446. This in turn leads to the unity of God's essence and existence and the link between the Economic and the Immanent Trinity. Finally with the eschatological consummation of the world we will "fully know God as the true Infinite who is not merely opposed by the world of the finite, and thus himself finite" (447).

119. Some scholars suggest that Pannenberg's claim to have found a paradox in Hegel's thought on infinity is part of a wider critique of Hegel. In their view Pannenberg's critique is leveled against Hegel's elevation of concrete history to the realm of spirit through the process of philosophical abstraction. Instead Pannenberg locates the concept of the true Infinite in a theological account of God's Trinitarian relation to the world. I will not pursue this question further here. Johanne Stubbe Teglbjaerg, private communication with the author, summer 2010.

120. The Enlightenment mechanistic philosophy associated with classical physics includes the rejection of Aristotle's notion of teleology as one of the four natural causes. As we will see, Pannenberg not only reinvokes teleology but challenges Aristotle's reduction of it to a present causal factor, or "germ," arguing instead for a future cause acting on the present as more faithful to Aristotle's true intentions.

121. Here it is irrelevant whether physics provides explicitly deterministic theories such as classical mechanics, or stochastic theories which assume that there are additional causal factors not yet included in these theories, such as statistical thermodynamics, or even theories such as quantum mechanics which can be interpreted as telling us that some events in nature lack a sufficient efficient cause. The point is that whatever causality there *is* is efficient—the past causes the present. (Classical mechanics can be formulated through a least action principle which suggests teleological rather than efficient causality, but it is a far cry from the full teleology offered by Aristotle and to which Pannenberg turns, as we will see shortly. Moreover it is an alternative explanation, not one in combination with efficient causality such as Pannenberg adopts.)

122. Pannenberg ultimately sets these eschatological themes in an even wider theological context, the relation of, and connection between, creation and eschatology. "Far from creation being at one end of the time spectrum and

eschatology at the other, creation and eschatology are partners in the formation of reality." TKG, 60.

123. See TKG, esp. chaps. 1 and 4.

124. TKG, 54. This will shortly lead Pannenberg to a related claim, one that is notably famous and controversial: since the being of God is God's rule, and since God's rule is still grounded in the future, "in a restricted but important sense, God does not yet exist. . . . God's being is still in the process of coming to be" (56).

125. Ted Peters originated the term "retroactive ontology" to make a very similar point. See, for example, his *Anticipating Omega: Science, Faith, and Our Ultimate Future* (Gottingen: Vandenhoeck and Ruprecht, 2006).

126. TKG, 57.

127. TKG, 59 (emphasis mine). He continues: "If we, in our anxiety and hope, contemplate the power of the future, we recognize both its breathtaking excitement and its invitation to trust. For those who accept the invitation, the world is widened with new possibilities for joy. In every present we confront the infinite future, and in welcoming the particular finite events which spring from that future, we anticipate the coming of God."

128. TKG, 67.

129. Pannenberg's view can be compared with the process philosophies of Alfred North Whitehead and Charles Hartshorne. For extended comments on their similarity and difference, see TKG, 62, 63, 67.

130. TKG, 139.

131. In a review of *Systematic Theology,* volume 2, Robert Jenson concludes that by prolepsis, Pannenberg is not urging an "ontological reconstruction of metaphysics" in which the resurrection of Christ provides "a structure of prolepsis throughout reality." Instead Pannenberg is working with "the usual understandings of time and eternity. A prolepsis . . . is simply a claim staked out in history, which, when and if history is fulfilled, will be verified or falsified." Robert W. Jenson, "Parting Ways?" *First Things: The Journal of Religion, Culture, and Public Life* (May 1995), http://www.firstthings.com/article.php3?id_ article=4034. I disagree with Jenson here. Of course a prolepsis is minimally an historical claim about the eschaton, but as I read Pannenberg it is much more the power of the eschaton to transform history as it is realized in the anticipatory present moment beginning with Easter. As Ted Peters writes, "What happened to Jesus on the first Easter was a prolepsis of what will happen to the entirety of the created universe at omega. What we see in the Easter resurrection of Jesus' incarnation is the first instantiation of that eschatological law of nature . . . [found] in the kingdom of God." Peters, *Anticipating Omega,* 46.

132. TKG, 133. Ted Peters has built upon and extended Pannenberg's arguments about prolepsis in very impressive and promising ways that compel further exploration. See in particular his nine theses in Peters, *Anticipating Omega,* chaps. 1, 2.

133. We see this claim once again in the following comment: "The primacy of the future and its novelty are guaranteed only when the coming kingdom is ontologically grounded in itself and does not owe its future merely to the present wishes and strivings of man." Wolfhart Pannenberg, *Basic Questions in Theology,* trans. George H. Kehm, 2 vols. (Minneapolis: Fortress Press, 1971), 2:239–40.

134. The third claim surfaces again when Pannenberg addresses it to the question of God's entering into history while distinguishing Godself from history: "The God who constitutes history has himself fully entered the process of history in his revelation. But he has done so in such a way that precisely as he is here transmitted in a process of tradition, he is at the same time the future of his history, the coming God who remains distinct from, or, better, who is always distinguishing himself in a new way from, what happens in this history" (Pannenberg, *Basic Questions,* 1:158). Pannenberg returns to this claim again in the *Systematics.* For example, in discussing revelation in volume 1, Pannenberg writes that "the final saving reality of the death-defeating resurrection life has come already in Jesus, but [it] is still future for us." The juxtaposition of provisional and definitive features resurfaces in "the Pauline tension between the Already of present salvation and the Not Yet of its consummation." ST1, 211.

135. Pannenberg, *Basic Questions,* 2:24–25.

136. JMG, 136. Pannenberg also returns to this claim in the *Systematics.* In his discussion of revelation in volume 1, Pannenberg writes that with Isaiah 40, prophecy becomes "a proleptic disclosure of what will be made universally manifest in the future. The experience thus became provisional, and its truth depended on the future self-demonstration of the truth of God" (ST1, 213). This is one of the few places in the *Systematics* where the term "proleptic" is still used and not its "replacement" in such terms as "anticipation," "appearance," etc. We find the hermeneutical move of assigning the meaning of the present to the eschatological future in the following remark: "God creates the world in the light of its latter end, because it is only the end which decides the meaning of the things and beings with which we have to do in the present." Wolfhart Pannenberg, *The Apostles' Creed in the Light of Today's Questions* (Louisville, KY: Westminster/John Knox Press, 1972), 39.

137. ST3, 145. Pannenberg picks up this theme briefly in ST3, chap. 13. He first locates the biblical and Patristic meaning of "prolepsis" in the context of the Pauline view of faith as assent in which Paul understands "faith as acceptance of the apostolic message of salvation." He then notes that Clement of Alexandria developed this concept with Hebrews 11:1 as a backdrop by using the Stoic concept of prolepsis: "an intelligent anticipation prior to secure comprehension" (145n145). Clement also refers to faith as "rational assent (*synkatathesis*)." Pannenberg also reminds us that Augustine and later Scholasticism defined faith as "thinking with assent (*cum assensione cogitare*)" (145n146).

138. M, 96. As Pannenberg tells us in the *Systematics,* the development of the apocalyptic view of revelation in the early church took into account "the debatability of the reality of God." See ST1, 213–14.

Chapter 2. Co-presence and Prolepsis in Light of Math, Physics, and Cosmology

A previous, shorter version of section C.2 appeared in Robert John Russell, "Scientific Insights into the Problem of Personal Identity in the Context of a Christian Theology of Resurrection and Eschatology," in *Personal Identity and Resurrection,* ed. Georg Gasser (Ashgate, 2010), 241–58.

1. With the focus strictly on Pannenberg's work it would not be appropriate here to offer a general introduction to the thematic of "time and eternity" in the scholarly literature. For a helpful point of entry and an extended bibliography on time and eternity, see Paul Helm, "Eternity," *Stanford Encyclopedia of Philosophy,* last modified February 4, 2010, http://plato.stanford.edu/entries/eternity/. See also Peter Manchester, "Eternity," *Encyclopedia of Religion,* 2nd ed., ed. Lindsay Jones, 15 vols. (New York: MacMillan-Thomson-Gale, 2005), 2853–57; J. J. C. Smart, Introduction, *Problems of Space and Time: From Augustine to Albert Einstein on the fundamentals for understanding our universe* (New York: Macmillan, 1964); Brian Leftow, *Time and Eternity* (Ithaca, NY: Cornell University Press, 1991); and Eleonore Stump and Norman Kretzmann, "Eternity," *Journal of Philosophy* 78, no. 8 (1981): 429–58.

2. It is fruitful to compare Pannenberg's interpretation of Boethius to that of Anselm. For a very helpful account of the latter, see Leftow, *Time and Eternity,* chap. 9. Leftow also provides a persuasive critique of the Stump-Kretzmann interpretation of Boethius. I would of course question his view of God as timeless, but that will have to wait for another occasion.

3. For a very instructive overview of the split between Anglo-American analytic philosophy and Continental philosophy, see Dean Zimmerman, "Three Introductory Questions," in *Persons: Human and Divine,* ed. Peter van Inwagen and Dean Zimmerman (Oxford: Oxford University Press, 2007), 1–32.

4. Michael Tooley, *Time, Tense, and Causation* (Oxford: Clarendon Press, 1997), chap. 1. Tooley provides a helpful bibliography to the philosophy of time. For extensive references, see Dean Zimmerman, "Temporary Intrinsics and Presentism with Postscript (2005)," in *Persistence,* ed. Roxanne Marie Kurtz and Sally Haslanger (Cambridge, MA: MIT Press, 2005), http://fas-philosophy .rutgers.edu/zimmerman/TempIntrinsicsPostscript.pdf, nn37, 40. See also William Lane Craig, *The Tensed Theory of Time: A Critical Examination* (Dordrecht: Kluwer, 200); and William Lane Craig, *The Tenseless Theory of Time:*

A Critical Examination (Dordrecht: Kluwer, 2000). For helpful online introductions with key references, see Bradley Dowden, "Time," last modified June 11, 2011, *Internet Encyclopedia of Philosophy,* http://www.iep.utm.edu/t/time.htm; and Ned Markosian, "Time," *Stanford Encyclopedia of Philosophy,* last modified November 10, 2010, http://plato.stanford.edu/entries/time/.

5. This terminology is due to Cambridge philosopher John Ellis McTaggart. His groundbreaking article, "The Unreality of Time," has stimulated a century of debate. See McTaggart, "The Unreality of Time," *Mind: A Quarterly Review of Psychology and Philosophy* 17 (1908): 456–73, http://www.ditext .com/mctaggart/time.html; also available from *Wikisource,* last modified November 3, 2010, http://en.wikisource.org/wiki/The_Unreality_of_Time.

6. Tooley, *Time,* 12. According to Zimmerman, some scholars associate an objective flow of time with temporal asymmetries in physics, such as increasing entropy in thermodynamics or the expansion of the universe according to general relativity, while others relate it to a postulated primitive temporal orientation in spacetime (Dean Zimmerman, private correspondence with the author, October 2008). For a brief and very readable defense of flowing time available online, see Conrad Robinson, "The Coherence of Tenses," paper presented at the Kentucky Philosophical Association, Fall 2005, http://www.wku.edu/~jan.garrett/ kpa/robitime.htm.

7. Philosophers often distinguish between two meanings of ontology: 1) the things that have to exist for a theory to be true, and 2) the fundamental terms or concepts that the theory uses. In general I use ontology in the second sense here.

8. Defenders of presentism include Arthur Prior, Roderick Chisholm, and more recently William Lane Craig, Thomas Crisp, Trenton Merricks, Ned Markosian, and Dean Zimmerman. See Zimmerman's defense of presentism in "The Privileged Present: Defending an A-Theory of Time," *Contemporary Debates in Metaphysics,* ed. John Hawthorne, Ted Sider, and Dean Zimmerman (Malden, MA: Blackwell, 2009), 211–25. I discuss William Lane Craig's arguments for presentism in relation to relativity in chapter 5.

9. A classic example of this view is found in C. D. Broad, *Examination of McTaggart's Philosophy,* 3 vols. (Cambridge: Cambridge University Press, 1933–38). For a current debate, see Peter Forrest, "The Real but Dead Past: A Reply to Braddon-Mitchell," *Analysis* 64 (2004): 358–62. Tooley offers a growing universe view in which "the past and present are real, [but] the future is not." He claims that his view "differs sharply, in certain fundamental ways, from traditional formulations." See *Time,* 13. As we will see in chapter 5, Tooley deals in detail with the challenge raised by relativity to a dynamic view of the world.

10. Robin Collins, private communication with the author, November 2009. Zimmerman assesses a hybrid version of eternalism with an A-theorist's ontology III ("eternalist A-theory") in "The A-Theory of Time, the B-Theory of Time, and 'Taking Time Seriously,'" *Dialectica* 59, no. 4 (2005): 401–75.

11. The B-theorist view that everything in the past, present, and future simply and equally exists is known as "eternalism." The "block universe" is a specific form of eternalism that arises most frequently in B-theorists' interpretation of special relativity. I refer to it here to help connect this discussion with that of chapters 4 and 5 where I turn specifically to relativity. See, for example, Michael Esfeld, "The Impact of Science on Metaphysics and Its Limits," *Abstracta* 2, no. 2 (2006): 86–101, http://www.abstracta.pro.br/revista/volume2 number2/1_esfeld.pdf.

12. Werner Heisenberg's interpretation of quantum mechanics provides a striking example of ontology III-2. Consider the spontaneous decay of a radioactive atom. Before decaying, the atom is in a superposition, indicated by the summation sign Σ, of several potential states ϕ_n with relative amplitudes a_n, as represented by the initial wavefunction $\psi_i = \Sigma\, a_n \phi_n$. The decay is an ontological transition to one actual final state represented by $\psi_f = \phi_f$. Thus, during radioactive decay, $\psi_i \rightarrow \psi_f$ and, out of the many initial possible states ϕ_n, one state ϕ_f is actualized (i.e., realized) in nature. In Heisenberg's interpretation, quantum indeterminism reflects the fact that nothing in nature determines which of the many initial states ϕ_n will be actualized as the final state ϕ_f. Of course, there are several other interpretations of quantum mechanics that differ from Heisenberg's. Nevertheless his interpretation shows how the future can be conceived as including a number of potential states reflective of ontology III-2, and not just a single potential state with such properties as futurity as in ontology. It also underscores the radical difference between quantum mechanics and classical, Newtonian mechanics, including chaos and complexity theory, in which our uncertainty about the future state of a system is purely epistemic, as in ontology III-1. In short, classical mechanics cannot be interpreted in terms of ontology III-2, whereas quantum mechanics can. For an interesting example of a more technical form of ontology III-2, see Storrs McCall's argument that the future can be viewed as holding many "branches" that are trimmed down to a single branch as the future becomes the present. McCall, "The Strong Future Tense," *Notre Dame Journal of Formal Logic* 20, no. 3 (July 1979): 489–504, http://projecteuclid.org/DPubS? service=UI&version=1.0&verb=Display&handle=euclid.ndjfl/1093882655.

13. Tooley, *Time,* 13. According to Craig, the static view of time is rooted in both tenseless theories of language (cf. Bertrand Russell, D. H. Mellor, and Jeremy Butterfield) and a realist view of ontology (cf. John Ellis McTaggart, Albert Einstein, Hermann Minkowski, Adolf Grünbaum, and Huw Price). Craig compares this with a dynamic view of time based on tensed theories of language (cf. Alvin Plantinga), on experience and phenomenology (cf. Edmund Husserl and A. N. Prior) and on a realist view of ontology (cf. C. D. Broad, Milič Čapek, Michael Tooley, and Craig). See William Lane Craig, *Time and Eternity: Exploring God's Relationship to Time* (Wheaton, IL: Crossway Books, 2001).

14. Zimmerman points out that "most B-theorists (though not all) will resist the labels 'static view of time' or 'timeless view of nature.' They will insist

that the world is not 'static' on their view, lots of things change within it, things are not timeless but in time, etc." Dean Zimmerman, private correspondence with the author, October 2008.

15. Tooley, *Time,* 16.

16. This is not to say that Pannenberg himself needs to attend to the debates in analytic philosophy. Nevertheless it is crucial that I do so precisely because many A-theorists draw enormous support from physics (specifically special relativity) and because my goal in this volume is to reformulate Pannenberg's work on time and eternity in light of physics—without opening the gates to an assault by A-theorists on his theology. It will be an additional benefit if, in the process, I can draw from Pannenberg's understanding of time and eternity when it has been reformulated in light of physics to suggest a new approach to flowing time that is consistent with relativity (see chapter 5).

17. I am reminded of Paul Tillich's philosophical analysis of the term "existence." The standard definition of the Latin word *existere* is "to stand forth, to arise, to appear." Tillich extends this to include "to stand out." But from what do things "stand out" of? In response, Tillich turned to the context of Greek philosophy, where things exist by "standing out" of nonbeing understood in two ways: First, to exist is to stand out of absolute non-being (*ouk on*). Finite being as finite only partially achieves this, always remaining a mixture of being and non-being. Second, to exist is to change from potential being into actual being. Thus in existing, actual being stands partially out of relative non-being (*me on*). Still since it never fully actualizes its potential, its character changes in time. See Paul Tillich, *Systematic Theology,* vol. 2 (Chicago: University of Chicago Press, 1957), 20–21.

18. The indeterminateness of the future is not equivalent to the causal indeterminism suggested by quantum mechanics. Even in a fully deterministic causal theory, such as classical mechanics, one can claim that the future is not actually determinate until it is the present, although the character of its determinateness— namely what particular properties it will have when it is present—is entirely a function of the deterministic causality in such a theory.

19. John Ellis McTaggart, *The Nature of Existence,* vol. 2 (Cambridge: Cambridge University Press, 1927), 20; reprinted as McTaggart, "Time: An Excerpt from *The Nature of Existence,*" in *Metaphysics: The Big Questions,* ed. Peter van Inwagen and Dean Zimmerman (Malden, MA: Blackwell, 1998), 67–74, quotation at 72.

20. For details of temporal issues in accounts referred to by the acronym NIODA, see Robert John Russell, Nancey Murphy, and William R. Stoeger, S.J., eds., *Scientific Perspectives on Divine Action: Twenty Years of Challenge and Progress* (Vatican City State: Vatican Observatory; Berkeley, CA: Center for Theology and the Natural Sciences, 2008), esp. chaps. 4–6; and various articles in additional volumes of the Scientific Perspectives on Divine Action Se-

ries, published by the Vatican Observatory and the Center for Theology and the Natural Sciences.

21. To be more precise, the relation of an event to itself (i.e., the "present") will be viewed as ontologically different from the relations of an event considered as present to events related to it as past and future, as we shall see below.

22. Interestingly, many scholars in theology and science hold such a view about the unreality of the future, including Arthur Peacocke, Ian Barbour, and John Polkinghorne.

23. The ontological status of the event at time B—actual versus potential, determinate versus indeterminate—refers to the event at time B, and not to the time of the event. In essence, B, as an event in the future of the event at time A, is only potentially real and indeterminate and characterized by at least two and perhaps an infinite set of possible states of affairs. (Think, for example, of the endless variety of choices and amounts of food one might have for lunch tomorrow, although when the time comes, something quite definite and finite is consumed!) Its "location" in time t, however, is *not* indeterminate or potentially real: it is definitely at the time B. Thus, while its temporal location does not change in this sense, what I am calling its "ontological status" does change as its time changes from future to present to past.

24. The "tangle" refers, roughly, to what McTaggart saw as conflated property ascriptions cited above.

25. Broad, *Examination of McTaggart's Philosophy,* 2:313; reprinted in C. D. Broad, "McTaggart's Arguments against the Reality of Time: An Excerpt from *Examination of McTaggart's Philosophy,*" in *Metaphysics: The Big Questions,* ed. Peter van Inwagen and Dean Zimmerman (Malden, MA: Blackwell, 1998), 74–79, quotation at p. 77.

26. To be correct, $R^{A, D, U}$ and $R^{P, I}$ will be assigned to the relations between different events and an event as present while $R^{A, D}$ will be assigned to an event as present.

27. While I occasionally refer to temporal relations as tensed, to be accurate "tense" is a feature of language by which one makes assertions about events in the world. So, the sentence, "C is future in relation to B" is a tenseless sentence and, if it is true, it is "eternally true." Hence, following Tooley's insight about a "dynamic conception of the world," in order to claim that mine is an A-theory of time I must add something about the nature of events and their relations in the world, which cannot be described by a tenseless sentence. It is this challenge that I am responding to here in terms of ontology III-2.

28. Again, my attempt to shift the concept of tense, primarily that of past/ present/future, from being seen as a property of an event to a relation between events, is not meant to suggest that all temporal features of the world can be expressed by tenseless sentences as can past/present/future, since this would transform my theory into a B-theory of time. Instead, I intend to anchor its standing

as an A-theory of time in Pannenberg's theological assertion that God is the Creator of flowing time: it is God's act of continuous creation that brings about the transition from future potentiality to present actuality.

29. See, for example, Dowden, "Time," sec. 6, http://www.iep.utm.edu/t/time.htm#H6.

30. An early example of this idea can be found in St. Augustine's *De Trinitate* 6.10. I am not arguing that such vestiges or "traces" in nature (or in culture, history, religion, or the human soul) are to be taken as an independent basis for the doctrine of the Trinity separate from and alongside of Scripture. Rather my point is that we should expect to find instances of three-foldness in nature because nature is the creation of the Trinitarian God. I thus agree with Barth in his rejection of the traditional *vestigium* argument when it takes the form of a purported second route to the Trinity, but I am adopting his remarkable inversion of the argument here. "The case then was not that men wished to explain the Trinity by the world, but on the contrary that they wished to explain the world by the Trinity in order to be able to speak of the Trinity in this world." He then referred to this as "*vestigia creaturae in trinitate*." Karl Barth, *Church Dogmatics,* 4 vols. in 12, ed. G. W. Bromiley and T. F. Torrance, trans. G. T. Thomson et al. (T and T Clark: Edinburgh: 1936–77), 1.1.8.3, 383–99, esp. 391–92.

31. Here I am drawing on a diagram by Jürgen Moltmann in which he describes his view of Augustine's analysis of time. See Moltmann, *God in Creation: A New Theology of Creation and the Spirit of God,* Gifford Lectures 1984–1985 (San Francisco: Harper and Row, 1985), chap. 5. Moltmann understands Augustine's language about memory, expectation, and perception to be purely subjective. Following my interpretation of Pannenberg (see chapter 1, note 36), I will suggest that it can also apply to objective physical time (see chapter 6).

32. A vector is a mathematical object with two properties: scale (i.e., size) and direction. An intuitive example is velocity. Imagine that at noon you are driving north on San Francisco's Golden Gate Bridge at 50 mph. A vector representing that velocity could be drawn at your location at noon. Its tail would lie at your location, its length would represent 50 mph, and its direction, indicated by an arrowhead, would point north. A comparison between velocity and temperature is helpful here. Temperature is merely a scalar quantity: its measure can vary but the term "direction" is meaningless. So it might be a warm 80°F in San Francisco, but it would make no sense to add that the temperature is "80° North." Vectors, as well as scalars, are ubiquitous in physics.

33. Actually, vectors typically do not lie within the space of the phenomena they represent and they do not literally connect different points in that space. Instead, they lie in an abstract mathematical space associated with the space of the phenomena. So, in the preceding example, the vector that represents a velocity of 50 mph north lies in a tangent space whose origin is mapped to the position of your car as you enter the Golden Gate Bridge at noon. Since

the physical space (x, y, z) in which your car moves is roughly "flat" and since the vector space is identically flat, it is easy to conflate them and project the vector "down" from the vector space onto the physical space at the point where your car is located. Thus vectors representing velocity are typically drawn in pictures such as the one showing your car, the bridge, etc. Nevertheless the vector space is an abstract, mathematical space separate from the physical space of the phenomena it describes.

34. For a readable introduction, see Martinus Veltman, "The Higgs Boson," *Scientific American* (November 1986): 88. See also Chris Quigg, "The Coming Revolutions in Particle Physics," *Scientific American* (February 2008): 46–53.

35. Karl Rahner, *The Trinity* (New York: Herder and Herder, 1970), esp. 76–79.

36. Thomas Aquinas says that the Trinitarian persons *are* the subsisting relations; see *Summa Theologica* I.40.

37. Moltmann makes a similar claim in his diagram of "the network of the times." He cites Augustine's *Confessions* 11.20, as the source of this claim, and he notes several scholars who have further differentiated Augustine's temporal modes.

38. It is important to distinguish between the flow of time and the direction of time's flow. Consider the wind as an analogy: we measure the speed of the wind's flow with an anemometer and its direction with a windsock or a weather vane. So, we could say that time flows at a rate of one second per second, although this is mostly a tautology and the underlying conceptual problem is vast, while its direction is universally one way, as the examples below suggest.

39. The equations of fundamental physics, such as quantum mechanics, are said to be reversible in time because they remain unchanged if we switch "t" and "–t" in them. This is because the time variable enters as a squared term, t^2. Their reversibility means that they allow solutions that include time running either from the past to the future or from the future to the past. These equations in themselves, therefore, do not offer a basis for the universal phenomenological fact that time for all observers runs the same way, as these examples of water waves diverging from their source and coffee cooling suggest. There is a minor exception, however, involving the decay of kaons into pairs of pions.

40. For a helpful introduction, see "Self-similarity," *Wikipedia,* last modified August 15, 2011, http://en.wikipedia.org/wiki/Self-similarity. See also Robert L. Oldershaw, "Nature Adores Self-Similarity," http://www3.amherst.edu/~rloldershaw/nature.html.

41. I say "nearly" because the iterative process cannot be endlessly repeated but these examples still convey the essential idea that the shape of the whole is congruent with the shape of its parts.

42. By "below" I mean that the copy of t_1 lies below the plane formed by the three axes t_1, t_2, and t_3 in figure 2.12.

43. Recall that for presentist A-theorists (ontology I) and for some growing universe A-theorists (ontology II), future events are not just indeterminate, they are not real. Here I am assuming what I call ontology III—growing universe A-theorists with future events as potential—with the qualification that these tenses are relations between events and not properties of events.

44. As we will see in chapter 3, "dense" means roughly that between any two points along the axis, no matter how close, there are an uncountable infinity of points. Compare this with the natural, discrete numbers. The continuum (including the time axis) is thus at least an order of infinity higher than the real numbers, according to the still debated "continuum hypothesis."

45. See Georg Cantor, *Contributions to the Founding of the Theory of Transfinite Numbers,* trans. Philip E. B. Jourdain (New York: Dover, 1895); see also the discussion of Cantor in chapter 3, particularly section A.

46. See chapter 1, note 45, where I point out that Pannenberg describes this separation as the "brokenness of our experience of time for which all life is torn apart by the separateness of past, present, and future." He relates this to the "structural sinfulness of our life" and tells us that it is to be overcome eschatologically (ST3, 561).

47. Again, creaturely time is not essentially broken by the separation of its present moments although it is experienced as broken. Johanne Stubbe Teglbjaerg, private communication with the author, summer 2010.

48. Recall that these temporal relations are abstractions; they do not lie in physical time. See the discussion related to figure 2.12 in section A.3 above.

49. In 1984 I first separated out several related aspects regarding time, including its directionality, the loss of the present, and the anticipation of the future by coining the metaphor "the talons of time":

> In religion, the direction of the passage of time is an underlying, though normally tacit, assumption without which most personal experience and community history would be meaningless. We watch our children grow; we sorrow over lost friends and broken promises; we mourn the death of our loved ones; we can be tormented by the anticipation of suffering; we believe in the promise of divine redemption; in short, we *know* the passage of time. Yet is the "arrow of time" a thoroughly established fact in physics? And if not, would our human experience of its *talons* be, ultimately, an illusion? For me, victory can only be true, and defeat only bearable, if time will not one day erase the stages of my living.

See Russell, *Cosmology from Alpha to Omega: The Creative Mutual Interaction of Theology and Science,* chap. 7, "Entropy and Evil: The Role of Thermodynamics in the Ambiguity of Good and Evil in Nature," 226–48, esp. 238, where the original 1984 article is revised and reprinted.

50. A somewhat amusing example of the difference would be the physics department library at University of California, Berkeley, which is closed stacks,

and the Graduate Theological Union's Hewlett Library, which is open stacks. If you have ever tried getting access to books from both libraries you will understand the difference between them!

51. To make this suggestion more concrete, consider the mathematics of ordering a set of N elements, here the N books of our life. A simple example will help us deduce the general rule. Let's consider three such books A, B, and C written about the whole of our life from the perspective of three different moments as in the example of the library (above). Suppose you first read book A, then book B and then book C. In this example B provides one context of interpretation for A, and C provides a context of interpretation for B and thus in turn a different context of interpretation for A. We can represent this symbolically as follows. For three books A, B, and C read in the order A, B, C, then B(A) and C(B(A)) (B is the context of A and C is the context of B which is the context of A). Thus A means something different when first read in light of B than when B is read in the context of C. Now consider all orders in which to read the three books:

C(B(A))
C(A(B))
B(A(C))
B(C(A))
A(C(B))
A(B(C))

From this we can infer how many ways there are to read N books: We first read each of N books, then we read each of N-1 books second (and these provide the first context of interpretation), then N-2 books third (and these provide the second context of interpretation), and so on until all the books are listed. The result is that for N books there are "N!" orders in which they can be read, where N! = N(N-1)(N-2) . . . 1. The result, N!, is called "N factorial." This number increases extremely rapidly as N increases. For example:

N	N!
1	1
2	2
3	6
4	24
5	120
6	720
7	5,040
8	40,320
9	362,880
10	3,628,800

Larger values of N and N!:

25	$> 10^{25}$
50	$> 10^{64}$
70	$> 10^{100}$
100	$> 10^{158}$
1000	$> 10^{2,567}$

To have even a clue as to how big these numbers are remember that the number of stars in the visible universe is approximately 10^{22} and the number of atoms approximately 10^{80}. So the number of ways you can read even 70 books is 10^{20} times the number of atoms in the visible universe (that is, one thousand million million million times the number of atoms). The upshot is clear: the combinations of N books always remain finite, and yet it very quickly becomes larger than an almost inconceivably large number. Clearly we could imagine reading the books of our lives endlessly in the New Creation and always find ever-increasing newness and insight as the order of their reading and interpretation undergoes endless exquisite permutations. For an introduction, see "Factorial," *Wikipedia,* last modified August 8, 2011, http://en.wikipedia.org/wiki/Factorial.

52. In other words, "It's hermeneutics all the way down." In his widely read book *A Brief History of Time* (1988), Stephen Hawkins recounts the story of a scientist (most likely Bertrand Russell) once lecturing on astronomy. After describing the structure of the solar system and the galaxy, the young scientist is confronted by an audience member: "What you have told us is rubbish. The world is really a flat plate supported on the back of a giant tortoise." The scientist replies, "What is the tortoise standing on?" "You're very clever, young man, very clever," said the questioner, "but it's turtles all the way down!" Hawkins, *A Brief History of Time: From the Big Bang to Black Holes* (New York: Bantam Books, 1988), 1.

53. There is no scholarly consensus on this point. Other locations of canonical decisions that differ from the Western Latin tradition in some details include those of the Byzantine East, Ethiopian, Armenian, and Syriac churches. Joshua Moritz, private communication with the author, spring 2010.

54. Martin Luther is a notable example from the sixteenth century, since he questioned the inclusion of the Epistles of James, Jude, Hebrews, and the Revelation of John in the New Testament canon.

55. On a personal note, growing up with training as a classical pianist, I took for granted that music is homophonic: a single melody line played in time against a well-harmonized progression of chords. As an undergraduate at Stanford University, though, in studying music history I came to realize that polyphony was the historical route to homophonic music. Still, a tantalizing question lingers for me: Is it preferable to consider a piece of music as the evolution of a series of chords by well-winnowed rules of harmonic progression with a lyrical melody floating in consonance above them, or as a simultaneous sequencing of multiple

melodies, each voice beautiful in itself but together producing an emergent, polyphonic harmony of irreducible and surpassing beauty?

56. We could embellish this concept further by thinking of the library of books as hyperlinked, the books at each level of the "fractal" hyperlinked to their subterranean labyrinth, making their availability for exploration that much simpler and immediate.

57. If we use Cartesian coordinates, distances between points are calculated as the square root of the sum of squares of the coordinates, following the Pythagorean Theorem. For example, a point P with coordinates x, y in a plane lies a distance r from the origin, where $r^2 = x^2 + y^2$.

58. As is well known, non-Euclidean spaces play a central role in Albert Einstein's general theory of relativity, where matter "curves" space and space "curves" the motion of matter.

59. Say we start at x = 0.5 and move incrementally toward increasing values of x. As we approach x = 1 we make successively smaller moves ad infinitum. In this way we never reach x = 1, a fact immortalized in Zeno's paradoxes.

60. A simple example is the x-axis. It is Hausdorff because any two points along it have neighborhoods which are disjoint. For example the points x = 1 and x = 2 have disjoint neighborhoods, 0.9 < x < 1.1 and 1.9 < x < 2.1.

61. See, for example, Mathieu Baillif and Alexandre Gabard, "Manifolds: Hausdorffness Versus Homogeneity," preprint, submitted September 4, 2006, http://arxiv.org/abs/math/0609098, 1–6. The difficulty in representing the topologies of non-Hausdorff manifolds is suggested by the figures in Baillif and Garbard's article and in the drawings in my text. In figure 2.23 below, for example, one way to express this difficulty is by noting that while the origins O_1 and O_2 are the "first points" along R_1 and R_2 which are not identified, and while R_1 and R_2 are each continuous for, say, the intervals $-1 < x_1 < +1$ and $-1 < x_2 < +1$, nevertheless there is no "last point" along either $x_1 < 0$ or $x_2 < 0$ to which the indicator of continuity " / " in figure 2.23c can be joined. This of course reflects one of the inner complications of the fact that the real line is dense. Unlike the common conception, the real line is composed of a series of points, which might suggest that there are gaps between them (however small), there are actually an uncountable infinity of points between any two points on the real line.

62. Baillif and Gabard, "Manifolds," 1.

63. Ibid. Baillif and Gabard credit the discovery of the complete feather to A. Haefliger and G. Reeb.

64. It is not sufficient to account for all aspects of the *flow* of time, particularly the "arrow" of time (see note 49 above). More important, as Pannenberg stresses, phenomenological features of nature such as this require a metaphysical argument for real becoming in relation to being, as his interactions with Hegel, Heidegger, Bergson, Whitehead, and Dilthey underscore.

65. I characterize these temporal features of the created world as unnecessary but inevitable intentionally to lay the grounds for proposing a close connection with Reinhold Niebuhr's rendering of Augustine's understanding of original sin as unnecessary but inevitable. In future work I want to explore the ways in which these temporal features reflect the "fallenness" of nature and possibility of their being dissolved in the New Creation. See Reinhold Niebuhr, *The Nature and Destiny of Man: I. Human Nature* (New York: Charles Scribner's Sons, 1941, repr. 1964), chap. 6. For an earlier discussion of the traces of these features at the biological and physical levels of nature, see Russell, *Cosmology from Alpha to Omega,* chaps. 7–8.

66. I am tempted to borrow language here from the ancient Calcedonian Christological formula to suggest in a highly metaphorical way what I hope to have represented here in mathematical terms: namely the two axes t_B and t_C in the region prior to event B_{pr} and C_{pr}, where they are identified, co-exist without confusion, division, or separation. Another way of saying this is that B as present along t_B remains connected to its past relations along t_B, and similarly for C as present along t_C, even while t_B and t_C are identified for all their events prior to B_{pr} and C_{pr}.

67. Unlike special relativity (SR), which I discuss in some detail in chapter 4, I will not offer an introduction to quantum mechanics (QM) here. There are several reasons for this. One is that there are a variety of very adroit introductions to QM available for the nonspecialist both in print and online (see below). In contrast, my approach to SR is relatively unique, relying almost entirely on space-time diagrams to lead the reader into the inner logic of the theory, and it is this approach that I find invaluable for the philosophical and theological appropriations of SR that I wish to make in this book. Another reason is that here I am only going to draw on one, though central, feature of QM: entanglement. A third consideration is the length of this volume. I hope to provide enough of an introduction that the nonspecialist can gain what is needed for my philosophical and theological arguments. Introductions to QM and its diversity of competing philosophical interpretations include John Polkinghorne, *Quantum Theory: A Very Short Introduction* (Oxford: Oxford University Press, 2002); Daniel F. Styer, *The Strange World of Quantum Mechanics* (Cambridge: Cambridge University Press, 2000); Robert John Russell, "Divine Action and Quantum Mechanics: A Fresh Assessment," in *Quantum Mechanics: Scientific Perspectives on Divine Action,* ed. Robert John Russell, Philip Clayton, Kirk Wegter-McNelly, and John Polkinghorne (Vatican City State: Vatican Observatory; Berkeley, CA: Center for Theology and the Natural Sciences, 2001), 293–328. For an online introduction, see "Introduction to Quantum Mechanics," *Wikipedia,* last modified August 8, 2011, http://en.wikipedia.org/wiki/Introduction_to_quantum_mechanics.

68. Clearly this statement needs much refinement given the "downfall" of simultaneity in special relativity. Nevertheless the basic idea is to make the measurements on the two particles such that nothing could propagate between

them even at the speed of light and influence the results. Such measurements are said to be "local." For details, see the references below. For an extended discussion of special relativity, see chapter 4 of this volume.

69. See "Double-Slit Experiment," *Wikipedia,* last modified August 5, 2011, http://en.wikipedia.org/wiki/Double-slit_experiment.

70. In a very rough way, an electron with spin is analogous to a top that rotates around its axis. But, as with all analogies between quantum mechanics and the classical world, this analogy breaks down rapidly.

71. Technically I need to multiple the expression on the right-hand side of the equation with the factor $1/\sqrt{2}$, but I have omitted this here and below for simplicity.

72. We can see this mathematically as follows: if state $\psi(+)_1$ equals state $\psi(-)_2$ then the combined state $\psi_T = \psi(+)_1\psi(-)_2 - \psi(-)_1\psi(+)_2$ vanishes: $\psi_T = 0$.

73. One way of explaining why we need to use the entanglement equation 3 instead of the classical equations 4a or 4b is to start with the breakdown of the classical ideal of distinguishable particles. When two electrons interact, there is simply no way to follow their individual trajectories to see if they pass by each other or recoil off each other. And thus, as with the double slit experiment, both possibilities must be included in calculating the results. When this more complex type of superposition is built into the mathematics of quantum mechanics, the result is the entangled wavefunction (3). Another way is to explore the connection between quantum entanglement and the statistical properties of fundamental particles. Now fundamental particles are divided into two types. Particles that carry the fundamental forces in nature (photons, gravitons, etc.) are called "bosons" and are represented by a symmetric version of equation 3: $\psi_{TS} = \psi(+)_1\psi(-)_2 + \psi(-)_1\psi(+)_2$. (We say that ψ_{TS} is "symmetric" because it is unchanged when we interchange particle labels 1 and 2.) Bosons are much more likely to be clumped together than to be randomly distributed, as the coherence of tightly knit photons in a laser beam demonstrates. Particles that are the basic "building blocks" of matter (electrons, protons, neutrons, etc.) are called "fermions" and are represented by an antisymmetric equation such as (3): $\psi_{TA} = \psi(+)_1\psi(-)_2 - \psi(-)_1\psi(+)_2$. (We say that ψ_{TA} is "antisymmetric" because its sign changes when we interchange particle labels 1 and 2.) It is easy to see why the Pauli exclusion principle—that no two electrons can be in the same state at the same time—is a result of this antisymmetry. If both electrons were to be spin up, the right-hand side of equation 3 would equal zero, and thus the square of the total state, which indicates the probability of finding such a state, would vanish. Since much of the chemical properties—indeed the shell structure of the orbits of electrons in complex atoms—results from the Pauli principle, one can say that the vast diversity of chemical properties that characterize our ordinary world of experience are rooted in the fermi statistics of electrons and thus in the quantum mechanics of entanglement. In essence, quantum entanglement underlies and gives rise to many of the far-reaching phenomena

that we take for granted in the ordinary world of experience and as reflected in classical physics and chemistry.

74. George Greenstein and Arthur G. Zajonc, *The Quantum Challenge: Modern Research on the Foundations of Quantum Mechanics* (Boston: Jones and Bartlett, 1997), 131; see chap. 5, esp. 124–30 for a very readable account of quantum mechanics and Bell's Theorem. Further introductions to the interpretive, philosophical, and theological issues of quantum mechanics include Nick Herbert, *Quantum Reality: Beyond the New Physics* (Garden City, NY: Anchor Press, 1985); David Z. Albert, *Quantum Mechanics and Experience* (Cambridge, MA: Harvard University Press, 1992); James T. Cushing and Ernan McMullin, eds., *Philosophical Consequences of Quantum Theory: Reflections on Bell's Theorem* (Notre Dame, IN: University of Notre Dame Press, 1989); Russell Clayton, Wegter-McNelly, and Polkinghorne, *Quantum Mechanics.*

75. Erwin Schrödinger, "Discussion of Probability Relations between Separated Systems," *Mathematical Proceedings of the Cambridge Philosophical Society* 31, no. 4 (1935): 555–63, quoted in Greenstein and Zajonc, *Quantum Challenge,* 132.

76. John S. Bell, "On the Einstein-Podolsky-Rosen Paradox," *Physics* 1 (1964): 195–200; J. Bell, "On the Problem of Hidden Variables in Quantum Mechanics," *Reviews of Modern Physics* 38 (1966): 447–52. For a careful and readable discussion of Bell's theorem, see Cushing and McMullin, *Philosophical Consequences*; and Greenstein and Zajonc, *Quantum Challenge.*

77. Alain Aspect, Philippe Grangier, and Gerard Roger, "Experimental Tests of Realistic Local Theories via Bell's Theorem," *Physical Review Letters* 47, no. 7 (1981): 460–63. For additional references to their work and to the work of others, see Greenstein and Zajonc, *Quantum Challenge,* chap. 6.

78. This challenge to realism is vibrantly communicated by David Mermin through a series of brilliant thought experiments. See, for example, N. David Mermin, "Is the Moon There When Nobody Looks? Reality and the Quantum Theory" *Physics Today* 38 (April 1985): 38.

79. For a careful analysis of the precise ways in which relativity is, and is not, violated, see Michael Redhead, "The Tangled Story of Nonlocality in Quantum Mechanics," in *Quantum Mechanics,* 141–58.

80. See, for example, Russell, *Cosmology from Alpha to Omega,* chap. 6.

81. See Robert John Russell, "The Bodily Resurrection of Jesus as a First Instantiation of a New Law of the New Creation: Wright's Visionary New Paradigm in Dialogue with Physics and Cosmology," in *From Resurrection to Return: Perspectives from Theology and Science on Christian Eschatology,* ed. James Haire, Christine Ledger, and Stephen Pickard (Adelaide: ATF Press, 2007), 54–94; and Russell, *Cosmology from Alpha to Omega,* introd., chaps. 1, 10.

82. Ted Peters calls the first *futurum* and the second *adventus.* See Peters, *God—the World's Future: Systematic Theology for a Postmodern Era* (Minneapolis: Fortress Press, 1992), 307–9.

83. The choice of specific elements is clearly rather personal, but I hope this list is at least suggestive of the possibility of elements of continuity within the eschatological transformation of the universe.

84. For definition and discussion of terms such as "natural evil," see Nancey Murphy, Robert John Russell, and William R. Stoeger, S.J., eds., *Physics and Cosmology: Scientific Perspectives on the Problem of Natural Evil* (Vatican City State: Vatican Observatory; Berkeley, CA: Center for Theology and the Natural Sciences, 2007); and Christopher Southgate, *The Groaning of Creation: God, Evolution, and the Problem of Evil* (Westminster/John Knox Press, 2008).

85. TKG, 133. Ted Peters has built upon and extended Pannenberg's arguments about prolepsis in very impressive and promising ways that compel further exploration. See in particular his nine theses in Peters, *Anticipating Omega: Science, Faith, and Our Ultimate Future* (Gottingen: Vandenhoeck and Ruprecht, 2006), chaps. 1–2.

86. One way of understanding what it means to say that a sphere is non-Euclidean is to realize that the sum of the interior angles in a triangle drawn on the sphere is greater than 180°. So, for example, start at the equator of the Earth and form a triangle out of two different longitudinal lines: they meet at an angle at the North Pole, yet they both lie perpendicular to the equator.

87. One kind of circle goes around the outside "equator" of the doughnut, the other goes around it like a ring.

88. See Charles W. Misner, Kip S. Thorne, and John Archibald Wheeler, *Gravitation* (San Francisco: W. H. Freeman, 1973), p. 921, fig. 34.4, for the Reissner-Nordstrom Penrose diagram. For the conformal structure of the Kerr solution, see Stephen Hawking and George F. R. Ellis, *The Large Scale Structure of Space-Time* (Cambridge: Cambridge University Press, 1973), 165.

89. See Andrei Linde, "Eternal Inflation and Sinks in the Landscape," powerpoint presentation given at Cosmo 2006: International Workshop on Particle Physics and the Early Universe, University of California, Davis, September 25–29, 2006, http://cosmo06.ucdavis.edu/talks/linde.ppt#546.

90. As before, these multiple prolepses should be understood not as paths lying on a background "space" that includes them and the timeline of the universe but rather as paths representing a topological "folding" of the timeline of the universe back onto itself (i.e., the lines on the paper *are* the spaces of prolepsis; there is no background "space" that includes them).

91. This is because the coordinates x, t of event B are related by $x = ct$, and thus $\tau^2 = t^2 - x^2/c^2 = x^2/c^2 - x^2/c^2 = 0$.

Chapter 3. From Hegel to Cantor

A version of this chapter was previously published as "The God Who Infinitely Transcends Infinity: Insights from Cosmology and Mathematics," in Robert John Russell, *Cosmology from Alpha to Omega: The Creative Mutual Interaction of*

Theology and Science (Philadelphia: Fortress Press, 2008), chap. 2. Material in this chapter is also revised from "God and Infinity: Theological Insights from Cantor's Mathematics," in *Infinity: New Research Frontiers,* ed. Michael Heller and W. Hugh Woodin (Cambridge: Cambridge University Press, 2011), 275–89.

1. For an insightful introduction, see Rudy Rucker, *Infinity and the Mind: The Science and Philosophy of the Infinite* (New York: Bantam, 1983). For a paper that explores themes in common with the present chapter, see Timothy J. Pennings, "Infinity and the Absolute: Insights in Our World, Our Faith and Ourselves," in *Christian Scholar's Review* 23, no. 2 (1993): 159–80. For an introductory online resource, see "Infinity," *Wikipedia,* last modified July 22, 2011, http://en.wikipedia.org/wiki/Infinity. Additional resources include John D. Barrow, *The Artful Universe Expanded* (Oxford: Oxford University Press, 2005); Eli Maor, *To Infinity and Beyond: A Cultural History of Infinity* (Princeton, NJ: Princeton University Press, 1987); and, online, query results for "infinity," *Stanford Encyclopedia of Philosophy,* http://www.seop.leeds.ac.uk/search/searcher.py?page=2&query=infinity.

2. See David Bentley Hart, "Notes on the Concept of the Infinite in the History of Western Metaphysics," *Infinity,* 255–74.

3. Hart also makes a crucial distinction, at the outset, between the physical/mathematical and the metaphysical/ontological meanings of the term "infinity." While accepting this distinction, the purpose of my paper is to begin with the ways in which the mathematical concept of infinity has changed with the work of Cantor and explore the implications of this change for the ways we implicitly use the mathematical sense of infinity in discussing metaphysical issues in the context of theology focused on revelation and on the divine attributes, following Pannenberg.

4. Aristotle, *Physics* 3.7.207b35, *The Basic Works of Aristotle,* ed. Richard McKeon (New York: Random House, 1941).

5. Aristotle, *Physics* 3.6.206a26–30.

6. Aristotle, *Physics* 3.4.204b1–206a8.

7. Aristotle, *Physics* 3.5.206a5b.

8. By a more positive conception of infinity, I mean that these early Christian writers were informed by Aristotle's understanding compared with that of Plato. In a recent essay, Denys Turner offers an important criticism of my reference to a "positive notion" of the infinite: "We need to distinguish between two sorts of predicates of God: those which can be said to be 'positive' (predicates) "such as 'goodness,' 'wisdom,' 'intelligence'—and are known 'by analogy' from what we know of such predicates as affirmed of creatures, and those which are . . . 'regulative'—such as 'infinity' and 'simplicity'—which are known to be true of God *only by denial* of what we know of creatures." See Denys A. Turner, "A (Partially) Skeptical Response to Hart and Russell," in *Infinity: New Research Frontiers,* ed. Michael Heller and W. Hugh Woodin (Cam-

bridge: Cambridge University Press, 2011), 294. I am appreciative of Turner's distinction between these two types of predicates of God. Using Turner's, together with Pannenberg's, terminology, the regulative predicates are appropriate for discussing those divine attributes that define the God who acts while the positive predicates are appropriate for discussing those divine attributes that describe God's actions. See my discussion of Pannenberg and the divine attributes in chapter 1, section E.3

9. Augustine, *City of God* 12.18.

10. Thomas Aquinas, *Summa Theologica* Ia, 7, 2–4. Most contemporary theists share this traditional understanding of infinity: God alone is infinite, and the world is strictly finite. God's infinity is a mode of God's perfection even while God is incomprehensible. We can know *that* God is perfect and infinite self-existence, but we cannot conceive of *how* God is perfect and infinite self-existence. Note the strict distinction continues to be held between divine infinity and created finitude.

11. For an excellent discussion of the history of the concept of infinity in ancient and modern philosophy, including the work of Nicholas of Cusa, see Wolfgang Achtner, "Infinity as a Transformative Concept in Science and Theology," in *Infinity: New Research Frontiers,* ed. Michael Heller and W. Hugh Woodin (Cambridge: Cambridge University Press, 2011), 19–53.

12. Galileo Galilei, *Dialogues Concerning Two New Sciences,* trans. Henry Crew and Alfonso de Salvio (New York: Dover, 1954), 30–33.

13. Rational numbers are numbers which can be expressed as a fraction, that is, as the ratio of two natural numbers (i.e., integers) such as 41/42. Irrational numbers are numbers which cannot be expressed as a fraction; a famous example is π, the ratio of the circumference of the circle to its radius. Rational numbers and irrational numbers together make up the real numbers. Real numbers can be notated as a decimal with an infinite sequence of integers, such as 0.9761904 . . . . Real numbers can be used to measure intervals along a linear continuum, such as the x-axis.

14. For a detailed biography of Cantor, see J. J. O'Connor and E. F. Robertson, "Georg Ferdinand Ludwig Philipp Cantor," MacTutor History of Mathematics Archive, October 1998, http://www-history.mcs.st-andrews.ac.uk/Biographies/Cantor.html.

15. This translation of Cantor's original text is from Abraham A. Fraenkel, Yehoshua Bar-Hillel, and Azriel Lévy, *Foundations of Set Theory,* 2nd rev. ed. (Amsterdam: North-Holland, 1973), 15. Jourdain suggests that "aggregate" is a better translation than "set" for Cantor's term, *Menge.* See Georg Cantor, *Contributions to the Founding of the Theory of Transfinite Numbers,* trans. Philip E. B. Jourdain (New York: Dover, 1895), 85. Nevertheless, "set" has become the standard term in mathematics.

16. Paul R. Halmos, *Naive Set Theory* (Toronto: D. Van Nostrand, 1960), 1.

17. Cantor, *Contributions,* 86.

18. Consider two sets X and Y. X is called a subset of Y if every element x_i of X is an element y_i of Y. (This entails that Y is a subset of itself.) X is called a *proper* subset of Y if there are elements y_i in Y which are not elements of X. So, for example, the set X = {2, 4, 6} is a proper subset of the set Y = {1, 2, 3, 4, 5, 6}.

19. For an introductory text on set theory, see Halmos, *Naive Set Theory.* For a technical text, see Abraham A. Fraenkel and Azriel Lévy, *Abstract Set Theory,* 4th rev. ed. (Amsterdam: North-Holland, 1976). Cantor's final essays on his theory of the transfinites were published in *Mathematische Annalen* in 1895–97 and translated into English in 1915 as Cantor, *Contributions.* For further references to Cantor's work as well as an extensive bibliography, see Fraenkel, Bar-Hillel, and Lévy, *Foundations of Set Theory.* See also Hans Hahn, "Infinity," *The World of Mathematics,* ed. James R. Newman, vol. 3 (New York: Simon and Schuster, 1956), 1593–1611; Rucker, *Infinity and the Mind,* chaps. 1, 2; and William Lane Craig, *The Kalam Cosmological Argument* (New York: Barnes and Noble, 1979). For introductory online resources, see "Set Theory," *Wikipedia,* last modified August 8, 2011, http://en.wikipedia.org/wiki/Set_theory; and Thomas Jech, "Set Theory," *Stanford Encyclopedia of Philosophy,* last modified July 11, 2002, http://plato.stanford.edu/entries/set-theory.

20. Technically, a one-to-one correspondence is a bijective map between sets A and B in which each element in A maps to a unique element in B, and all elements in B are mapped onto. By comparison, an injection is a map between sets A and B in which at least one element in B is not mapped onto, and a surjection is a map between sets A and B in which at least one element in B is mapped onto from at least two elements in A.

21. Halmos, *Naive Set Theory,* 66, 75. For example, when we say that an essay is twenty pages long, the number 20 is the cardinal number designating the length of the essay. If you are reading page 15, the number 15 is the ordinal number designating the fact that you are reading the fifteenth page in the essay.

22. We can put this more formally, and perhaps more intuitively, as follows: Start with the set of natural numbers {1, 2, 3, . . .} with cardinal number $\aleph_0$ and ordinal number ω. Then consider sets with the following three properties: a) they contain at least one element; b) each element has a predecessor and a successor, and c) they contain no last element. Such sets have the same ordinal number ω as the set of natural numbers. Now we are ready to distinguish between the set $1 + \omega$ and the set $\omega + 1$. In the case of $1 + \omega$, we start with the element 1 and add all the natural numbers to it, thus forming the set {1, 1, 2, 3, . . .}. This set shares the three properties that characterized the set of natural numbers, and thus, $1 + \omega = \omega$. On the other hand, if we start with the set of natural numbers and then add a new element to this set, thus forming the set {1, 2, 3, . . . , 1}, the new set no longer has property c, since unlike the set of natural numbers, there is a last element, 1, lying beyond the unending series 1, 2, 3, ... Thus its ordinal number, $\omega + 1$, cannot be equal to the ordinal number of the natural set, ω.

23. The best-known of Cantor's proofs for the uncountability of the real numbers is his diagonalization argument published in 1891. The proof starts by assuming the converse, namely that the real numbers are countably infinite. For convenience we represent the real numbers $0 < x < 1$ by infinite decimals such as 0.1234. . . . If the real numbers are countable we can list them and count them using the natural numbers as follows:

(1) 0.1234 . . .
(2) 0.2345 . . .
(3) 0.3456 . . .
(4) 0.4567 . . .
etc.

Now consider a diagonal that lies along the list:

(1) 0.1234 . . .
(2) 0.2345 . . .
(3) 0.3456 . . .
(4) 0.4567 . . .
etc.

If we produce a number y by changing the digit lying under the diagonal, that is, if we produce the first digit in y by changing the first digit in number (1), the second digit in y by changing the second digit in number (2), the third digit in y by changing the third digit in number (3), etc., the number so generated is a real number lying between zero and one which cannot be a member of the preceding list. Thus, contrary to our assumption, the real numbers are uncountable. See Lillian R. Lieber, *Infinity: Beyond the Beyond the Beyond,* ed. Barry Mazur (Philadelphia: Paul Dry Books, 2007), 109–24 (chap. 8). An online version of the diagonalization argument can be found at "Cantor's Diagonal Argument," *Wikipedia,* last modified August 12, 2011, http://en.wikipedia.org/wiki/Cantor%27s_diagonal_argument.

24. In 1963 Paul Cohen seemed to establish that the continuum hypothesis is neither provable nor disprovable in terms of the axioms of standard Zermelo-Fraenkel set theory together with the axiom of choice. Paul Cohen, "The Independence of the Continuum Hypothesis," *Proceedings of the National Academy of Sciences, U.S.A.* 50 (1963): 1143–48; and Cohen, "The Independence of the Continuum Hypothesis, II," *Proceedings of the National Academy of Sciences, U.S.A.* 51 (1964): 105–10. Hugh Woodin, however, has given reasons for believing that the continuum hypothesis might be false. See W. Hugh Woodin, "The Continuum Hypothesis, Part I," *Notices of the American Mathematical Society* 48, no. 6 (2001): 567–76. For a readable overview, see Erica Klarreich, "Infinite Wisdom: A New Approach to One of Mathematics' Most Notorious Problems," *Science News* 164, no. 9 (August 30, 2009), http://www.phschool.com/science/

science_news/articles/infinite_wisdom.html. For a recent technical report, see Patrick Dehornoy, "Recent Progress on the Continuum Hypothesis (after Wooden)," http://www.math.unicaen.fr/~dehornoy/Surveys/DgtUS.pdf.

25. We refer to the "class" or "collection" of all sets instead of "the set of all sets" for a very important reason. Such a class cannot be considered a set in the strict sense unless we are careful to avoid self-contradictions involving self-reference and unless we clarify what we mean by its properties. This discussion is based in part on Rucker, *Infinity and the Mind,* 53; see also the online resource, "Reflection Principle," *Wikipedia,* last modified June 2, 2011, http://en.wikipedia.org/wiki/Reflection_principle.

26. Georg Cantor, *Gesammelte Abhandlungen Mathematischen und Philosophischen Inhalts* (Springer-Verlag, 1980), 378, trans. Rucker, *Infinity and the Mind,* 10.

27. See, for example, William Lane Craig's argument against the existence of an actual infinite temporal regress based on his argument against an actual infinite in relation to Cantor's work. His objective is to defend the *kalam* cosmological argument for the existence of God. William Lane Craig and Quentin Smith, *Theism, Atheism, and Big Bang Cosmology* (New York: Oxford University Press, 1993), esp. 3–30.

28. "A certain theory contains an antinomy when each of two contradictory statements . . . has been proved within the theory, though the axioms of the theory seem to be true and the rules of inference valid." Fraenkel, Bar-Hillel, and Lévy, *Foundations of Set Theory,* 1; for a technical overview of the antinomies, see chap. 1; for discussion of the logical antinomies (Bertrand Russell's antinomy, Cantor's antinomy, and Burali-Forti's antinomy), 5–8; for discussion of semantical antinomies (Richard's antinomy, Grelling's antinomy, and the Liar), 8–10. For an online introduction, see José Ferreirós, "The Early Development of Set Theory," *Stanford Encyclopedia of Philosophy,* last modified July 6, 2011, http://plato.stanford.edu/archives/fall2011/entries/settheory-early. See also the helpful introduction, "Paradoxes of Set Theory," in *Wikipedia,* last modified February 25, 2011, http://en.wikipedia.org/wiki/Paradoxes_of_set_theory.

29. Douglas R. Hofstadter, *Godel, Escher, Bach: An Eternal Golden Braid* (New York: Basic Books, 1979), 20.

30. The antinomy arises as follows: The well-ordered set S of all ordinal numbers should define an ordinal number O. Hence, S should contain O, being the set of all ordinal numbers, and yet S should not contain O, since an ordinal number O is defined by the set of smaller ordinal numbers S. Cantor was aware of the problem of antinomies in his work, writing in 1905 to Dedekind that "a multiplicity can be such that the assumption that *all* its elements 'are together' leads to a contradiction, so that it is impossible to conceive of the multiplicity as a unity." Quoted in Rucker, *Infinity and the Mind,* 51.

31. Fraenkel, Bar-Hillel, and Levy, *Foundations of Set Theory,* 2. Fraenkel, Bar-Hillel, and Lévy report the depth of the impact of Russell's paradox on

the work of both Richard Dedekind and Gottlob Frege on the foundations of mathematics. Dedekind believed his arguments had been "shattered." Frege, who was about to publish over a decade of work, acknowledged that Russell's paradox had shaken the foundations of his argument. And they quote Hermann Weyl as writing nearly a half century later [in 1946] that "we are less certain than ever about the ultimate foundations of [logic and] mathematics" (2–4).

32. A famous example is Russell's "barber paradox": A barber shaves all men who do not shave themselves. Does the barber shave himself? If he does not shave himself than he has to shave himself since he is the kind of man he shaves—but he cannot shave himself because he only shaves men who do not shave themselves. See "Barber Paradox," *Wikipedia,* last modified July 11, 2011, http://en.wikipedia.org/wiki/Barber_paradox.

33. What we have been discussing so far, therefore, is called Cantor's "naive set theory."

34. Russell pursued a different tack, attempting to find a new logical solution to these paradoxes that led him, in collaboration with Alfred North Whitehead, to write the *Principia Mathematica* (1910–13).

35. See, for example, "Zermelo-Fraenkel Set Theory," *Wikipedia,* last modified August 4, 2011, http://en.wikipedia.org/wiki/ZFC; and "Axiom of Choice," *Wikipedia,* last modified August 13, 2011, http://en.wikipedia.org/wiki/Axiom_of_choice.

36. X is a proper subset of Y if there are elements y_i in Y that are not members of X.

37. Fraenkel and Lévy, *Abstract Set Theory,* 30.

38. Ernst Snapper, "The Three Crises in Mathematics: Logicism, Intuitionism and Formalism," *Mathematics Magazine* 52, no. 4 (September 1979): 207–16.

39. Kurt Gödel proved two now famous theorems that apply to any computable axiomatic system that can treat the arithmetic of the natural numbers (e.g., ZFC and the Whitehead/Russell program): 1. If the system is consistent, it cannot be complete (the incompleteness theorem); 2. The consistency of the axioms cannot be proved within the system. Gödel's theorems thus challenge ZFC, the Whitehead/Russell program, and other axiomatization approaches in fundamental ways. See Fraenkel, Bar-Hillel, and Lévy, *Foundations of Set Theory.*

40. Snapper, "Three Crises in Mathematics," 216.

41. In section C, I will suggest that Pannenberg's understanding of the concept of infinity underlying and structuring the divine attributes implies, and can be enhanced by, Cantor's claim that the transfinites *are realized in nature.* The antinomies not withstanding, this claim might lead to an interesting direction for research in the foundations of mathematics: can the ontological perspective of the doctrine of creation *ex nihilo* point to a new type of response to this crisis? This idea would formally come under part two: TRP → SRP. While I will not explore this question further in this book, I hope to return to it in future research.

42. Joseph W. Dauben, "Georg Cantor and Pope Leo XIII: Mathematics, Theology, and the Infinite," *Journal of the History of Ideas* 38, no. 1 (Jan.–Mar. 1977): 100.

43. Georg Cantor, quoted in Dauben, "Georg Cantor and Pope Leo XIII," 94n31.

44. See Augustine, *City of God* 12.18.

45. Dauben, "Georg Cantor and Pope Leo XIII," 94–95.

46. A striking example, considering that Cantor had been his student, is the finitist Leopold Kronecker, who argued that mathematics should be limited to finite numbers and to a finite set of operations on them.

47. A detailed comparison between Leo's encyclical and the statement in 1988 by John Paul II on theology and science would be particularly intriguing, but I must set it aside for a future occasion. See Pope John Paul II, "The Church and the Scientific Communities: A Common Quest for Understanding," in *John Paul II on Science and Religion: Reflections on the New View from Rome,* ed. Robert J. Russell, William R. Stoeger, S.J., and George V. Coyne, S.J. (Vatican City State: Vatican Observatory, 1990), M1–M14.

48. *Aeterni Patris* is available on the Vatican website, http://www.vatican .va/holy_father/leo_xiii/encyclicals/documents/hf_l xiii_enc_04081879_aeterni patris_en.html, paras. 29, 30.

49. Dauben, "Georg Cantor and Pope Leo XIII," 98–101.

50. Ibid., 100.

51. Ibid., 103.

52. ST1, chap. 6.

53. ST1, 400.

54. ST1, 446

55. As I noted in chapter 1, Pannenberg both uses and criticizes Hegel's concept of True Infinity. He even claims to have found a paradox in Hegel's thought on infinity, but he does not specify what this paradox is. In my opinion, this claim can be seen as part of a wider critique of Hegel by Pannenberg, one leveled against Hegel's speculative move from concrete history to the abstract, philosophical realm of spirit. Instead Pannenberg wants to return the theological focus to God's relation to the world. One might argue that the use of mathematics in discussing a theological concept of infinity would raise similar concerns for Pannenberg. My reply is that I am not suggesting we stop with the abstract concepts of mathematics, as Hegel did with those of philosophy, in discussing infinity. Instead I want to place Cantor's mathematical concept of infinity within the context of Pannenberg's theology and then rely on Pannenberg to carry out the task of keeping that theology grounded in the concreteness of history as revelation. Johanne Stubbe Teglbjaerg, private communication with the author, summer 2010.

56. ST1, 397n126. Pannenberg also refers to the *Science of Logic* in the context of discussing Hegel's critique of Schleiermacher's understanding of the relation of the finite and the infinite. See ST1, 166, esp. note 126. Hegel's

The Science of Logic is available online, Hegel by HyperText, http://www
.marxists.org/reference/archive/hegel/works/hl/hl000.htm.

57. One of the few exceptions is in Pannenberg's discussion of the traditional proofs of God (ST1, 91–92; 106–7). There, after a careful discussion of
natural theology and its Enlightenment critique, Pannenberg moves to Hegel's
version of the proofs of God and the role of infinity in them. According to Pannenberg, Hegel moved beyond the traditional arguments for the existence of
God as "the expression of the elevation of the human spirit above sensory data,
and above the finite in general, to the thought of the infinite and the universality
of the concept." While such philosophical reflection on the elevation of thought
to the infinite no longer offers a theoretical proof of God, it still plays a crucial
role in "imposing minimal conditions for talk about God. . . . In this sense it is
possible to have a philosophical concept which acts as a framework for what
deserves to be called God."

58. Hegel, *Science of Logic,* §275, 278, 280.

59. Ibid., §283–84, 286.

60. Ibid.,§272.

61. Ibid., §288, 290.

62. ST1, 440.

63. The decision regarding the relative importance of "difference" and
"similarity" here depends, in part, on one's assessment of Pannenberg's reading
of Hegel. Johanne Stubbe Teglbjaerg, private communication with the author,
summer 2010.

64. ST1, 337.

65. See Robert John Russell, "The God Who Infinitely Transcends Infinity," in *How Large Is God? Voices of Scientists and Theologians,* ed. John
Marks Templeton (Philadelphia: Templeton Foundation Press, 1997), 69–72.

66. I want to acknowledge two potential problems here, one new and one
that we have seen previously. First, Cantor's mathematical concept of Absolute
Infinity is unqualifiedly monistic and it refers to God as "a fully independent
other-worldly being." Both of these facts suggest that his concept would provide
an inadequate basis for a detailed Trinitarian doctrine of God such as Pannenberg deploys. Nevertheless I would not want to dismiss it as being contradictory
to Pannenberg's Trinitarian theology. Instead I believe it provides a new insight
into what Pannenberg calls the "logical structure" of infinity in his Trinitarian
theology. To decide if this is in fact the case we will move directly to Pannenberg's
discussion below. The second problem, which we saw in section A.4 above, involves the profound problems caused by the antinomies associated with set
theory and the eventual series of crises that developed in the foundations of
mathematics. If we treat the transfinites as an actual part of creation we would
need to address the implications of these antinomies for our concept of creation.
My response above was that I am seeking to use Cantor's mathematics as a conceptual tool or language to shed insight into the way Pannenberg explicates the

role of the concept of infinity in the doctrine of God. I will imply below that Pannenberg himself seems to assume that something like Cantor's transfinites actually exist in creation. That, in turn, leads to a further discussion of Pannenberg's doctrine of creation in light of the problem of the antinomies, and thus to a task beyond the limitations of this book.

67. ST1, 400: "God gives existence to the finite as that which is different from himself, so that his holiness does not mean the abolition of the distinction between the finite and the infinite."

68. As I stated above in section 4, the profound problems caused by the antinomies associated with set theory and the eventual series of crises that developed in the foundations of mathematics need not block our using Cantor's work theologically if we are seeking to use it as a conceptual tool to enhance the way Pannenberg explicates the role of the concept of infinity in the doctrine of God.

69. In chapter 2 the *vestigium* issue arose over the question of time being relational. There I stated that one of my reasons for viewing tense as relational is that the world is created by the Trinitarian God. I immediately noted that while this might *seem* like an argument from creation to God, that is, a traditional *vestigium* argument, it was not. Instead it was an argument from God the Creator, being Trinitarian, to the world as creation. I noted, there, Karl Barth's rejection of the traditional *vestigium* argument and yet his acceptance of this distinct and limited *vestigium* argument. He referred to this as "*vestigia creaturae in trinitate.*" Barth, *Church Dogmatics,* 4 vols. in 12, ed. G. W. Bromiley and T. F. Torrance, trans. G. T. Thomson et al. (T and T Clark: Edinburgh: 1936–77), 1.1.8.3, 383–399, esp. 391–92. I make a similar case here.

70. It should go without saying that this use of the doctrine of creation to establish the ontological transcendence of God in no way undermines God's immanence in creation as Logos and Spirit.

71. ST1, 408.

72. ST1, 414.

73. ST1, 446. The concept of Spirit "not in its fusion with thought" seems a clear reference to Hegelian idealism, and its rejection suggests that Pannenberg views the paradox he sees in Hegel's understanding of the Infinite as not only a *logical* paradox but even a *metaphysical* paradox—one that Hegel's dialectic between thesis and antithesis, and their synthesis through *Geist,* cannot solve without drawing explicitly on Trinitarian theology.

74. ST1, 400, 347.

Chapter 4. Covariant Correlation of Eternity and Omnipresence in Light of Special Relativity

1. Albert Einstein, quoted in Ilya Prigogine, *From Being to Becoming: Time and Complexity in the Physical Sciences* (San Francisco: W. H. Freeman, 1980), 203.

2. Russell Stannard, *Relativity: A Very Short Introduction* (Oxford: Oxford University Press, 2008); Delo E. Mook and Thomas Vargish, *Inside Relativity* (Princeton, NJ: Princeton University Press, 1987); Martin Gardner, *Relativity Simply Explained* (Mineola, NY: Dover, 1997); and Bruce Bassett, *Introducing Relativity: A Graphic Guide,* 3rd. ed. (Lanham, MD: Totem Books, 2005).

3. See, for example, "Introduction to Special Relativity," *Wikipedia,* last modified September 1, 2011, http://en.wikipedia.org/wiki/Introduction_to_special_relativity. For a helpful online resource, see PBS.org, http://www.pbs.org/wgbh/nova/einstein/. The reader might also find the following online resource helpful, "Overview Visualization of Special Relativity," *Space Time Travel,* last modified July 6, 2011, http://www.spacetimetravel.org/ueberblick/ueberblick1.html.

4. Albert Einstein, "Zur Elektrodynamik Bewegter Korper," *Annalen der Physic* 17 (1905): 891–921. There is a vast literature discussing the scientific and philosophical implications of SR. For a helpful guide, see Arthur I. Miller, *Albert Einstein's Special Theory of Relativity: Emergence (1905) and Early Interpretation (1905–1911)* (Reading, MA: Addison-Wesley, 1981). For introductory texts, see Joseph Schwartz and Michael McGuinness, *Einstein for Beginners* (New York: Pantheon Books, 1979); William L. Burke, *Spacetime, Geometry, Cosmology* (Mill Valley, CA: University Science Books, 1980); Leo Sartori, *Understanding Relativity: A Simplified Approach to Einstein's Theories* (Berkeley: University of California Press, 1996); and Sander Bais, *Very Special Relativity: An Illustrated Guide* (Cambridge, MA: Harvard University Press, 2007). For works more specifically treating philosophical issues, see Lawrence Sklar, *Space, Time, and Spacetime* (Berkeley: University of California Press, 1974); and James T. Cushing, *Philosophical Concepts in Physics* (Cambridge: Cambridge University Press, 1998). For a very readable online overview, see Michio Kaku, "The Theory behind the Equation," *NOVA,* PBS.org, October 2005, http://www.pbs.org/wgbh/nova/einstein/kaku.html.

5. To be more precise, we focus on a set of observers who are at rest with respect to each other and at various distances from each other. They construct a common coordinate grid, or frame of reference, by which events are described. Other sets of observers moving at various uniform velocities with respect to the first set do the same, and together all these form a set of "inertial frames." These observers and their frames of reference are distinguished from accelerating, or non-inertial, observers, such as rotating observers or observers moving along curves in space. To precisely define the distinction between inertial and non-inertial observers and their frames of reference, Isaac Newton introduced the metaphysical idea of absolute space and absolute time.

6. The first postulate is rooted in the Galilean/Newtonian principle of the relativity of motion. In SR Einstein extended this principle to include electromagnetism and, thus, light. See section C.1 below, where I describe Einstein's

objective in constructing special relativity: to unify Maxwell's electromagnetism and a modified form of Newton's mechanics.

7. This means that light does not obey the "addition of velocities" typical of classical physics and ordinary experience. Instead, whether it is emitted or received by observers in relative motion, the speed each measures for it will be the same. See section A.6 below.

8. For an excellent treatment with special attention to the many paths to the Lorentz transformations, see J. R. Lucas and P. E. Hodgson, *Spacetime and Electromagnetism* (Oxford: Clarendon Press, 1990), esp. p. 152, fig. 5.1.1.

9. In their 1963 publication, *Spacetime Physics,* Taylor and Wheeler report that there were at least one million tests of relativity per year with an accuracy of one part in ten thousand or better. See Edwin F. Taylor and John Archibald Wheeler, *Spacetime Physics* (San Francisco: W. H. Freeman, 1963, rev. 1992). For recent tests, see Peter Wolf, Sebastien Bize, Michael E. Tobar, Frederic Chapelet, Andre Clairon, Andre N. Luiten, and Giorgio Santarelli, "Recent Experimental Tests of Special Relativity," *Lecture Notes in Physics* 702 (2006): 451–78, preprint, submitted June 21, 2005, http://arxiv.org/abs/physics/0506168; and Tom Roberts and Siegmar Schleif, "What Is the Experimental Basis of Special Relativity?" Physics FAQ, October 2007, http://www.phys.ncku.edu.tw/mirrors/physicsfaq/Relativity/SR/experiments.html.

10. I will be following closely the presentations by William L. and Peter L. Scott Burke in "Special Relativity Primer" (Santa Cruz: Department of Physics, University of California, Santa Cruz—photocopied manual, 1978); and Burke, *Spacetime, Geometry, Cosmology.* Fortunately the first seven sections of the (unpublished) Primer are available online at http://physics.ucsc.edu/~drip/SRT/. In 1972, as a physics graduate student at the University of California, Santa Cruz, I worked with Burke on background material for this text and I am deeply grateful for his unique vision in formulating SR in this way. For a detailed exposition, particularly of the paradoxes of SR, see Taylor and Wheeler, *Spacetime Physics.* See also the elegant treatment by Charles W. Misner, Kip S. Thorne, and John Archibald Wheeler, *Gravitation* (San Francisco: W. H. Freeman, 1973).

11. SR does not give us an "arrow of time" distinguishing between future and past. For that we must turn to other areas in physics, such as thermodynamics and/or cosmology. This in turn leads to a very technical discussion about whether these areas do indeed break the symmetry (reversibility) of time that holds in classical physics, SR, quantum mechanics, etc. Here I simply presuppose the irreversibility of time and indicate the future along the t-axis by the arrowhead at the top of the axis.

12. Note that measuring time in seconds and distance in light-seconds is equivalent, mathematically, to setting $c = 1$. Often c is dropped out of the equations of SR for simplicity. I will continue to use it explicitly to avoid confusion with presentations that do not follow the "$c = 1$" convention.

13. For a helpful online resource, see "Einstein's Big Idea," NOVA, June 2005, http://www.pbs.org/wgbh/nova/einstein/hotsciencetwin/.

14. Technically, all muons do not decay in the same time but rather decay at a precise statistical rate characterized by their "half-life": the period in which half of an initial collection of muons will have decayed. Quantum mechanics predicts their half-life but no known physics can explain why a particular muon decays when it does.

15. For details, see, for example, Taylor and Wheeler, *Spacetime Physics* (1992), 89. Time dilation is as empirically well confirmed as the rotation of the earth on its axis and its orbiting the sun.

16. A hyperbola is one of several "conic sections": shapes obtained by slicing through a cone at various angles. The other conic sections are the circle, the ellipse, and the parabola. Conic sections also represent, to a good approximation, the shapes of the orbits of planets, comets, meteors, and other objects revolving around our sun. In the seventeenth century René Descartes showed how to express the geometric shapes of conic sections using algebra such as in equation 1 above.

17. Caveat 1: The attentive reader might ask whether these results are consistent with Einstein's first postulate: namely, if we shift to the coordinates (x', t') measured by any of these moving observers, will their plot of the one-second events lie along the same hyperbola as ours? The answer is yes, if we use the right mathematics to describe how we move from the coordinates (x, t) of our rest observer to a friend moving through the origin O at velocity v and measuring (x', t'): namely the Lorentz transformations (see below, section D).

18. Caveat 2: Again this requires a bit more untangling. We are using Einstein's second postulate, that is, light travels in all directions at the same finite speed for all observers regardless of relative motion. But is it an empirical fact? Most scientists would answer yes, viewing it as a direct result of the famous Michelson-Morley null experiment (1887), though Einstein probably did not rely on the results of the Michelson-Morley experiment in constructing SR. Instead, Einstein was able to retain Newton's principle of the relativity of motion, now including the constancy of the speed of light, but only by replacing the Galilean transformations with the Lorentz transformations. See Gerald Holton, "Einstein, Michelson, and the 'Crucial' Experiment," *Thematic Origins of Scientific Thought: Kepler to Einstein* (Cambridge, MA: Harvard University Press, 1973, rev. ed. 1980), 279–370. Meanwhile the equivalence of the speed of light *in all directions* is an assumption that has been challenged by scholars wishing to explore alternative, neo-Lorentzian formulations of SR because of the problems with "flowing time" such as those we have been discussing. We will return to this discussion in chapter 5 below. Wolfgang Rindler gives an interesting account of what would happen to SR without the constancy of c and without its being finite. See Rindler, *Essential Relativity: Special, General, and Cosmological,* 2nd ed. (New York: Springer-Verlag, 1977), esp. 2.17.

Caveat 3: Notice the similarity and difference between viewing light as a classical particle or as a classical wave. If it were a wave, then it makes sense that the speed of propagation is independent of the emitter or receiver, though the receiver and emitter would measure different speeds than c for it, which they do not—and there is no evidence of a wave medium (as the Michelson-Morley null experiment implies). It if were a particle, then it would not need a medium but its speed would depend on the speed of the emitter or receiver, which it does not. So, given SR, light is really neither a classical wave nor a classical particle, suggesting a subtle analogy with the famous "wave-particle duality" accorded it by quantum mechanics for rather different reasons.

19. In 1905 Henri Poincaré named these transformations after the Dutch physicist and mathematician Hendrik Lorentz (1853–1928). In his 1905 paper, Einstein derived them in the context of his treatment of the invariance of electromagnetism for observers in uniform relative motion.

20. Caveat: The Lorentz transformations assume $c' = c$, that is, the speed of light measured in the vacuum by O' is the same as that measured by O, as Einstein's second postulate requires and experiments demonstrate. We will see explicitly below that the Lorentz transformations also preserve the value of the speed of light in vacuum for observers in relative motion, thus justifying this assumption.

21. First, the ruler's right end crosses the x-axis at event B, then a moment later its left end passes the origin. At that instant the right end of the ruler lies at event A along its x'-axis, marking off the proper length, L_0, of the ruler.

22. To reiterate, SR does not give us an "arrow of time" distinguishing between future and past. Here I simply presuppose it.

23. "Elsewhen" and "elsewhere" are used interchangeably in the literature.

24. For example, see Taylor and Wheeler, *Spacetime Physics,* for a discussion of dozens of "paradoxes."

25. For an online resource, see Mark L. Irons, "Pole in the Barn Paradox," website of Mark L. Irons, last modified August 10, 2007, http://www.rdrop.com/~half/Creations/Puzzles/pole.and.barn/. For an animated version, see Mario Belloni and Wolfgang Christian, "Physics 220/230 Lab 10: Relativity; Exercise 3: Pole in the Barn," *WebPhysics,* Davidson College, 2004, http://webphysics.davidson.edu/course_material/Py230L/relativity/relativity-ex3.htm, 2004. For additional helpful discussions, see section 5 on the pole-barn paradox in "Special Relativity/Simultaneity, Time Dilation and Length Contraction," *Wikibooks,* last modified May 9, 2011, http://en.wikibooks.org/wiki/Special_Relativity/Simultaneity,_time_dilation_and_length_contraction.

26. From the definition of γ it is easy to show that $v = ((\gamma^2 - 1/\gamma^2))^{1/2}$ c. Thus $\gamma = 2$ implies that $v = (3/4)^{1/2}$ c.

27. My presentation of the pole-in-the-barn paradox dates back to 1972 when I was an entering physics doctoral student at the University of California,

Santa Cruz. There I worked with Professor Bill Burke on developing what we then called a "generalized spacetime diagram." What I take to be my own contribution to this approach is the construction described here.

28. On a final note, contrary to common tendencies to use SR as a justification for a "relativistic" and even nihilistic view of truth, SR does not allow for an indefinite number of distinct perspectives on reality. For example, there is no way to "sew together" pieces of the consecutive views of the present from either the barn's or the pole's perspective such that the barn doors remain closed, or remain open, or break the pole, or the pole goes around the barn instead of through it, etc. In essence, SR incorporates two distinct, factual accounts of reality (that of the pole and of the barn) into a single spacetime perspective while ruling out of bounds an endless, perhaps infinite, series of possible mis-accounts of reality. I summarize this very important philosophical argument with the catchphrase: "relativity is not relativistic."

29. Hermann Minkowski, Lecture at the 80th Assembly of German Natural Scientists and Physicians, September 21, 1908, quoted in Miller, *Albert Einstein's Special Theory of Relativity,* see references on pp. 238–43.

30. To be more precise, I will not challenge the geometric interpretation of spacetime but I will challenge the additional ontological assumption that all events in spacetime must be equally "present," that is, the block universe interpretation of the geometric interpretation of spacetime.

31. The view of spacetime as a geometry helps explain the SR paradoxes as merely perspectival. According to this view they arise because we naturally look at the world as "3 + 1" instead of as "4", that is, as a three-dimensional spatial universe changing in time, a perspective lodged in both ordinary human experience and the classical physics of Galileo and Newton. From the perspective of spacetime and the Lorentz metric, distances in space and intervals in time are like the shadows cast by a rotating ruler in ordinary Euclidean space. The ruler's length is invariant while the shadows vary in size, but the sum of the square of their sizes is always equal to the square of the ruler's length. In spacetime, the interval of proper time τ is invariant between observers in relative motion. The measures these observers make of the interval in space x and time t vary, but the difference in the squares of these measurements is always equal to the square of the proper time, τ. Because the geometry of spacetime is both like Euclidean three-space (for example, both are flat spaces) and different from Euclidean three-space (due to the difference in their metrics), spacetime is often called pseudo-Euclidean.

32. The Lorentz transformations are, in turn, analogous to the Euler transformations for rotations in three-dimensional Euclidean space. When t is replaced by imaginary time the analogy is sharpened.

33. I am thinking explicitly here about Rudolf Bultmann's interpretation of the New Testament cosmology with its three-storied universe populated with demons as "mythological" to use his well-known phrase.

34. ST2, 91.

35. ST2, 91.

36. ST2, 90–91.

37. Pannenberg returns to many of these themes in recent writings. Most important to the present book is Pannenberg's adamant assertion that "space-time is not eternity." The geometric description of spacetime may suggest a similarity between spacetime and the traditional concept of eternity as timeless. However, this only holds if spacetime entails a spatialization of time in which "the differences of tense—the distinctions between present, past and future—are removed" (ETS, 167–68). In contrast, by eternity I claim that Pannenberg means a flowing-time form of simultaneity that preserves these distinctions. It is this claim that I have tried to capture by the term "co-presence" and that I have added to Pannenberg's use of "duration" to suggest that duration in time is intrinsically structured by the co-presence of past and future within each extended present (chapter 2).

38. Pannenberg is certainly not alone in persisting in this traditional approach. Cosmologists William R. Stoeger and George F. R. Ellis have developed a highly complex critique of multiverse theory and its basis in superstring theory. A key element in their critique is their rejecting the claim that the multiverse is actually infinite. While they offer a variety of robust reasons for this claim, the most telling one seems to be the rejection of an actual infinity, much like what is found in the traditional, pre-Cantorian, literature where infinity and finitude are understood as radically different. However even the open model in standard big bang cosmology can include an actual physical infinity: namely its size which is infinite—and growing! So if one wants to reject infinity in cosmology, one best not wait until the multiverse to do so! See William R. Stoeger, George F. R. Ellis, and Ulrich Kirchner, "Multiverses and Cosmology: Philosophical Issues," preprint, submitted January 19, 2006, http://arxiv.org/abs/astro-ph/0407329; George F. R. Ellis, "Issues in the Philosophy of Cosmology," preprint, submitted March 29, 2006, http://arxiv.org/abs/astro-ph/0602280, esp. 9.3.2.

39. Recall both similarities and differences between the rules for transfinite cardinality ($\aleph_0 + \aleph_0 = \aleph_0$) and transfinite ordinality ($\omega + \omega \neq \omega$) on the one hand and the rules for finite cardinality ($\aleph_0 + \aleph_0 \neq \aleph_0$) and finite ordinality ($\omega + \omega \neq \omega$) on the other hand.

42. Recall that in the four dimensions of spacetime the light cone is a continuous volume, as suggested already in the three dimensions of diagram 4.3 for x, y, and t. In the two-dimensional spacetime diagram of diagram 4.16 the light cone appears misleadingly as two separate regions.

41. Of course we can still use the classical worldview to frame theology within it if (1) we are mindful that it is radically different in principle from the relativistic worldview and (2) the differences in practice between the relativistic and classical views are for all intents and purposes irrelevant. As an example

of this distinction, I basically adopted the classical view in chapter 2 where I discussed time and eternity independently of SR.

42. ST2, 86n221.

43. ST1, 414.

Chapter 5. A New Flowing Time Interpretation of Special Relativity Based on Pannenberg's Eternal Co-presence and the Covariant Theological Correlation of Eternity and Omnipresence

1. Their classic text is Edwin F. Taylor and John Archibald Wheeler, *Spacetime Physics* (San Francisco: W. H. Freeman, 1963, rev. 1992); see chap. 1, "The Geometry of Spacetime."

2. Here and throughout this chapter I refer the reader to chapter 4 for the relevant definitions and discussion of terms such as "Lorentz invariance."

3. Here I have again used the convention $c = 1$; cf. chapter 4, section A.

4. One way to understand this is to picture a ruler rotating around the x-axis, with lights projecting its shadows on the y-z plane. One could then discover the existence of the ruler's invariant length by playing with the values of the y, z coordinates as they change in time: According to the Euclidean metric, the square root of the sum of the squares of the coordinates, $(y^2 + z^2)^{1/2}$, renders a constant, d, regardless of the specific values of the coordinates y, z. From the perspective of philosophical realism, it refers to the length of an object rotating in the y-z plane. Similarly we picture a spacetime event P with coordinates (t, x, y, z) seen from the perspective of two observers in uniform relative motion. According to the Lorentz metric, the square root of the difference of the squares of the coordinates, $(t^2 - x^2 - y^2 - z^2)^{1/2}$, yields another constant, τ, which, again from the perspective of realism, refers to the "length" of the interval between the origin and the event P in spacetime. Just as the constant length of the ruler implies the existence of a ruler with a property called length, even if only its changing shadows are seen, so the constant interval between spacetime events implies the existence of a spacetime "object," even if its separate space and time measurements are registered differently by the observers in relative motion.

5. Taylor and Wheeler, *Spacetime Physics* (1992), 1–5.

6. J. T. Fraser, ed., *The Voices of Time: A Cooperative Survey of Man's Views of Time as Expressed by the Sciences and by the Humanities* (Amherst: University of Massachusetts Press, 1966, repr. 1981).

7. Olivier Costa de Beauregard, "Time in Relativity Theory: Arguments for a Philosophy of Being," in *Voices of Time* (1981), 417–33; Milič Čapek, "Time in Relativity Theory: Arguments for a Philosophy of Becoming," in *Voices of Time* (1981), 434–54. For a detailed exposition, see Milič Čapek, *The Philosophical Impact of Contemporary Physics* (Princeton, NJ: D. Van Nostrand, 1961).

8. Costa de Beauregard, "Arguments for a Philosophy of Being," 429.

9. The terms "elsewhere" and "elsewhen" are used interchangeably in the literature.

10. Costa de Beauregard, "Arguments for a Philosophy of Being," p. 428, fig. 3.

11. Ibid., 429.

12. Although Costa de Beauregard does not say explicitly why this should be the case, the idea is that an object lying at rest along the x-axis of one observer will be seen as lying along a spacelike interval, that is, as extending through both space and time, by a moving observer.

13. Costa de Beauregard, "Arguments for a Philosophy of Being," 428–30.

14. Čapek, "Arguments for a Philosophy of Becoming," 439.

15. Ibid., 441–42.

16. Ibid., 452.

17. Ibid., 443–47.

18. C. J. Isham and J. C. Polkinghorne, "The Debate over the Block Universe," in *Quantum Cosmology and the Laws of Nature: Scientific Perspectives on Divine Action,* ed. Robert John Russell, Nancey Murphy, and C. J. Isham (Vatican City State: Vatican Observatory; Berkeley, CA: Center for Theology and the Natural Sciences, 1993), 135–44. In the published text, there is no explicit reference to who took which position. Polkinghorne returns to these issues in *Exploring Reality: The Intertwining of Science and Religion* (New Haven, CT: Yale University Press, 2007). His rather quick dismissal of the philosophical implications of special relativity's challenge to the global present—"so much the worse for physics"—seems to contradict his well-known adoption of "critical realism," particularly his very strong form of realism in which "phenomenology models ontology" (*Exploring Reality,* 114–16).

19. Isham and Polkinghorne, "Block Universe," 135.

20. Ibid., 136–37.

21. Ibid., 138.

22. Ibid., 141–43.

23. Ibid., 147.

24. Ibid., 145–46. I attempt to respond positively to this challenge in chapter 6, not by proposing a change in SR but by pointing to ways in which the reformulated theology of the divine attributes, discussed in chapter 4, can inform choices between current research programs in physics or lead to new research directions in physics.

25. In a private communication Polkinghorne reiterates this view: "the cosmic frame of simultaneity . . . is undetectable in local physics but experienced in human consciousness" (John C. Polkinghorne to the author, November 2009). A similar view is taken by Alan G. Padgett in *God, Eternity and the Nature of Time* (Eugene, OR: Wipf and Stock, 1992). While Padgett defends

what he calls a "relatively timeless" view of God and so differs fundamentally from Polkinghorne, he makes a similar claim about the existence of an ontological global present. So, while the finite speed of light places an epistemic limit to what we can know about reality, "it does not follow from this epistemic limitation that there is no ontologically genuine simultaneity." What we would need to settle the matter is a signal which, unlike light, would be capable of arbitrarily fast speeds. While we have no such signals physically, "there is nothing illogical or impossible about such a signal from a philosophical point of view" (89).

26. Isham and Polkinghorne, "Block Universe," 142.

27. Ibid., 144.

28. In 1924 Hans Reichenbach introduced the parameter ε in his analysis of Einstein's assumption that the speed of light is the same in either direction along a given axis. Suppose a light ray is emitted by observer A at time t_1 along the positive x-axis. It is reflected by a mirror held by observer B at time t_2 and returns to observer A along the negative x-axis at time t_3. If the speed of light is the same as it moves from A to B and then from B to A, it is clear that $t_2 = t_1 + \frac{1}{2}(t_3 - t_1)$. As Reichenbach notes, "this definition [of t_2] is essential for the special theory of relativity, but it is not epistemologically necessary." Instead Reichenbach offers a definition of t_2 as follows: $t_2 = t_1 + \varepsilon(t_3 - t_1)$ for $0 < \varepsilon < 1$. For special relativity $\varepsilon = \frac{1}{2}$ but other values reflect the possibility that light propogates at different speeds between A and B. The modified, neo-Lorentzian transformations include this revised definition of t_2. See Hans Reichenbach, *The Philosophy of Space and Time,* trans. Maria Reichenbach and John Freund (New York: Dover, 1958), 126–27. The details of the construction of the ε-Lorentz transformations are given in John A. Winnie, "Special Relativity without One-Way Velocity Assumptions," *Philosophy of Science* 37.1 (March 1970): 81–99, and *Philosophy of Science* 37.2 (June 1970): 223–38; see p. 234 for a summary of the modified Lorentzian transformations. A helpful introduction to this work, with extensive references, is found in Max Jammer, *Concepts of Simultaneity: From Antiquity to Einstein and Beyond* (Baltimore: Johns Hopkins University Press, 2006), chap. 14. Michael Tooley is another leading supporter of the neo-Lorentzian generalization of SR. See Tooley, *Time, Tense and Causation* (Oxford: Clarendon Press, 1997), who describes the neo-Lorentzian generalization on pages 346–56 and then offers an assessment in its favor over standard SR on pages 356–73.

29. William Lane Craig, *The Tensed Theory of Time: A Critical Examination* (Dordrecht, Holland: Kluwer Academic Publishers, 2000); William Lane Craig, *The Tenseless Theory of Time: A Critical Examination* (Dordrecht: Kluwer, 2000); William Lane Craig, *Time and the Metaphysics of Relativity* (Dordrecht: Kluwer, 2001); and William Lane Craig, *God, Time, and Eternity* (Dordrecht: Kluwer, 2001).

30. William Lane Craig, *Time and Eternity: Exploring God's Relationship to Time* (Wheaton, IL: Crossway Books, 2001); chapter 5 provides a helpful overview on dynamic versus static concepts of time. For further details, see Craig's more technical works listed in note 29 above.

31. For an extensive analysis and critique of Craig's work, see Yuri Balashov and Michael Janssen, "Presentism and Relativity," preprint, January 9, 2002, http://philsci-archive.pitt.edu/525/.

32. Michael Tooley also adopts a form of the neo-Lorentzian interpretation. As we saw in chapter 4, Tooley in general defends a "growing universe" interpretation of flowing time in which the past and present are real but the future is not real. "[This] tensed view . . . is committed to the existence of absolute simultaneity—a relation that does not enter into the Special Theory of Relativity, and for which, on the face of it, there is no empirical evidence." And so, to defend his tensed view of time, Tooley argues that "the Special Theory of Relativity does not provide a complete account of the spatiotemporal relations that obtain between events." See Tooley, *Time, Tense, and Causation,* 339–40. In its place, Tooley supports a modified version of SR in which absolute simultaneity *can* be defined via the ε-Lorentz transformations. Tooley then argues that this modified theory is preferable to the standard version of SR, and thus that absolute simultaneity is real. See Tooley, *Time,* 342–73, sec. 11.4.3 and, for his response to objections, sec. 11.5. For the purposes of this book, I stay with Craig's discussion of a neo-Lorentzian interpretation of SR because of the connection he makes between it and the general theological concerns he and I share.

33. Craig, *Time and Eternity,* 169. "According to presentism, future times do not yet exist and past times no longer exist. Therefore, there literally are no times which have the properties of pastness or futurity. When a time becomes past, it does not exchange the property of presentness for the property of pastness; rather it just ceases to exist altogether . . . . [A] dynamic or tensed theory of time implies a commitment to *presentism,* the doctrine that the only temporal entities that exist are present entities" (ibid). Craig uses this argument in an attempt to defeat McTaggart's paradox, which is due, he claims, to a heterodox combination of dynamic and static views of time.

34. Craig, *Time and the Metaphysics of Relativity,* 77–83. Craig often uses the terms "co-present" and "co-exist" in discussing simultaneous events. I will follow him in using "co-exist" for events that are simultaneous but I wish to reserve the term "co-present" for the specifically theological context of Pannenberg's understanding of the divine eternity.

35. Lawrence Sklar, "Time, Reality, and Relativity," in *Reduction, Time and Reality,* ed. Richard Healey (Cambridge: Cambridge University Press, 1981), 140.

36. Craig, *Time and Eternity,* 169–73.

37. Ibid., 172.

38. Ibid., 173–80. Craig usually refers to it as the Lorentzian interpretation, but in the wider literature it is usually called the neo-Lorentzian interpretation as I indicated above. I will continue to follow that convention here.

39. Ibid.

40. Clearly much more needs to be said about the ontology that I have only begun to describe here, and the resources of Continental philosophy will play an increasingly important role in this discussion. Nevertheless I believe the beachhead has been established for a genuinely robust flowing time interpretation of the spacetime interpretation of SR.

41. Not all of these points are explicitly discussed by all four scholars but their agreement is, in my view, a given.

42. Defenders of endurance contend that objects have no temporal parts, only spatial parts. They are three-dimensional entities moving in time. This means that objects are present as a whole in each moment of time. They may change properties in time, and they may come into existence or go out of existence in time, but they do not extend in time as they do in space. Defenders of perdurance believe objects in nature have both temporal and spatial parts. In essence, objects are four-dimensional, not three-dimensional; they are extended in time as well as space. See Harold Noonan, "Identity," in *Stanford Encyclopedia of Philosophy,* last modified November 7, 2009, http://plato.stanford.edu/entries/identity/; Neil McKinnon, "The Endurance/Perdurance Distinction," *Australasian Journal of Philosophy* 80, no. 3 (2002): 288–306; Curtis Brown, "Persistence: Some Arguments," Phil. 3330: Metaphysics, last modified October 18, 2007, http://www.trinity.edu/cbrown/metaphysics/persistence-arguments.html.

43. This claim will take a great deal of unpacking in order to show its difference from a simplistic form of realism. I hope for an opportunity to do so in the future.

44. Actually the choice between competing metaphysics is never forced by science, although one version might be highly indicated, perhaps by an "inference to the best explanation."

45. For a readable account, see Gerald B. Cleaver, "Before the Big Bang: String Theory, God, and the Origin of the Universe," paper presented at Valparaiso University, Valparaiso, IN, January 2006; Metanexus 2006; and the Faith and Faithfulness: Christianity in an Age of Science conference, Seguin, Texas, July 2007. http://www.metanexus.net/conferences/pdf/conference2006/Cleaver.pdfard. His paper includes technical references.

46. Isham actually makes this point in Chris J. Isham, "Creation of the Universe as a Quantum Process," in *Physics, Philosophy, and Theology: A Common Quest for Understanding,* ed. Robert J. Russell, William R. Stoeger, S.J., and George V. Coyne, S.J. (Vatican City State: Vatican Observatory, 1988), 375–408.

47. John Lucas goes even further in challenging SR because of his commitment to flowing time. See John R. Lucas, *The Future: An Essay on God, Temporality, and Truth* (New York: Blackwell, 1989).

48. The work that several of us, including Nancey Murphy, Tom Tracy, and George Ellis, have already done on what I call NIODA (non-interventionist objective divine action) may help address this challenge, but that remains to be seen in future research. I hope the work of this book in suggesting how God can act in relation to a variety of axes of simultaneity precisely because of the richness of the elsewhen will provide additional insights into NIODA, whose focus to date has been primarily ontological indeterminism in quantum mechanics. See, for example, Robert John Russell, Nancey Murphy, and William R. Stoeger, S.J., eds., *Scientific Perspectives on Divine Action: Twenty Years of Challenge and Progress* (Vatican City State: Vatican Observatory; Berkeley, CA: Center for Theology and the Natural Sciences, 2008), chaps. 4–6.

49. For a lengthier discussion, see "God and Time: A New Flowing Time Interpretation of Special Relativity and Its Importance for Theology" in the International Society of Science and Religion's publication *God and Physics: An Exploration of the Work of John Polkinghorne,* Fraser Watts and Chris Knight, eds. (Ashgate: forthcoming in 2012), sec. 6a, "God and Time: Do John and I Agree or Differ on Key Theological Questions?"

50. Craig, *Time and Eternity,* 233.

51. Ibid.

52. See, for example, Padgett, *God, Eternity and the Nature of Time,* 93; John Polkinghorne, *Belief in God in an Age of Science* (New Haven, CT: Yale University Press, 1998), 71; and Polkinghorne, *Exploring Reality,* 116–17.

53. For example, in the spacetime of "black holes" the spatial and temporal coordinates exchange places in the region within the horizon of the hole.

54. I discussed the relation between big bang cosmology with its absolute beginning of time ("t = 0") and the doctrine of creation in the Introduction to this volume. What is germane here is the role big bang cosmology has played in discussing God's *temporal* relation to the universe.

55. The spacetime interval along a curved geodesic for the FLRW models can be written as $ds^2 = -dt^2 + a^2(t)[d\chi^2 + \Sigma^2 (d\theta^2 + \sin^2\theta \, d\phi^2)]$. Here ds^2 plays an analogous role to τ^2 in equation 5, although in differential form and with the sign reversed (i.e., here timelike intervals have a negative length). The term $a(t)$ is the "expansion factor" which in a very rough sense is a measure of the "size" of the model universe; it is clearly a function of time, t. The term in brackets $[d\chi^2 + \Sigma^2 (d\theta^2 + \sin^2\theta \, d\phi^2)]$ represents the global spacelike "slice" or "hypersurface" through spacetime at a given time t and value of $a(t)$. Finally, depending on the definition of Σ, the hypersurface can represent any one of the three standard closed and open big bang models with constant curvature. For details see Misner, Thorne, and Wheeler, *Gravitation,* sec. 27.6, esp. eq. 27.24 and box 27.2.

56. While defending a timeless view of the divine eternity, Padgett appeals to GR for a cosmic global present. See Padgett, *God, Eternity and the Nature of Time,* 93–95.

57. The argument for the role of big bang cosmology in establishing a global present goes like this: We begin by focusing on the clusters of galaxies that "float" on the surface of the expanding universe. In this sense, they are at rest in spacetime. Observers at rest in relation to these clusters of galaxies are, in turn, at rest on the surface of the expanding universe. The axes of simultaneity for these observers become the physically significant axes that SR cannot provide.

58. See, for example, George Ellis, Antonio Lanza, and John Miller, eds., *The Renaissance of General Relativity and Cosmology: A Survey to Celebrate the 65th Birthday of Dennis Sciama* (Cambridge: Cambridge University Press, 1994); W. R. Stoeger, S.J., Stanley D. Nel, and G. F. R. Ellis, "Observational Cosmology. IV. Perturbed, Spherically-Symmetric Dust Solutions," *Quantum Gravity* 9 (1992): 1711–23; D. R. Matravers and R. Maartens, "Comments on Standard Cosmologies," *Lecture Notes in Physics* 455 (1995): 135–36, http://www.springerlink.com/content/f5tg4m575205q64p/. See also Chris Clarkson and Roy Maartens, "Inhomogeneity and the Foundations of Concordance Cosmology," *Classical and Quantum Gravity* 27, no. 12 (2010): 124048, http://iopscience.iop.org/0264-9381/27/12/124008; preprint, revised June 2010, http://arxiv.org/abs/1005.2165.

59. An interesting rough analogy exists with the relation between quantum mechanics (QM) and classical physics. If we assume that classical physics is a key to physical reality and then encounter QM, we are quickly led into such seeming paradoxes as wave-particle complementarity, Bell correlations between once coupled particles, etc. But the real question is *not* posed by classical physics to QM, but by QM to classical physics: since nature in fact is quantum mechanical, why is there an ordinary, macroscopic, classical world?

Chapter 6. Duration, Co-presence, and Prolepsis: Insights for New Research Directions in Physics and Cosmology

1. It should go without saying that methodological naturalism does *not* commit one to metaphysical naturalism and thus atheism. Unfortunately, the conflation of methodological and metaphysical naturalism is frequent within current popular literature, especially among "militant atheists," who claim that science strongly supports atheism, and by some of the leading voices of Intelligent Design.

2. This is the equivalent of saying that "emergence" is not the correct philosophical category for the theology of "resurrection as transformation" since it assumes the new elements of discontinuity presuppose an underlying and permanent set of elements of continuity, as, say, biological novelty (i.e., discontinuity) during the evolution of diverse species of life on earth presupposes the constancy of the underlying laws of physics (i.e., continuity).

3. See the introduction and appendix to this volume. See also Robert John Russell, *Cosmology from Alpha to Omega: The Creative Mutual Interaction of Theology and Science* (Philadelphia: Fortress Press, 2008), esp. introd. and chap. 10.

4. Recall from chapter 2 that "past" and "future" are here meant as temporal relations between events and not as properties of events.

5. At the recent "Fundamental Questions in Science FQ(x)" conference (August 2011), a rich variety of topics related to the role of time in fundamental physics and philosophy were presented and explored that potentially bear on the theme drawn here from Pannenberg's theology. These clearly deserve attention in the future. See http://www.fqxi.org/conference/talks/2011.

6. See appendix to the introduction, background material, section E.2.

7. See Pannenberg's *Metaphysics and the Idea of God* (M).

8. Alfred North Whitehead, *Process and Reality,* corr. ed., ed. David Ray Griffin and Donald W. Sherburne (New York: Free Press, 1978). Actual occasions do not "exist" in time and space as one might expect. Instead, in a very subtle way, they are the core reality out of which the "extensive continuum" arises, that which we know as time and space and which we take—wrongly, from Whitehead's perspective—as foundational.

9. Ibid., 208.

10. Ibid., 209. The lines are taken from the hymn written by Anglican priest Henry Francis Lyte (1793–1847) for his congregation at All Saints Church at Lower Brixham, Devonshire. Lyte, dying of tuberculosis, finished the text on the Sunday evening in which he gave his farewell sermon before departing for Italy in hopes of recuperating. He never made his destination, dying instead just weeks later in Nice, France. Its most popular tune was written by William Henry Monk (1823–89) in 1861. For over a century the bells of Lyte's All Saints have rung out "Abide with Me" daily. The first verse is probably the most well known: "Abide with me; fast falls the eventide; the darkness deepens; Lord, with me abide. When other helpers fail and comforts flee, Help of the helpless, O abide with me." The text comes from Luke 24:29: "They urged Him strongly, 'Stay with us, for it is nearly evening; the day is almost over.'"

11. Whitehead, *Process and Reality,* 209. I am continually amazed at the fact that Whitehead uses an explicitly religious quotation to capture in a single metaphor the essence of one of the most complex parts of his overall philosophical system, the double temporality of the actual occasion.

12. Ibid. On p. 210 of *Process and Reality* Whitehead cites Lock's *Essay* for the origin of the phrase "perpetual perishing."

13. Writing a decade later, Whitehead further clarifies this concept: "a duration [involves] a definite lapse of time, and not merely an instantaneous moment." Whitehead, *Science and the Modern World* (New York: Free Press, 1925), 124.

14. Miliĉ Ĉapek, *The Philosophical Impact of Contemporary Physics* (Princeton, NJ: D. Van Nostrand, 1961), 231.

15. Ibid.

16. David Ray Griffin, *Physics and the Ultimate Significance of Time: Bohm, Prigogine, and Process Philosophy* (Albany: State University of New York Press, 1986).

17. David Bohm, "A Suggested Interpretation of the Quantum Theory in Terms of 'Hidden Variables' I," *Physical Review* 85 (1952): 166–79; David Bohm, "A Suggested Interpretation of the Quantum Theory in Terms of 'Hidden Variables' II," *Physical Review* 85 (1952): 180–93.

18. George Greenstein and Arthur G. Zajonc, *The Quantum Challenge: Modern Research on the Foundations of Quantum Mechanics* (Boston: Jones and Bartlett, 1997), 148. For a comparison of classical mechanics and Bohm's formulation, see Robert John Russell, "Divine Action and Quantum Mechanics: A Fresh Assessment," in *Quantum Mechanics: Scientific Perspectives on Divine Action,* ed. Robert John Russell, Philip Clayton, Kirk Wegter-McNelly, and John Polkinghorne (Vatican City State: Vatican Observatory; Berkeley, California: Center for Theology and the Natural Sciences, 2001), 325–28. For an excellent assessment of the historical factors that led to the dominance of the Copenhagen over the Bohmian tradition in physics, and the associated philosophical issues in Bohm's approach, see James T. Cushing, *Quantum Mechanics: Historical Contingency and the Copenhagen Hegemony* (Chicago: University of Chicago Press, 1994); see also Don Howard, "Who Invented the 'Copenhagen Interpretation'? A Study in Mythology," *Philosophy of Science* 71, no. 5 (December 2004): 669–82, http://www.jstor.org/pss/10.1086/425941. For a helpful online introduction, see "De Broglie–Bohm Theory," *Wikipedia,* last modified August 8, 2011, http://en.wikipedia.org/wiki/De_Broglie%E2%80%93 Bohm_theory.

19. David Bohm, *Wholeness and the Implicate Order* (London: Routledge and Kegan Paul, 1980).

20. For more recent work, see David Bohm and Basil J. Hiley, *The Undivided Universe: An Ontological Interpretation of Quantum Theory* (London: Routledge, 1993).

21. Hua Wu and D. W. L. Sprung, "Quantum Chaos in Terms of Bohm Trajectories," *Physics Letters A* 261, nos. 3–4 (1999): 150–57.

22. B. J. Hiley and R. E. Callaghan, "Delayed Choice Experiments and the Bohm Approach," *Physica Scripta* 74 (2006): 336–48; a version of this article is available online, Theoretical Physics Unit, Birbeck College, University of London, last modified April 15, 2010, http://www.bbk.ac.uk/tpru/BasilHiley/ DelayedChoice.pdf.

23. Wheeler's "delayed choice experiment" illustrates the radically non-classical behavior of a particle in an interferometer: Consider a narrow beam of

light entering an interferometer at point A where it is split into two beams (1 and 2). These beams are then reflected back by two mirrors where they meet at point B. The pattern of light at B shows interference effects due to small differences in the distances each beam traversed from A to B. Now consider the light to be so weak that at any moment only an individual photon is moving within the interferometer. The standard explanation, drawing on quantum mechanics, is that the photon interferes with itself, that is, that in some sense it traverses both the paths of beams 1 and 2 between points A and B. Wheeler then introduces a switch along the path of beam 2. The switch can either let the photon continue along this path, or it can divert the photon out of the interferometer. Experiments have shown that in the first case, the interference pattern at B is maintained, but in the second case it is lost and light is recorded only 50% of the time at B. Thus, while the first case suggests that the photon has *wave-like* properties leading to self-interference, the second suggests that the photon is *particle-like* and traverses either the path of beam 1 and is detected at B or the path of beam 2, where it is diverted away by the switch. Now for the startling new result: Wheeler suggested that we delay the decision as to how to set the switch until after the photon has already started traveling from A to B. The result is that even though the choice is delayed, it affects the behavior of the photon prior to the choice, namely when the photon is first split at point A. This sounds like a choice made in the present determines a fact about the past, and thus the phrase "delayed choice experiment." See John A. Wheeler, "The Past and the Delayed-Choice Double-Slit Experiment," in *Mathematical Foundations of Quantum Theory,* ed. A. R. Marlow (New York: Academic Press, 1978), 9–47. See also Greenstein and Zajonc, *Quantum Challenge,* 37–42. For a nontechnical introduction, see "Wheeler's Delayed Choice Experiment," *Wikipedia,* last modified June 21, 2011, http://en.wikipedia.org/wiki/Wheeler%27s_delayed_choice_experiment; and John Polkinghorne, *Quantum Theory: A Very Short Introduction* (Oxford: Oxford University Press, 2002), 64–66.

24. Wheeler, "The Past and the Delayed-Choice Double-Slit Experiment," 14, 41.

25. Hiley and Callaghan, "Delayed Choice Experiments," *Physica Scripta* 74 (2006): 336–48.

26. Roderick I. Sutherland, "Causally Symmetric Bohm Model," preprint, last modified October 7, 2010, http://philsci-archive.pitt.edu/3606/.

27. Henry Pierce Stapp, "Quantum Mechanics, Local Causality, and Process Philosophy," *Process Studies* 7, no. 4 (Winter 1977): 173–82; Henry P. Stapp, *Mind, Matter and Quantum Mechanics,* 2nd ed. (Berlin: Springer, 2004); Henry P. Stapp, *Mindful Universe: Quantum Mechanics and the Participating Observer* (Berlin: Springer, 2007).

28. B. J. Hiley and F. David Peat, eds., *Quantum Implications: Essays in Honor of David Bohm* (Routledge, 1991).

29. See the Bohmian Mechanics website at the Institute for Theoretical Physics, University of Innsbruck, last January 20, 2006, http://bohm-c705.uibk .ac.at/. This website includes excellent visualizations of Bohmian mechanics. See also "Introduction to Bohmian Mechanics," Bohmian Mechanics, December 2010, http://www.bohmianmechanics.org/.

30. See, for example, Detlef Dürr, Sheldon Goldstein, Roderich Tumulka, and Nino Zanghì, "Bohmian Mechanics," January 6, 2008, http://arxiv.org/ PS_cache/arxiv/pdf/0903/0903.2601v1.pdf.

31. See Detlef Dürr and Stefan Teufel, *Bohmian Mechanics: The Physics and Mathematics of Quantum Theory* (Berlin: Springer, 2009), http://www .springer.com/physics/quantum+physics/book/978-3-540-89343-1.

32. Think of a deck of playing cards ordered perfectly from ace to deuce. Throw them down on a table and the result is a relatively disordered set of cards. The entropy has increased in the process. Again, make a fire in your fireplace. The energy contained in the burning logs is lost to the environment during the fire, leaving only soot.

33. The earth is a fine example of an open system. Earth absorbs energy from the sun whose light is emitted at roughly 6000°C. It in turn exhausts energy as microwave light at 2.7°C emitted from Earth's night sky. The difference in the temperatures of the absorbed and emitted light powers the nonlinear thermodynamics underlying and, in part, making possible the biological evolution of life.

34. So, for example, in thermodynamics hot coffee left sitting in a cup on a table spontaneously cools. More to the point, five hundred cups—or five thousand or five million—all filled with hot coffee would all cool, and none get hotter, suggesting a direction of time. But in classical mechanics a pendulum alternately swings left, then right, then left again, suggesting that time in dynamics, as compared with thermodynamics, could be reversed and the physical process would look the same.

35. For a technical introduction, see Ilya Prigogine, *From Being to Becoming: Time and Complexity in the Physical Sciences* (San Francisco: W. H. Freeman, 1980). For a readable introduction, see Ilya Prigogine and Isabelle Stengers, *Order Out of Chaos: Man's New Dialogue with Nature* (New York: Bantam Books, 1984).

36. For a simple example, the phase space of a pendulum is two-dimensional: one dimension represents the angle of inclination of the bob from vertical while the other represents the velocity of the bob.

37. Peter V. Coveney, "The Second Law of Thermodynamics: Entropy, Irreversibility and Dynamics," *Nature* 3332 (June 2, 1988): 409–15, http://www .nature.com/nature/journal/v333/n6172/abs/333409a0.html.

38. R. M. Kiehn, "Prigogine's Thermodynamic Emergence and Continuous Topological Evolution," 2007, http://www22.pair.com/csdc/download/ ecosud07.pdf.

39. Giorgio Sonnino, "Nonlinear Closure Relations Theory for Transport Processes in Nonequilibrium Systems," *Physics Review E* 79, no. 5 (2009), http://scitation.aip.org/getabs/servlet/GetabsServlet?prog=normal&id=PLEEE8000 07900000505112600001&idtype=cvips&gifs=yes.

40. See the Center for Complex Quantum Systems, http://order.ph.utexas.edu/.

41. Key contributors to string theory include Michael Green, Stephen Hawking, Yoichiro Nambu, Holger Nielsen, Joel Scherk, Leonard Susskind, Gabriele Veneziano, and Edward Witten. Critics include Richard Feynman and Sheldon Lee Glashow. For an engaging introduction to string theory, see Brian Greene, *The Elegant Universe: Superstrings, Hidden Dimensions, and the Quest for the Ultimate Reality* (New York: Random House, 1999); Stephen Hawking, *The Universe in a Nutshell* (Bantam Books, 2001). For online resources, see Official String Theory WebSite, http://www.superstringtheory.com/index.html; "Introduction to String Theory," ThinkQuest, http://library.thinkquest.org/27930/stringtheory1.htm; and Superstrings! website, http://www.sukidog.com/jpierre/strings/intro.htm.

42. The strong and weak forces are only effective over ranges the size of the atomic nucleus. The strong force holds together the nucleus of the atom composed of nucleons, protons and neutrons. The weak force accounts for radioactivity and related phenomena. Compared with these forces, the electrodynamic and gravitational forces extend to arbitrarily large distances, making distant galaxies visible to us (since photons are particles of the electrodynamic force) and binding distant galaxies within the expanding universe by gravity. See "Strong Interaction," *Wikipedia,* last modified August 9, 2011, http://en.wikipedia.org/wiki/Strong_interaction; and "Weak Interaction," *Wikipedia,* last modified August 10, 2011, http://en.wikipedia.org/wiki/Weak_interaction.

43. See chapter 2 section C.3 for references to introductory texts on quantum mechanics and a brief discussion of quantum entanglement.

44. The electromagnetic force and the weak force were a single, unified force for an incredibly brief time in the early universe, 10^{-36} seconds after the big bang. By then the temperature of the universe had cooled enough to allow the electroweak force to split into electromagnetism and the weak force. See "Timeline of the Big Bang," *Wikipedia,* last modified August 8, 2011, http://en.wikipedia.org/wiki/Timeline_of_the_Big_Bang.

45. See "Standard Model," *Wikipedia,* last modified August 10, 2011, http://en.wikipedia.org/wiki/Standard_Model.

46. Greene, *Elegant Universe,* 129.

47. According to Noether's rule (1918), symmetries in physics are connected with conserved quantities. So if a physical system behaves the same anywhere in space, the laws of physics that describe it are the same in all places. In this case the momentum of the system is conserved. If the behavior of the sys-

tem and the laws which represent it are the same for all times then energy is conserved. Finally if the same laws hold for all angular orientations of the system in space then angular momentum is conserved.

48. A telephone cable hanging between poles is often used as an analogy to compactness. From a distance the cable looks one-dimensional, but up close its thickness shows that it is actually two-dimensional.

49. Gerald Cleaver, email message to the author, October 5, 2009. Obviously the Planck time is orders of magnitude (at least 44!) below what Augustine and Pannenberg are thinking about in terms of duration. On the other hand, *any* finite value for the duration of a temporal event is qualitatively different from an event of absolutely zero duration. Whether this insight will prove fruitful awaits further research.

50. Gerald Cleaver, email message to the author, October 5, 2009.

51. S. Roy, "Planck Scale Physics, Pregeometry and the Notion of Time," in *The Nature of Time: Geometry, Physics and Perception,* ed. Rosolino Buccheri, Metod Saniga, and William Mark Stuckey (Dordrecht: Kluwer Academic, published in cooperation with NATO Scientific Affairs Division, 2003), http:// www.chronos.msu.ru/EREPORTS/nature_of_time.html; preprint, arXiv.org., http://arxiv.org/abs/gr-qc/0311012v1.

52. For Roy's discussion, see Roy, "Planck Scale Physics," sec. 5.

53. Gerald Cleaver, email message to the author, October 5, 2009.

54. See Itzhak Bars, Research Interests, USC Physics and Astronomy, accessed August 2011, http://physics.usc.edu/~bars/research.html.

55. See Richard Healey's "Holism and Non-separability in Physics," *Stanford Encyclopedia of Philosophy,* last modified December 10, 2008, http://plato .stanford.edu/entries/physics-holism/. His article provides an excellent overview, in-depth analysis, and wealth of references to the general problem of nonseparability. See also Vassilios Karakostas, "Nonseparability, Potentiality, and the Context Dependence of Quantum Objects," *Journal for General Philosophy of Science* 38, no. 4 (2007): 279–97; preprint, last modified October 7, 2010, http:// philsci-archive.pitt.edu/4357. Tim Maudlin gives a detailed and challenging account of the problems involved in reconciling SR and nonlocality/non-separability in QM when attempting to embed QM in Minkowski spacetime. Bohm's theory, for example, seems to require a preferred space-like hypersurface (but recall Sutherland's "Causally Symmetric Bohm Model," cited in note 26 above). Alternatives, such as "many minds theories," seem only to worsen the problem. According to Maudlin, it is even conceivable that we will be forced "to reject Relativity as the ultimate account of spacetime structure." See Tim Maudlin, *Quantum Non-Locality and Relativity: Metaphysical Intimations of Modern Physics,* 2nd ed. (Wiley-Blackwell, 2002), 220.

56. For an introduction, see "Non-Hausdorff Manifold," *Wikipedia,* last modified April 2, 2011, http://en.wikipedia.org/wiki/Non-Hausdorff_manifold.

57. Mark F. Sharlow, "A New Non-Hausdorff Spacetime Model for Resolution of the Time-Travel Paradoxes," *Annals of Physics* 263, no. 2 (March 1998): 179–97, http://www.ingentaconnect.com/content/ap/ph/1998/00000263/00000002/art05772;jsessionid=n4go9dams0ih.alexandra.

58. Mark F. Sharlow, "The Quantum Mechanical Path Integral: Toward a Realistic Interpretation," website of Mark Sharlow, http://www.eskimo.com/~msharlow/path.pdf, abstract; preprint, submitted January 10, 2008, http://philsci-archive.pitt.edu/3780/. See also Mark F. Sharlow, "What Branching Spacetime Might Do for Physics," preprint, submitted January 21, 2008, http://philsci-archive.pitt.edu/3781/.

59. Bertram Yood, "Banach Algebras with Non-Hausdorff Structure Spaces," *Proceedings of the American Mathematical Society* 123, no. 2 (February 1995): 411–15; P. Hajicek, "Causality in Non-Hausdorff Space-times," *Communications in Mathematical Physics* 21, no.1 (March 1971): 75–84, http://www.springerlink.com/content/u313g61386q43p6n/; in the abstract to his paper, Hajicek reports that the properties of certain non-Hausdorff spacetimes include causal anomalies.

60. Additional effects in physics and cosmology that might be suggestive of the physical preconditions for Pannenberg's concept of the "causality of the future" are listed in Robert John Russell, "Eschatology and Physical Cosmology: A Preliminary Reflection," in *The Far Future: Eschatology from a Cosmic Perspective,* ed. George F. R. Ellis (Philadelphia: Templeton Foundation Press, 2002), 307–8.

61. While this approach might suggest that the force between particles at a distance is instantaneous, this is incorrect. In Newton's theory of gravity the interaction is instantaneous but in the Wheeler-Feynman theory the interactions propagate at the speed of light (see fig. 6.1 below).

62. As far as I am aware, George Murphy is the only other scholar who has explored such a possibility. See George L. Murphy, "Prolepsis and the Physics of Retrocausality," *Theology and Science* 7, no. 3 (August 2009): 213–23; Murphy, *The Cosmos in Light of the Cross* (New York: Continuum, 2003); and Murphy, "Hints from Science for Eschatology—and Vice Versa," in *The Last Things: Biblical and Theological Perspectives on Eschatology,* ed. Carl E. Braaten and Robert W. Jenson (Grand Rapids, MI: Eerdmans, 2002), 146–68.

63. Here letters in bold, such as **E** and **B**, stand for vector functions of the variables x, y, z, t with components along the x-axis, y-axis, and z-axis. The object ∇ (pronounced "del") represents both a vector and an operator that involves the three spatial derivatives of scalars and vectors. For introductory details, see Frank S. Crawford Jr., *Waves* (Berkeley: University of California, 1965), 589–90.

64. Instead, magnetic fields are produced entirely by the motion of electric charges (i.e., by electric current).

65. For an introductory discussion, see Edward M. Purcell, *Electricity and Magnetism* (New York: McGraw-Hill, 1963), 263–64. What Einstein showed in his 1905 paper is that the form of Maxwell's equations, given above, is unchanged (i.e., "invariant") when we use the Lorentz transformations, instead of the Galilean transformations of classical mechanics, for observers in uniform relative motion.

66. Basically, we differentiate equation 1d with respect to time. Then, using the vector identity $\nabla \times (\nabla \times C) = \nabla(\nabla \cdot C) - (\nabla \cdot \nabla) C$ and the fact that $\nabla \cdot E = 0$ we obtain equation 2.

67. Unlike the electromagnetic wave equation which is second-order in time (i.e., the time variable is squared in the δt^2 term), thermodynamic equations are typically first-order in time (i.e., the time variable is not squared).

68. J. A. Wheeler and R. P. Feynman, "Interaction with the Absorber as the Mechanism of Radiation," *Review of Modern Physics* 17 (1945): 156; J. A. Wheeler and R. P. Feynman, "Classical Electrodynamics in Terms of Direct Interparticle Action," *Review of Modern Physics* 21 (1949): 424. Technically, Wheeler and Feynman produced an "action at a distance" or "direct interparticle action" (DIA) form of electromagnetism which is relativistically correct and which is essentially equivalent to the predictions of Maxwell's theory. Their formulation, following the conceptual trajectory laid out by Schwarzschild, Tetrode, and Fokker, offers "the natural and self-consistent generalization of Newtonian mechanics to the four-dimensional space of Lorentz and Einstein" (Wheeler, "Classical Electrodynamics in Terms of Direct Interparticle Action," 425–26). For a helpful online introduction, see "Wheeler-Feynman Absorber Theory," *Wikipedia,* last modified August 15, 2011, http://en.wikipedia.org/wiki/Wheeler%E2%80%93Feynman_absorber_theory. For a more technical introduction, see Richard Fitzpatrick, "Classical Electromagnetism: An Intermediate Level Course," website of Richard Fitzpatrick, University of Texas at Austin, 2006, http://farside.ph.utexas.edu/teaching/em/lectures/lectures.html. See also F. Hoyle and J. V. Narlikar, *Action at a Distance in Physics and Cosmology* (San Francisco: W. H. Freeman, 1974), chap. 3.

69. See the original figure 40 in Fitzpatrick, "Classical Electromagnetism." Dr. Joshua Moritz has redrawn it in figure 6.1 to depict the direction of time as upward and he has relabeled the backward and forward moving waves for clarity.

70. To be correct, the backward and forward moving waves emitted from the source and the absorber are only half the amplitude of the usual forward moving waves. This allows the combination of the two forward moving waves to equal the classical amplitude of the forward moving wave.

71. In this sense the reaction of the past absorber to the backward moving wave is "instantaneous" but not in a way that violates special relativity since it involves the time of propagation $t = r/c$ for both "B" and "f" waves.

72. Figure 6.1 can be confusing because it is two-dimensional: one spatial dimension and time. It is better to think of all the waves mentioned here as spherical waves in space's three dimensions, which expand or contract in time from the source at the origin.

73. For an introductory discussion see Fred Hoyle and Jayant Narlikar, *The Physics-Astronomy Frontier* (San Francisco: W. H. Freeman, 1980), appendix D. See also "Steady State Theory," *Wikipedia,* last modified August 9, 2011, http://en.wikipedia.org/wiki/Steady_State_theory.

74. Without this assumption an infinitely old universe would be essentially empty of matter contrary to the obvious fact that we and the visible universe are here.

75. With the development of inflationary/hot big-bang models in the 1980s and eventually the inclusion of quantum effects in such approaches as quantum gravity / quantum creation of the universe models, domain models and eternal inflation, the problem of t=0 was, arguably, overcome. In these models our 13.7 billion year old universe is merely as a tiny part of, and emerges out of, an overwhelmingly vast, probably infinite "multiverse" of one kind or another. I will not follow up on the details of this long account here as my primary goal is to motivate more recent work in light of the early steady-state model.

76. As recently as 1980, Hoyle and Narlikar wrote that "if it were not for the existence of the microwave background . . . the steady-state model might reasonably be judged to stand well at the present time." Hoyle and Narlikar, *Physics-Astronomy Frontier,* 462.

77. Mach's Principle can be described as claiming that the inertia of an object, and thus its mass, is not an intrinsic property of matter, as Newton assumed. Instead it is the effect of the rest of the universe on an object. See Hoyle and Narlikar, *Physics-Astronomy Frontier,* 409.

78. Hoyle and Narlikar start with the insight that we can express all of the basic units of physics in terms of length L, mass M, and time T. Using special relativity we can reduce these three units to two if we define the speed of light, c, as c = 1 by equating the unit of length with the unit of time. From quantum mechanics we know that the product of the uncertainties in position and momentum can never be less than Planck's constant, h. If we set h = 1 we can express length L in terms of the inverse of mass M: $L \sim 1/M$. Hoyle and Narlikar then chose mass as the fundamental unit of measurement. The point here is the following: Normally we follow Newton's assumption that mass is a constant and intrinsic property of matter. But if we adopt Mach's assumption then the mass of a particle is an extrinsic property of matter arising from the existence and motion of the rest of the universe. Now if the structure of the universe changes in time, so will the mass of a particle, and this will lead to the idea of fundamental rulers as shrinking in time instead of space as expanding in time following the insight that $L \sim 1/M$.

79. Hoyle and Narlikar, *Physics-Astronomy Frontier,* chap. 14.

80. Wheeler and Feynman's original calculation had assumed a static Euclidean universe in which the future and past absorbers, i.e., the rest of the universe, were identical. For discussion see F. Hoyle and J. V. Narlikar, *Action at a Distance in Physics and Cosmology* (San Francisco: W. H. Freeman, 1974), chap. 3. See also Fred Hoyle, Geoffrey Burbidge, and Jayant V. Narlikar, *A Different Approach to Cosmology: From a Static Universe through the Big Bang towards Reality* (Cambridge: Cambridge University Press, 2000).

81. Hoyle and Narlikar, *Action at a Distance,* chap. 9.

82. S. W. Hawking, "On the Hoyle-Narlikar Theory of Gravitation," *Proceedings of the Royal Society of London A* 286 (July 20, 1965): 313–19; and P. C. W. Davies, "Hoyle-Narlikar Theory of Gravitation," *Nature* 228 (1970): 270–71.

83. G. F. R. Ellis, "Alternatives to the Big Bang," *Annual Review of Astronomy and Astrophysics* 22 (1984): sec. 4.2. Ignazio Ciufolini and John Archibald Wheeler, *Gravitation and Inertia* (Princeton, NJ: Princeton University Press, 1995); here Mach's principle is built into the concept of the curvature of spacetime as a response to the distribution of matter in spacetime.

84. Helge Kragh, *Cosmology and Controversy: The Historical Development of Two Theories of the Universe* (Princeton, NJ: Princeton University Press, 1996), 358–73.

85. Davide Fiscaletti and Amrit Srecko Sorli, "Non-Locality and the Symmetrized Quantum Potential," *Physics Essays* 21, no. 4 (2008), http://www.fqxi.org/data/forum-attachments/1_Non_locality_and_the_Symmetrized_Quantum_Potential_.pdf.

86. Yakir Aharonov and Jeff Tollaksen, "New Insights on Time Symmetry in Quantum Mechanics," preprint, submitted June 8, 2007, http://arxiv.org/abs/0706.1232. Their research is supported in part by a grant from STARS (Science and Transcendence: Advanced Research Series), a program of the Center for Theology and the Natural Sciences (CTNS) funded in part by a grant from the John Templeton Foundation. For information on STARS, see www.ctnsstars.org. For information on the Aharonov/Tollaksen team's research visit to CTNS, see http://www.ctnsstars.org/enews/news_team5.html.

87. For a careful and balanced treatment of the role of Hoyle's atheism in his search for an alternative to Einstein's big-bang cosmology, see Kragh, *Cosmology and Controversy.*

88. Revised Standard Version.

89. From the Heidelberg Disputation (1518): "19. That person does not deserve to be called a theologian who looks upon the invisible things of God as though they were clearly perceptible in those things which have actually happened [Rom. 1:20]"; "20. He deserves to be called a theologian, however, who comprehends the visible and manifest things of God seen through suffering and the cross" (http://www.catchpenny.org/heidel.html#E020).

90. Ted Peters, *God—The World's Future: Systematic Theology for a Post-modern Era* (Minneapolis: Fortress Press, 1992), 74.

91. Martin Luther, *Luther's Works,* 55 vols., ed. Jaroslav Pelkian (St. Louis: Concordia, 1955–67), 14:342.

92. Jürgen Moltmann, *God in Creation: A New Theology of Creation and the Spirit of God,* Gifford Lectures 1984–1985 (San Francisco: Harper and Row, 1985), 133. Peters, *God—The World's Future,* 307–9; Peters also proposes the term *venturum* to emphasize the proleptic coming of the future.

93. John Hick, *Evil and the God of Love,* rev. ed. (San Francisco: Harper and Row, 1978), see esp. chap. 13.

94. Ibid., 281. Hick cites Bonhoeffer as the source of the phrase *etsi deus non daretur* in the endnotes on p. 373 of *Evil and the God of Love,* but he does not refer to him when introducing the term in the main text on p. 281.

95. I part company with Hick's central claim that ours is a "soul-making" world for various reasons not all that relevant here, but I value his idea of epistemic distance and use it in my own theodicy.

96. Russell, *Cosmology from Alpha to Omega,* chap. 8, esp. 260–62.

INDEX

Absolute Infinity. *See* Cantor:
  infinity, Absolute
Aeterni Patris, 20, 208, 209, 404n48
Aharonov, Yakir, 27, 346, 347, 429n86
Albert, David, 246
anthropic principle, 2, 3, 32–34, 37
  fine-tuning, 3, 33
  many worlds, 33, 367n109
  multiverse, 33, 59, 179, 186, 189,
    306, 321, 412n38, 428n75
antinomies. *See under* Cantor: set
  theory
apeiron. See under Cantor: infinity
apocalyptic eschatology. *See*
  eschatology: apocalyptic
Aquinas, Thomas, 111, 112, 130,
  197, 389n16, 399n10
Aristotle, 108, 119, 129, 196, 197,
  200, 371n9, 380n120, 398n8
arrival of the future. *See under*
  Pannenberg
Athanasius, 110
A-theories of time. *See* special
  theory of relativity: *subentries
  for* interpretation, flowing time;
  time: flowing dynamic
Augustine, 16, 36–39, 77, 81–82,
  95–94, 104, 106–7, 183, 197,
  372n19, 373n35, 376n68,
  377n79, 382n137, 388nn30, 31,
  389n37, 394n65, 425n49

Bars, Itzhak, 335
Barth, Karl, 8, 37, 50–51, 54, 96–97,
  359n34, 372n25, 374n36,
  380n114, 388n30, 406n69

Big Bang cosmology. *See* cosmology:
  subentries Big Bang *and*
  inflation
bodily resurrection. *See* resurrection
  of Jesus: objective
  interpretation, bodily
block universe. *See* time: static;
  special theory of relativity:
  interpretation, block universe
Boethius, 95–97, 106, 108, 372n20,
  383n2
Bohm, David, 27, 322, 326–29, 346,
  421n18, 425n55
Broad, C. D., 133–34
Brower, L. E. J., 206
B-theories of time. *See* time: static
Bultmann, Rudolf, 40–41, 50–51,
  54, 359n38, 411n33

Callaghan, R. E., 328
Cantor, Georg, 19–21, 194–223,
  267, 398n3, 399n15, 401n23,
  402n27, 30, 403n41, 404n46,
  55, 405n66, 406n68, 412n38
  infinity
  —Absolute, 19–21, 203–22, 267,
    390, 405n66
  —actual, as grasped by the
    human mind and found in
    the world, 210
  —actual, in the mind of God, 210
  —actual, the transfinite as an
    endless series taken as a whole,
    200
  —*apeiron,* 19, 110, 195, 196, 210,
    215, 218, 220, 221, 267

ROBERT JOHN RUSSELL

is the Ian G. Barbour Professor of Theology and Science in Residence

at the Graduate Theological Union, Berkeley. He is the author or co-editor

of seventeen books, including *Cosmology From Alpha To Omega:*

Theology and Science in Creative Mutual Interaction

and *Resurrection: Theological And Scientific Assessments.*

He is also the founder and director of the Center for Theology

and the Natural Sciences.

Lightning Source UK Ltd.
Milton Keynes UK
UKHW022015181220
375483UK00008B/297

9 780268 040598